OF SOUTHERN BRITISH COLUMBIA

Bill Mathews and Jim Monger

2005
Mountain Press Publishing Company
Missoula, Montana

First Printing, July 2005

Cover image constructed using
Canadian Digital Elevation Data obtained from GeoBase
Cover by Kim Ericsson and James Lainsbury

Library of Congress Cataloging-in-Publication Data

Mathews, William Henry, 1919–2003
Roadside geology of southern British Columbia / Bill Mathews and Jim Monger.—1st ed.
p. cm. — (Roadside geology series)
Includes bibliographical references and index.
ISBN 0-87842-503-9 (pbk. : alk. paper)
1. Geology—British Columbia—Guidebooks. 2. Automobile travel—Guidebooks. I. Monger, J. W. H. II. Title. III. Series.
QE187.M29 2005
557.11—dc22
2005011929

PRINTED IN CANADA

MOUNTAIN PRESS PUBLISHING COMPANY
P.O. Box 2399 • Missoula, Montana 59806
(406) 728-1900

Roads and sections of Roadside Geology of Southern British Columbia

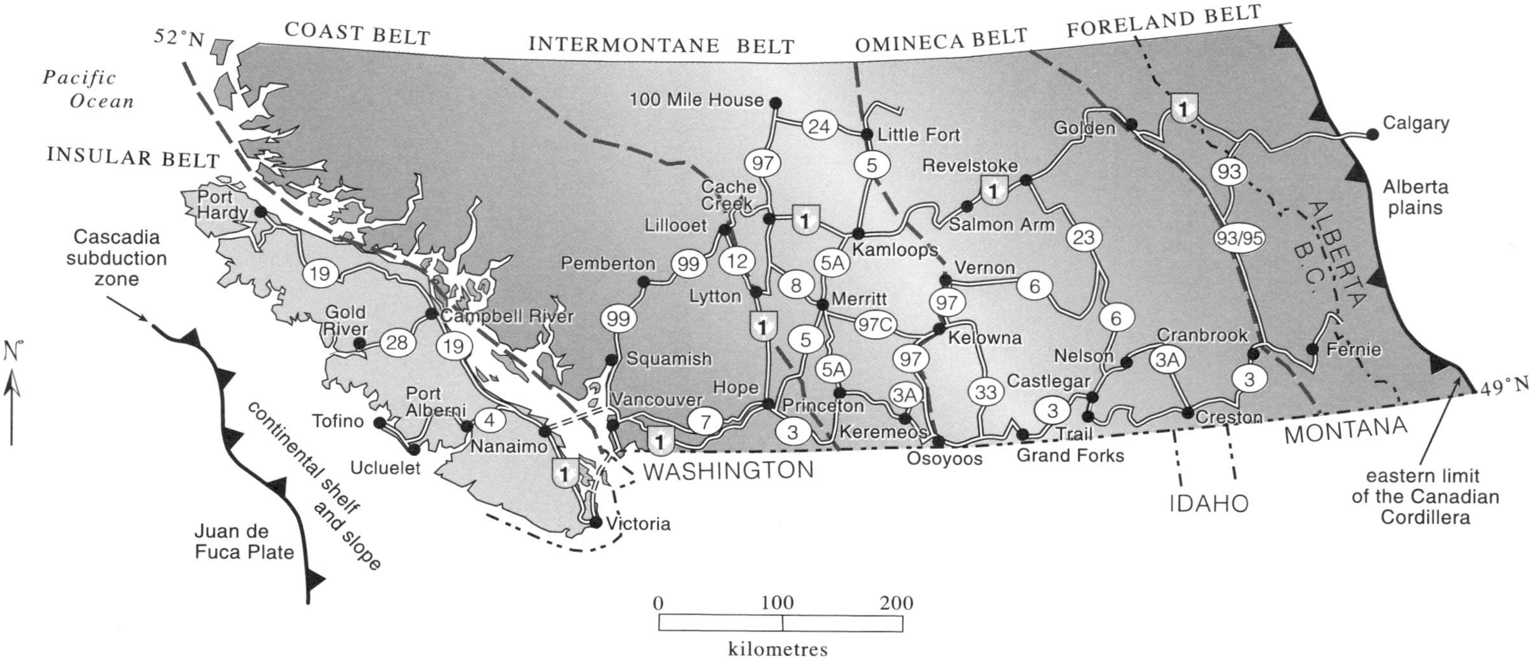

COAST BELT
INTERMONTANE BELT
OMINECA BELT
FORELAND BELT
INSULAR BELT
52°N
49°N
Pacific Ocean
100 Mile House
Little Fort
Golden
Calgary
Revelstoke
Cache Creek
Lillooet
Kamloops
Salmon Arm
Port Hardy
Pemberton
Lytton
Merritt
Vernon
Alberta plains
ALBERTA
B.C.
Cascadia subduction zone
Gold River
Campbell River
Kelowna
Cranbrook
Fernie
Squamish
Nelson
Hope
Castlegar
Port Alberni
Tofino
Vancouver
Princeton
Creston
Nanaimo
Keremeos
Trail
Ucluelet
Osoyoos
Grand Forks
MONTANA
WASHINGTON
IDAHO
eastern limit of the Canadian Cordillera
continental shelf and slope
Juan de Fuca Plate
Victoria
N
0 100 200
kilometres

CONTENTS

PREFACE

British Columbia is roughly 1,200 kilometres from north to south and nearly 800 kilometres wide. Much of it is sparsely populated and accessible only by boat, floatplane, or helicopter, or as in the old days, by slogging through the bush with or without a pack train. Most people live in the southern part of the province, and we've focused on this region because its relatively dense road network provides ready access to the geology from major population centres. The guidebook covers the region that extends up to 300 kilometres north of the boundary with the conterminous United States, and the 800 kilometres between the Pacific coast of Vancouver Island and the Continental Divide, which marks the provincial boundary between southeastern British Columbia and southwestern Alberta.

The main part of the book consists of descriptive road guides and anecdotal material pertaining to the geology along different highways in southern British Columbia. The book spans almost the entire width of the mountainous region of western North America, within which the geology and scenery vary widely from place to place. The region is divided into five geologic belts, each of which contains geology and topography that are generally distinct from those featured in the other belts. The guides to the roads in any one belt are grouped together in a single chapter with a general introduction to the belt, although we've lumped the two eastern belts together. Where one highway, such as Trans-Canada Highway 1, crosses more than one belt, road guide descriptions are split up into the different chapters.

For each chapter, an introduction is followed by descriptive material presented generally in order of highway number from lowest to highest. For example, on Vancouver Island, which is part of the Insular Belt, the material is in order of Highways 1, 4, 19, and 28. A simplified geological map in the introduction to each chapter shows how the rocks along the different routes in that chapter relate spatially to one another.

We comment on the geology of a few ferry routes even though the sea bottom is hidden from view. Although this is perhaps not strictly "roadside geology," it falls within the spirit of this book, and to omit it would leave a big gap in the geology. In addition, although ferry travellers are denied the opportunity to lay hands on the rocks, they have as good a chance of observing distant outcrops or landforms as does someone in a car and can do so without the distractions of passing traffic.

Note that the use of hammers is not allowed in provincial or national parks. Most important of all, please be extremely careful when stopping to examine rocks in roadside outcrops. Fast and heavy traffic and the possibility of rocks falling from steep outcrops make stopping hazardous in some places.

Many people, from the first Geological Survey of Canada (GSC) geologists assigned in 1871 to work in British Columbia, to the present community of academic, federal, and provincial survey, and industry geologists, have contributed directly and indirectly to this book. We've modified several of their figures to use in this book. Jim Monger's colleagues in the former Cordilleran section of the GSC unselfishly contributed information and for thirty years helped form his views on Cordilleran geology. Since 1981, Ray Price has taught him most of what he (Jim) knows about the eastern Cordillera, in the course of jointly running Trans-Cordilleran geology field trips. Earl Dodson and Jim Sears read early versions of the manuscript. Bob Turner of the GSC found money to support a joint photo-taking trip. Bob, Cathie Hickson, and John Clague contributed photos. Jennifer Carey, James Lainsbury, and their colleagues at Mountain Press patiently wrestled the text and figures into their present form. Finally, our wives, Laura Lou and Jackie, have unstintingly supported our efforts to complete the book.

Geologists, geological engineers, and physical geographers universally respected Bill Mathews as a scientist. He had an unrivalled knowledge of matters pertaining to the landscape, geological engineering, and past mining activities in British Columbia. Bill's peers have formally recognized his contributions to the geosciences. He was a Fellow of the Royal Society of Canada, a recipient of the Willet G. Miller Award of that society for outstanding research in earth sciences in 1989, and the winner of the first C. J. Westermann Memorial Award of the Association of Professional Engineers and Geoscientists of British Columbia in 1995.

In 1985, after his retirement from the University of British Columbia, Bill started work on the *Roadside Geology of Southern British Columbia*. In the mid-nineties, Bill recognized that his failing health would not allow him to complete the book. At that time, Jim Monger left GSC employment (although he retains emeritus scientist status there) and Bill, with typical gruff humour, asked him if he "was interested in cooperating with him on the book, now that he (Jim) was not working for a living!" Completion of the book has taken a long time, as other things have intervened, but we hope that *Roadside Geology of Southern British Columbia* provides an overview and details of the province's complex geology, and conveys the sense of how important geology has been, and is, to everyone who lives in and passes though the region.

Mountains, waterways, and some towns of British Columbia

Introduction to the Geology of Southern British Columbia

The southern part of British Columbia extends across almost the entire width of the North American Cordillera, the mountainous part of western North America. *Cordillera* is the Spanish word for a system of interconnected mountain ranges, plateaus, valleys, and basins. Geographers and geologists have used the name in western North America since the 1860s, and that part in Canada is known as the Canadian Cordillera.

The southern Canadian Cordillera contains a tremendous range of scenery. Rocky, densely forested islands at the province's western edge give way to high, rugged mainland coastal mountains indented by deep, glacially carved saltwater inlets. In the rain shadow east of the coastal mountains, rolling, semiarid interior plateaus merge eastward into more mountain ranges, culminating in the spectacular Canadian Rockies. Active tectonics and volcanic eruptions, an enormous variety of rock types and structures, and a landscape carved by both running water and ice make the region a vast natural geological laboratory from which we can learn much about the origins of mountains and continents.

Topography and Geology: The Five Belts of British Columbia

British Columbia can be divided into several topographic regions that parallel the north-northwest trend of the Canadian Cordillera for the length of the province. From west to east these are the Insular Mountains (which include the Vancouver Island Ranges); the topographic depression containing the Strait of Georgia; the Coast and Cascade Mountains; the Interior Plateaus; the Shuswap and Okanagan Highlands; the Columbia Mountains; the great valley called the Rocky Mountain Trench; and the Rocky Mountains. The latter extend east of southern British Columbia for about 100 kilometres into southwestern Alberta, where they are flanked by the plains of the continental interior.

Geologists in British Columbia have traditionally used a slightly different set of divisions in order to emphasize the distinctive nature of the bedrock of these major topographic features. These divisions are called *belts.* From west to east they are the Insular, Coast, Intermontane, Omineca, and Foreland Belts. For the most part, but not always precisely, the boundaries of the belts correspond with those of the topographic divisions. For example, the Coast Belt in southwestern British Columbia includes the Coast Mountains and much of the Cascade Mountains, which have many rock units in common.

The five belts of the Canadian Cordillera. Most rocks in the Omineca and Coast Belts were once buried to much greater depths than rocks in the other three belts.

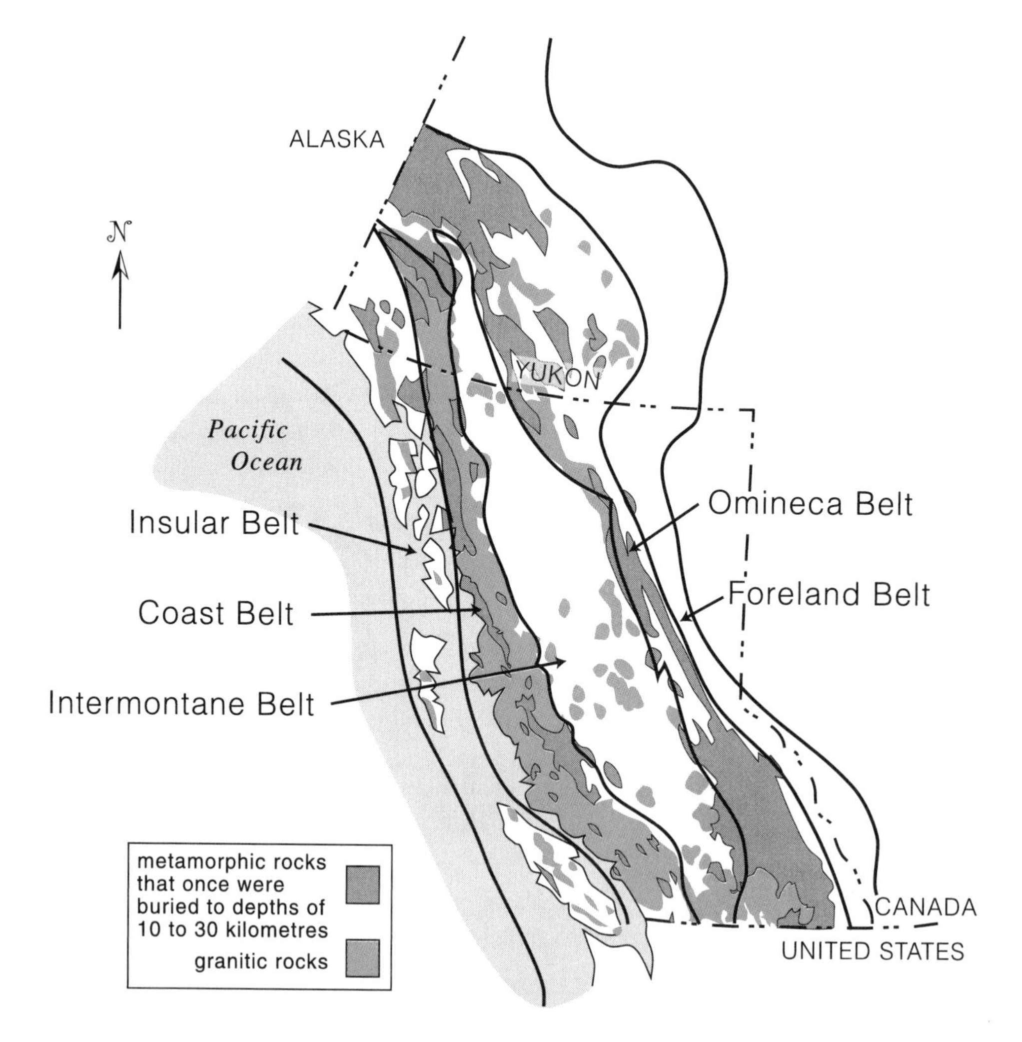

The geological nature of the belts is most clearly seen on a map that shows the surface distribution of rocks that once were at Earth's surface but later were deeply buried and changed by heat and pressure to form metamorphic rocks. From laboratory studies, we know the conditions of temperature and pressure at which different metamorphic rocks are made. From this we can say with some confidence that many rocks now at the

Predominant bedrock of the belts of southern British Columbia.

INSULAR BELT
(Vancouver Island, Strait of Georgia, modern continental shelf and slope)
Magmatic arcs and accretionary complexes welded to the continent about 95 million years ago; 90-million-year-old to present sediments eroded from uplifted Cordilleran mountains.

COAST BELT
(Coast and Cascade Mountains)
Terranes accreted to the continent about 95 million years ago, including 170- to 110-million-year-old granitic rock of an island arc; granitic rock of the Coast Mountains belong to a continental arc that formed 95 to 45 million years ago; young volcanic rock of the Cascade continental arc.

INTERMONTANE BELT
(Interior Plateaus)
Rocks of three terranes (Quesnel island arc, Cache Creek accretionary complex, and Slide Mountain ocean basin) were welded to the continent about 185 to 170 million years ago and intruded by granites of continental arcs; plateau basalts spread across the region 10 million years ago.

OMINECA BELT
(Okanagan and Shuswap Highlands; Monashee, Purcell, Selkirk, and Cariboo Mountains, collectively called the Columbia Mountains)
Precambrian to Paleozoic sedimentary rocks deposited on or near the ancient North American continent and metamorphosed during Cordilleran mountain building between 180 and 60 million years ago; granitic rock formed in magmatic arcs; normal faulting during crustal stretching began 55 million years ago.

FORELAND BELT
(Rocky Mountains)
Precambrian through early Mesozoic rocks deposited on or near the margin of the ancient North American continent and thrust over the edge of the ancient continental platform during Cordilleran mountain building between 100 and 60 million years ago.

surface in the Omineca and Coast Belts once were buried to depths between 10 and 30 kilometres but later were vertically uplifted, and the rocks above them were removed by erosion; they are now exposed at Earth's surface. By contrast, rocks in the Insular, Intermontane, and Foreland Belts are little metamorphosed and most were never buried to depths greater than 10 kilometres.

The three belts with the little-metamorphosed rocks can be further distinguished by rock types. Rocks in the Foreland Belt are mainly sedimentary, whereas those in the Intermontane and Insular Belts are a mixture of volcanic and sedimentary rocks as well as granitic rocks that formed at shallow depths in the crust.

How did the complex pattern of British Columbia geology come to be the way it is, with belts of once deeply buried and now greatly uplifted rocks alternating with belts of little-metamorphosed volcanic and sedimentary rocks? To answer this question, we first must know something about the nature of the Earth and the theory of *plate tectonics.*

Plate Tectonics

The Earth is layered like an egg: shell, white, and yolk. The "eggshell" we walk around on is the rigid *lithosphere*, which means "rock sphere." It varies in thickness from 0 to more than 150 kilometres. In places, such as near chains of volcanoes or mid-ocean ridges where the rocks are hot, they are weak and the lithosphere is thin. Where rocks are relatively cold and strong, as they are below the plains of the continental interior, the lithosphere is thick. The lithosphere includes continental crust, which is mainly made of light, silica-rich granitic rock, and oceanic crust, which consists of basalt, a heavy, dark volcanic rock. In addition, the uppermost rigid part of the underlying mantle, made of very heavy dark rock low in silica and high in iron and magnesium, may be included in the lithosphere where the latter is thick.

The "egg white" is most of the mantle that extends down from the base of the crust to a depth of 2,900 kilometres and makes up about four-fifths of the volume of the planet. The mantle is made of rock that behaves in a plastic manner and can flow very slowly like glass; glass in ancient windowpanes sometimes is thicker at the bottom than the top. That part of the mantle directly below the lithosphere is particularly plastic and is called the *asthenosphere*, which means "weak sphere." The "egg yolk" is the Earth's core, which consists of outer liquid and inner solid parts that probably are made mainly of iron.

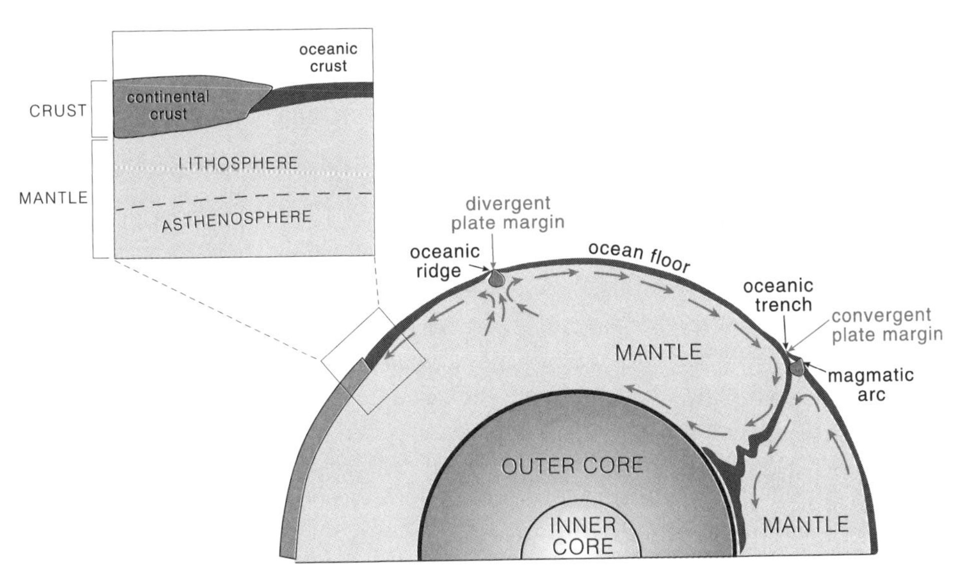

A cross section of the Earth showing its crust, mantle, and core; lithosphere and asthenosphere; divergent and convergent plate margins; and convection within the mantle.

Like a cracked eggshell, the lithosphere is not continuous across Earth's surface but instead is broken into about seven big and several small pieces, which are called *tectonic plates*. Some plates, such as the enormous Pacific Plate that underlies most of the Pacific Ocean, are made entirely of oceanic lithosphere. Others, such as the North American Plate, contain oceanic and continental crust: the western part of the plate is the North American continent and the eastern part is the floor of the northwestern Atlantic Ocean.

Plates move about continuously because the Earth is constantly losing heat, mostly through the ocean floors. The heat comes from the mantle and is lost through the ocean floor to seawater, then to the atmosphere, and ultimately to space. Convection cells circulate within the plastic mantle and continually bring molten basaltic lava to the surface at cracks, or rifts, along the crests of oceanic ridges. The oceanic lithosphere moves laterally away from the ridges, and as it cools by conduction at the rock-water interface, it becomes denser and gradually sinks, so that the ocean basins are deeper away from the ridges. Eventually the oceanic lithosphere becomes so dense that it sinks back into the mantle at the oceanic trenches along *subduction zones*, which are the descending limbs of convection cells. Most oceanic lithosphere returns to the mantle, and because of this, the ocean floors are constantly renewed—none are more than 180 million years old.

In contrast, the continents are made of lighter rock and are not subducted because they float on the underlying mantle; the convection cells move them continually across Earth's surface. By analogy with a pot of boiling dirty water, continents can be thought of as the "scum of the earth." Because they don't disappear into the mantle, the continents contain some very old rocks. The oldest rocks in Canada, located in the northwestern part of the Canadian Shield, are just over 4 billion years old and are some of the oldest known on Earth.

Plates move over the face of the Earth at rates of up to 20 centimetres per year, but mostly the rates are much less (for comparison, fingernails grow at rates of about 4 centimetres per year). These rates don't seem like much, but combine them with geological time measured in millions of years, and it is easy to see why plates may travel great distances across the Earth's surface. A rate of 10 centimetres per year for 100 million years translates to 10,000 kilometres of movement, a distance equal to about one-quarter of the circumference of Earth.

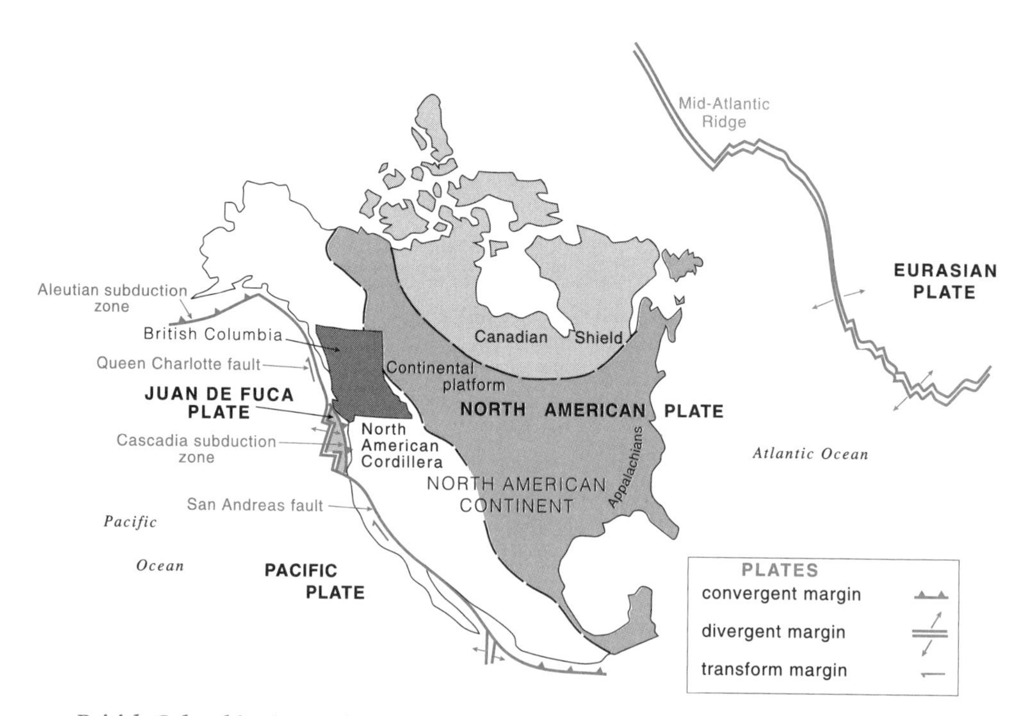

British Columbia sits on the western edge of the North American Plate, which extends eastward as far as the Mid-Atlantic Ridge, where the North American Plate is diverging from the adjacent plate. Its western margin is either converging toward adjacent oceanic plates, or moving laterally with respect to them.

Although tectonic plates are internally strong and rigid, their continual movements in different directions mean they interact with one another at their edges, at what are called *active plate margins.* There are three kinds of active plate margin. First, plates move apart at *divergent margins.* A local example is the boundary between the small oceanic Juan de Fuca Plate and the enormous oceanic Pacific Plate, located about 150 to 500 kilometres off the coast of southern British Columbia. Second, plates come together at *convergent margins,* where one plate dives beneath another to disappear into the mantle. This is happening at the boundary between the Juan de Fuca Plate and the large North American Plate, about 50 to 150 kilometres off the southern British Columbia coast. Third, *transform margins* are places where plates grind past one another and are marked by great faults, such as the well-known San Andreas fault in California or the Queen Charlotte fault submerged west of the Queen Charlotte Islands off the coast of central British Columbia.

A fourth kind of geologically important boundary lies within plates that contain both continental and oceanic crust and is called a *passive continental margin.* The modern example within the North American Plate is the boundary between the continental crust of eastern North America and the oceanic crust of the western North Atlantic Ocean.

Rock Formed at Plate Margins

Geologists recognize that different and distinctive kinds of rocks are forming today in association with the three types of active margin and with passive margins. If we are able to identify similar associations in the geological record, and if we apply the old adage "the present is the key to the past," used by geologists for the past two hundred years, we can hope to decipher something of the record of ancient plate tectonic activity and to gain an idea of how the Canadian Cordillera came to be the way it is today.

At divergent margins, a rift first forms within a plate, splitting it into two or more pieces, and molten lava wells up from the asthenosphere along the rift. If rifting continues, the plate separates and the pieces—new plates—drift apart. The first-formed lava cools to form basalt, and the new plates grow incrementally by addition to the plate margins of fresh lava extruded at the rift. Eventually an ocean basin forms, which is floored by basalt and overlain by any sediments deposited in the ocean, with an elevated (but still submerged) mid-ocean ridge marking the active rift. Thus, rocks that typify ocean floors include basalt, sedimentary rocks deposited in deep water, and, rarely, scraps of *ultramafic* rock, a very heavy iron- and magnesium-rich rock that originates in the upper mantle.

divergent plate margin

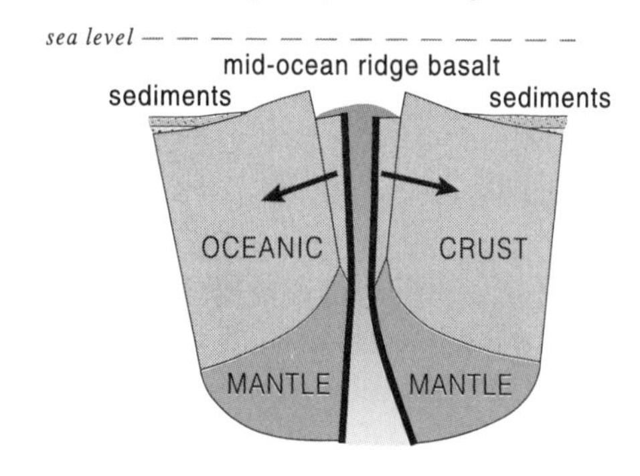

convergent plate margin

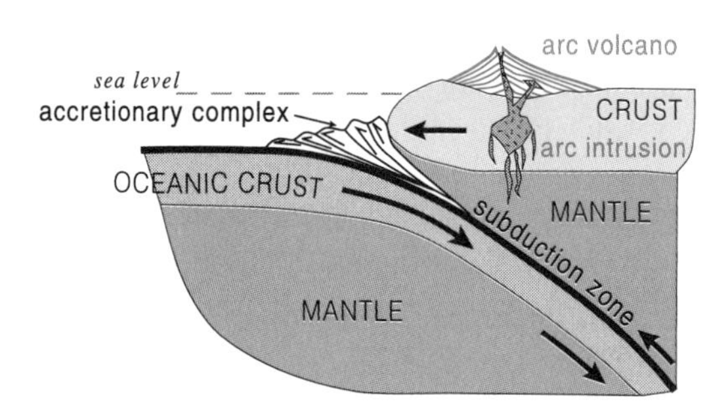

transform plate margin

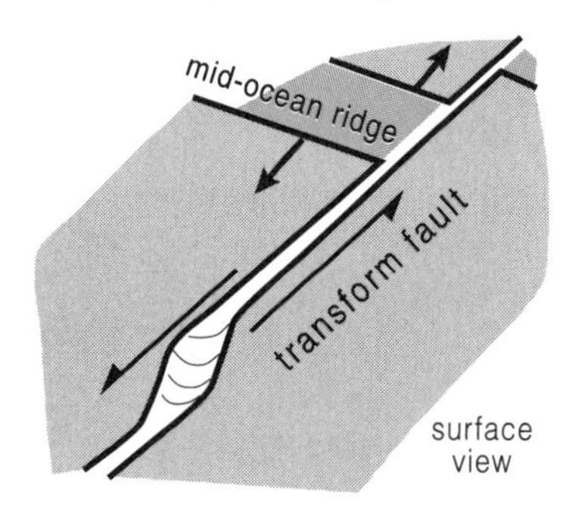

passive continental margin

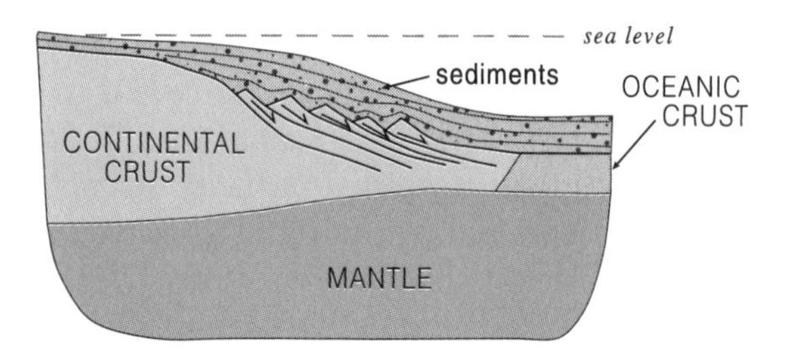

The three types of plate margins and the passive continental margin.

Two different rock associations form at convergent plate margins. First, where one plate dives beneath another to disappear into the mantle, places typically marked by deep oceanic trenches, some of the sediment and volcanic rock on the ocean floor may be scraped off squeegee-fashion and plastered to the edge of the upper plate to form the distinctively messy rocks called *accretionary complexes.* Second, in the upper plate about 100 kilometres above the top of the subducting plate, arc-shaped chains of volcanoes are fed by molten rock, or magma, intruded into the crust beneath the volcanoes, some of which may cool in the crust. These chains are called *magmatic arcs.*

There are two contrasting types of magmatic arc. *Island arcs*, such as those of the Japanese islands, Philippines, and elsewhere in the western Pacific Ocean, are separated from continents by *back-arc basins* floored mainly by oceanic crust. Island arcs form where the subduction zone dips steeply into the mantle and the crust of the arc is being stretched. *Continental arcs* lie on the margins of continents and are exemplified locally by the chain of active and dormant Cascade volcanoes that extends from southwestern British Columbia to northern California, and farther south by the Andes of South America. In continental arcs, the subduction zones

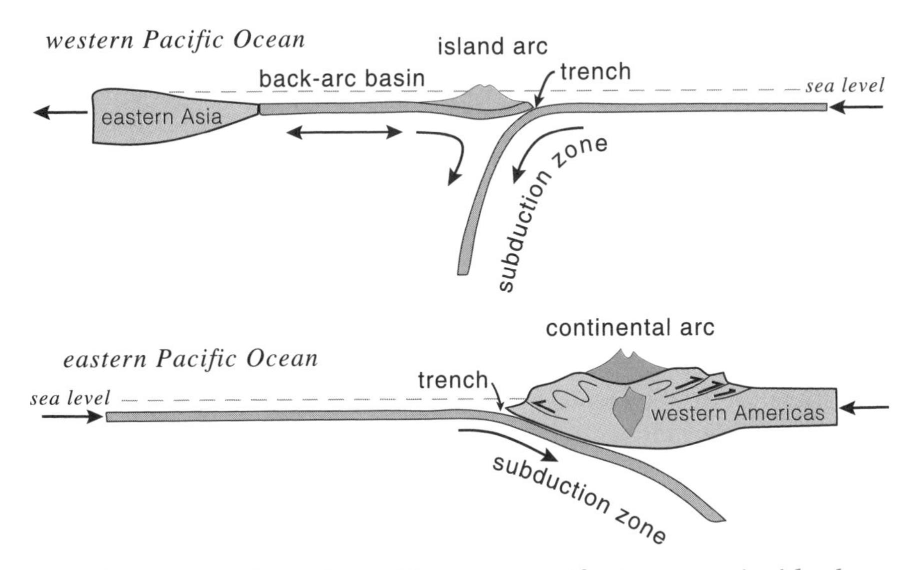

Island arcs and continental arcs. The western Pacific Ocean contains island arc chains, such as the Japanese, Philippine, and Mariana island chains, that are separated from continents by back-arc basins. The eastern margin of the Pacific Ocean features continental arcs, such as those of the Cascades and Andes.

dip at a less steep angle than those beneath island arcs, and the lithosphere of most continental arcs is under lateral compression. It seems that the difference between arc types is largely due to plate motions. In places where the continent retreats from the oceanic trench, an island arc and back-arc basin form. In places where the upper, continental plate advances rapidly toward the trench, the lower plate is unable to subduct fast enough to get out of the way, and the two converging plates collide. The two converging plates form a "tectonic vise" that squeezes and thickens the hot, weak rocks of the arc and eventually raises them into mountains capped by continental arc volcanoes.

Ancient arc rocks are widespread in the Intermontane, Coast, and Insular Belts and in places in the Omineca Belt and provide a record of plate convergence that extends back in time for nearly 400 million years (600 million years in southeastern Alaska). Arc activity was intimately linked to Cordilleran mountain building. Ancient arcs can be recognized by a range of volcanic rocks whose composition, like that of modern arcs, may range from dark, heavy basalt, through andesite, to light, silica-rich rhyolite. The volcanic rocks may be interlayered with sedimentary rocks deposited either in the sea or on land. In close time and space association with the volcanic rocks are bodies of intrusive rock, such as granite, that intruded the crust below the volcanic chains when they were active, later cooled and crystallized, and now are exposed by erosion. In places, such as the southern Intermontane Belt, remnants of old accretionary complexes are preserved near arc rocks of the same age.

Transform boundaries may leave little trace in the geological record. They are recorded mainly by linear tracts of ground-up and highly faulted rock and in some cases by sediments deposited in deep narrow basins and by volcanic rocks that escaped to the surface along the faults.

The present passive continental margin that lies along the eastern seaboard of North America has an ancient analogue in the Foreland and Omineca Belts of the eastern Canadian Cordillera. There, enormously thick deposits of sedimentary rock were laid down between about 550 and 160 million years ago. These thick deposits were laid down in part on top of late Precambrian sedimentary and volcanic rocks formed when Rodinia, an ancient supercontinent, rifted apart, and in part directly on the crust of the old continent.

Terranes of Southern British Columbia

There are two very different kinds of older rock associations in the Canadian Cordillera, with a boundary between them that lies within the Omineca Belt. East of the boundary are thick deposits of sedimentary rock (and their metamorphosed equivalents) laid down along the ancient continental margin. Because we can trace these rocks eastward into the continental interior, we know they formed close to where they are today. West of the boundary, the Intermontane, Coast, and Insular Belts contain a chaotic jumble of fragments of former island arcs, accretionary complexes, and back-arc basins, whose original geographic relationships to one another and to the old continental margin are very uncertain. These rocks were added, or accreted, to the continental margin starting about 180 million years ago, a process that coincides with the initiation of Cordilleran mountain building.

Geologists attempt to sort out the chaotic jumble by identifying regions in which the rocks within them have so much in common that they probably always were together. Such regions are called *terranes*, and any one terrane has a geologic record that is different from those of adjacent terranes. For example, late Paleozoic and early Mesozoic rocks on Vancouver Island are remarkably similar to those in the Queen Charlotte Islands and southern Alaska but differ in many ways from those of the same age in all other parts of the Canadian Cordillera; these rocks form the Wrangellia terrane. In addition, terranes are bounded by big faults on which there are mostly unknown but probably large amounts of lateral displacement, although in many places the faults are hidden by younger intrusions or by rocks deposited across them. Geologists also find evidence, based on fossils and ancient magnetic directions in the rocks, that leaves them to suspect a terrane formed somewhere far removed from its present position relative to the ancient continental margin. This has led to the term *suspect terrane*. The Insular, Coast, and Intermontane Belts all contain suspect terranes; most of the rocks in the Omineca Belt, and all rocks in the Foreland Belt have not moved distances of more than a few hundred kilometres with respect to rocks in the continental interior.

Terrane names are taken from some geographic feature or settlement within them. Some terranes, such as the Quesnel, Stikine, and Wrangellia, formed mainly in island arcs, whereas others, such as the Cache Creek and Slide Mountain terranes, are the remains of ancient ocean floors.

We learn how and when the terranes came together and how the geology of British Columbia was assembled by observing several kinds of geological relationships between the terranes and younger rocks. First, where sedimentary rocks have been deposited on top of two or more terranes, we know that the terranes were together at the time of deposition. Second, if a granitic rock has intruded two or more terranes, we know that the terranes were together before the rock intruded. Third, if deposits of conglomerate and sandstone on one terrane contain cobbles of rocks that were eroded off another terrane, we know they were not far apart at the time the sandstone or conglomerate was deposited. Finally, if we can date the youngest rock in the terranes and the oldest rock that links the terranes, then we can bracket the time at which they came together. By making these observations we see how the components of the crust of the Cordillera in British Columbia were gradually assembled.

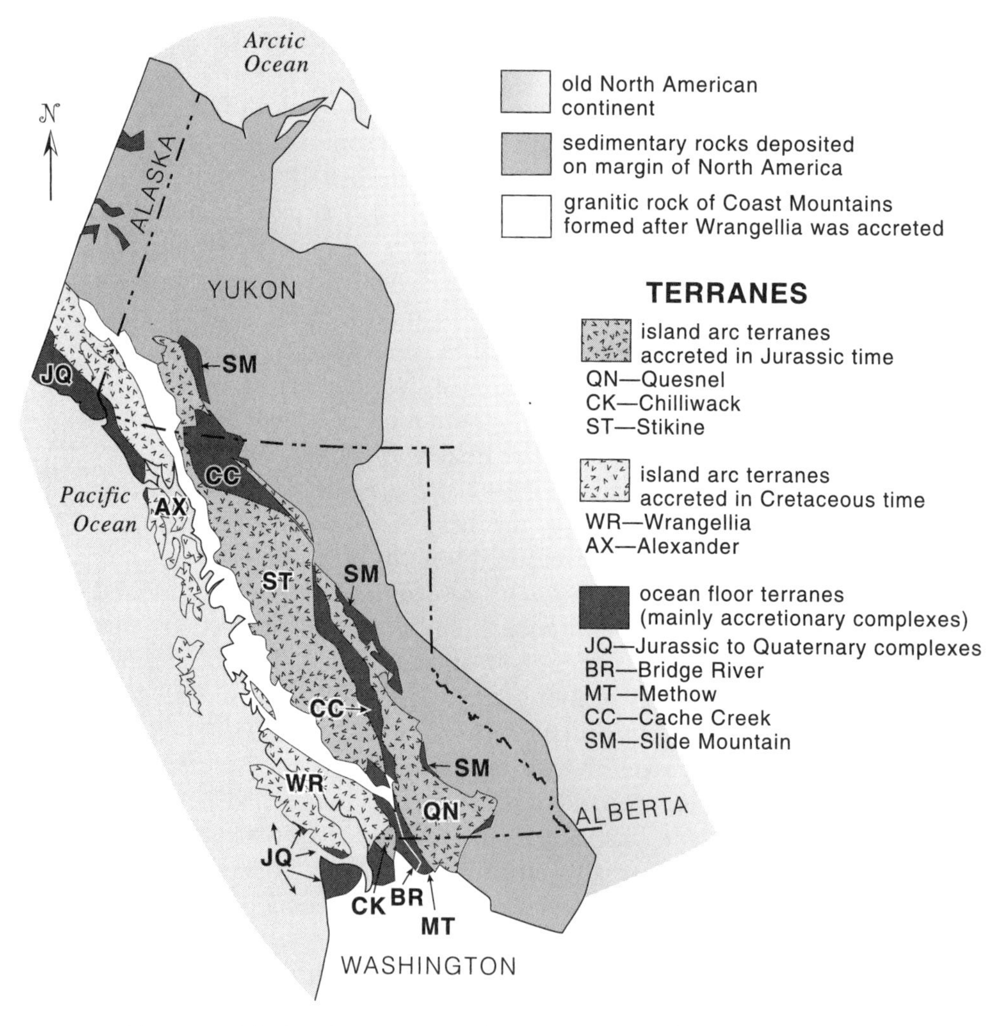

Terranes added to the ancient continental margin of North America in the past 185 million years. The region where terranes overlapped the ancient continental margin by accretion in Jurassic time is along the Omineca Belt; the region where terranes accreted in Cretaceous time overlap those accreted earlier is in the Coast Belt.

Terranes of Southern British Columbia (listed from west to east)

JQ	**Jurassic-Quaternary.** Accretionary complexes that have been welded to the western side of Vancouver Island along the Cascadia subduction zone and its predecessors from Jurassic to Quaternary time; the younger complexes are mainly submerged below the continental shelf and slope.
WR	**Wrangell terrane (Wrangellia).** Devonian to middle Jurassic island arc with conspicuous late Triassic basalt; welded to North America 95 million years ago in Cretaceous time.
CK	**Chilliwack terrane.** Devonian to Permian island arc; fossils in Permian limestone similar to those in the Quesnel and Stikine terranes, and to fossils in the southwestern United States.
BR	**Bridge River terrane.** An accretionary complex of Carboniferous through middle Jurassic age sandwiched between Wrangellia and the Cretaceous continental margin of North America. Includes the Hozameen group.
MT	**Methow terrane.** Ocean basin rocks of Permian age overlain by early Jurassic through early Cretaceous mainly sedimentary rocks, mostly eroded from volcanic arcs. Includes the Jackass Mountain group, Ladner group, and Pasayten group.
CC	**Cache Creek terrane.** Accretionary complex that accompanied the Quesnel and Stikine terranes and contains Carboniferous to early Jurassic ocean basin rocks; fossils in Permian limestone similar to fossils in rocks of the same age in eastern and central Asia.
QN	**Quesnel terrane.** Devonian to Permian island arc rocks overlain and intruded by early Mesozoic island arc rocks; fossils similar to those in rocks of the same age in the western conterminous United States; welded to North America about 185 to 170 million years ago. Includes the Nicola and Rossland groups and the Mount Lytton complex and Eagle plutonic complex.
SM	**Slide Mountain terrane.** Late Paleozoic to Triassic rocks that originated in a back-arc ocean basin to the east of the Quesnel terrane. Includes the Fennell formation.

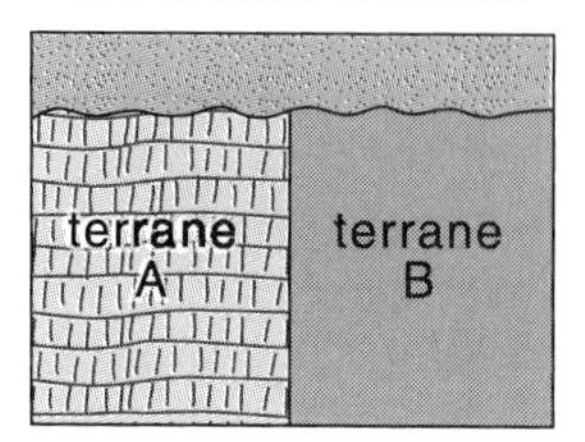

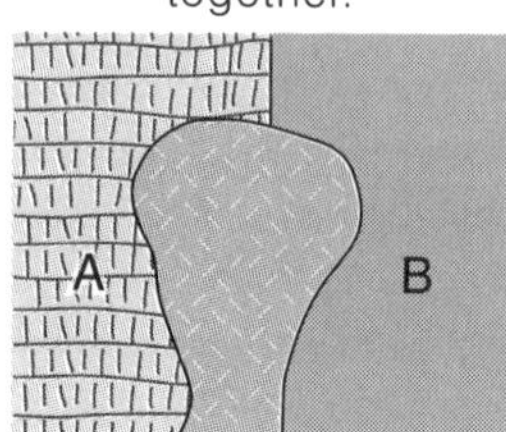

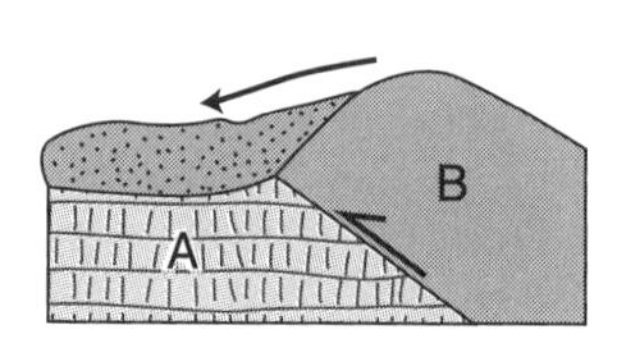

Spatial relationships between rocks can show how and when terranes came together.

Geological Time

Before discussing how the geology of British Columbia probably came to be the way it is today, we first need to mention geological time, whose enormous span is very difficult to comprehend in terms of a human lifetime. Geological time is measured in terms of relative age and absolute age. The *relative age* is based both on the order in which layers, or strata, of sedimentary or volcanic rocks were deposited one on top of another, younger on older, and on the changes in fossil content of the strata. The long-established, familiar names attached to rocks of different ages, such as Cambrian and Mesozoic, are derived from locations and fossil content: *Cambrian* is from *Cambria*, the old name for Wales, where rocks of this age were first studied, and *Mesozoic* means "middle animals," the time when the land was dominated by dinosaurs. The *absolute age* is the number of years before the present and is based on "isotopic clocks." The method measures in the laboratory the rates of decay of relatively long-lived isotopes in a mineral or rock and also the quantity of decay or "daughter" isotopes that have accumulated in the mineral or rock since it formed. Thus, the length of time it has taken for the daughter products in the rock to accumulate can be determined. The isotopic clock starts when the mineral or rock forms and cools below a temperature called the *blocking temperature*, which locks the isotopic system within it. The age quoted is the time, generally given in millions of years before the present, when the mineral or rock cooled through the blocking temperature. This method is most useful for rocks that have cooled from magma, such as lava or granitic rock, but it also can be used to determine the time when minerals in a metamorphic rock formed. Today, the relative and absolute times scales are well calibrated, and geologists use both sets of terms interchangeably.

British Columbia through the Ages

The tectonic plates that form the outer layer of the Earth continually move about as they grow, collide, or slide past one another. The effects of a plate interaction in one part of the Earth, such as a collision between continents, may be transmitted through the entire global system of plates to influence tectonic events thousands of kilometres away. This means we must have some idea of what happened globally if we wish to understand the geological evolution of our region.

Twice in the past billion years, mantle convection has brought all of the continents together into what is called a *supercontinent.* About 1 billion years ago, the continents were amalgamated into the supercontinent called Rodinia. About 750 million years ago, in late Precambrian time, Rodinia started to break up into smaller, continent-sized fragments, and by about

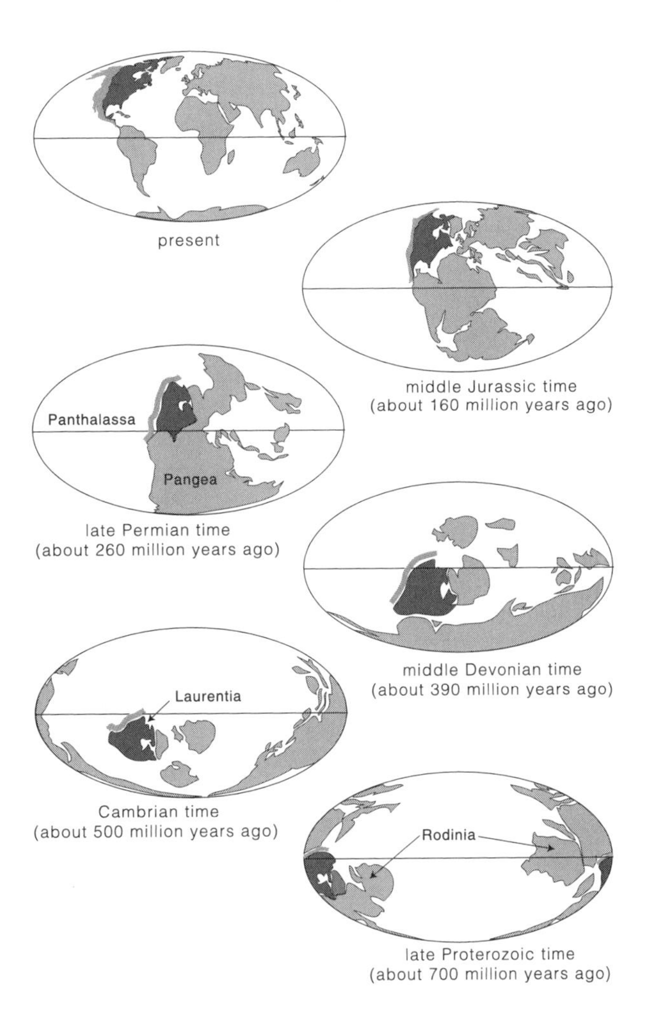

The global distribution of continents at selected times over the past 700 million years. The present North American continent is dark grey; other continents are in light grey; the site of the North American Cordillera is the brown area or line.

600 million years ago the fragments were drifting apart. As they separated, a new ocean basin formed on what was then the northern side of a continent called Laurentia, composed of North America and Greenland. The continent-ocean boundary along the margin of Laurentia eventually became the birthplace of the Canadian Cordillera.

For the next 300 million years, the new ocean expanded enormously and eventually came to occupy more than a hemisphere. Called Panthalassa, this ocean was the distant ancestor of the present Pacific Ocean. By about 300 million years ago, in late Paleozoic time, all of the dispersed continental fragments re-amalgamated in a different configuration to form a new supercontinent called Pangea.

About 200 million years ago, in early Mesozoic time, Pangea started to break up and separate into the continents we know today. The "superocean" Panthalassa gradually shrank and eventually formed the Pacific Ocean. The North American continent finally separated from the rest of Pangea as the Atlantic Ocean opened on its eastern side, and at about the same time the

Continents, oceans, and plate tectonic features that contributed to the evolution of the Canadian Cordillera.

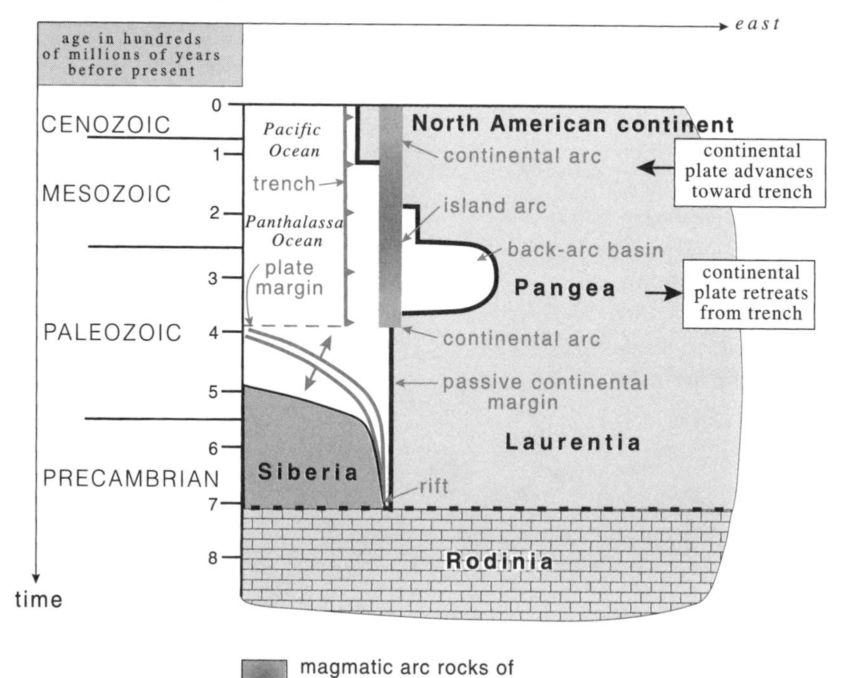

North American Cordillera started to form on its western side. The timing similarity suggests that the breakup of Pangea and migration of the North American Plate toward the ocean on its western side was the ultimate cause of Cordilleran mountain building.

Oldest Rocks in the Canadian Cordillera

—2 Billion Years Ago—

The exposed foundation of much of the North American continent, called the Canadian Shield, is centred on Hudson Bay. It consists mainly of metamorphic and granitic rocks that formed in the deep roots of Precambrian mountains raised between 1 and 3 billion (and very rarely 4 billion) years ago. By about 700 million years ago, erosion had leveled these old mountains to a surface of low relief. The old rocks exposed in the Canadian Shield are buried beneath the blanket of flat-lying, Paleozoic and younger sandstone, shale, and limestone that underlies the plains from near Lake Winnipeg in Manitoba to as far west as the eastern front of the Canadian Cordillera in southwestern Alberta. This part of the continent, where the shield is covered by a relatively thin sheet of sedimentary rock, is called the *continental platform.* At the eastern front of the Cordillera, the top of the Canadian Shield dives westward beneath the piled-up sedimentary strata of the Foreland Belt. A few small areas of 2-billion-year-old rocks occur in the southern Omineca Belt, but no Canadian Shield rocks have been recognized anywhere else in the Canadian Cordillera.

Oldest Sedimentary Rocks in Southern British Columbia

—1.5 to 1.4 Billion Years Ago—

A great thickness of sandstone and shale, and minor limestone, basalt, and related intrusions, was deposited between about 1.5 and 1.4 billion years ago in an enormous basin within an old continent whose western part is now missing. Called the Belt-Purcell supergroup, the rocks are exposed in the southernmost Omineca and Foreland Belts.

Oldest Specifically Cordilleran Rocks

—750 to 550 Million Years Ago—

About 750 million years ago the supercontinent Rodinia began to rift. The rifting event is recorded by deposits of coarse sandstone, shale, conglomerate, and some basalt, referred to in British Columbia as the Windermere supergroup. The rocks are found in the Omineca and Foreland Belts, and similar rocks are known along the Cordillera from Alaska to southeastern California.

Passive Margin Deposits
—540 to 390 Million Years Ago—

Rifting that began about 750 million years ago culminated in complete continental separation by about 540 million years ago, at the beginning of Cambrian time, when deposition started along the passive continental margin of Laurentia. The thick deposits of limestone, sandstone, and shale that were deposited along the continental margin during this time interval make up much of the Foreland Belt and can be traced eastward into thin deposits on the continental platform, and westward into shale, sandstone, and some basalt deposited in deep water, in what is now the southern Omineca Belt. The lateral transition from shallow-water deposits to deepwater deposits is characteristic of a passive margin.

Oldest Evidence of Plate Convergence
—390 to 355 Million Years Ago—

In middle Devonian time, about 390 million years ago, a volcanic arc developed along the western Laurentian margin from northern Alaska to California. It is represented by volcanic and granitic rocks in the Omineca Belt that overlie and intrude westernmost, deepwater deposits of the old passive margin. We can only speculate as to why there was a change from a passive continental margin to a convergent plate boundary. However, oceanic crust must have begun to form about 540 million years ago immediately after the breakup and dispersal of parts of Rodinia. The oldest oceanic crust was close to the continental margin where rifting began, and the older oceanic crust is, the cooler, heavier, and less buoyant it becomes. It may be that the oceanic crust was old enough and dense enough by middle Devonian time to start sinking into the mantle and subducting beneath the North American Plate. The convergent plate margin initiated in middle Devonian time has persisted until today.

Offshore Island Arcs and Back-Arc Basins
—355 to 185 Million Years Ago—

Between about 355 and 185 million years ago, a period when all continents were amalgamated into the supercontinent Pangea, a succession of island arcs replaced the arc that formed along the Laurentian continental margin in Devonian time. The arcs were separated from the old continental margin by deepwater basins of unknown width, floored in part by oceanic lithosphere. In southern British Columbia, the arcs form the Quesnel terrane, and the basins between the arc and the old continental margin, called back-arc basins, are represented by the Slide Mountain terrane and by the

Triassic shale and fine-grained sandstone that overlie it. The accretionary complex that accompanied at least the early Mesozoic arc of the Quesnel terrane is the Cache Creek terrane, which lies west of the Quesnel terrane. In early Mesozoic time, the Quesnel arc and its accompanying accretionary complex marked the western, offshore edge of the Pangean Plate.

In late Paleozoic and early Mesozoic time, British Columbia probably resembled the present western Pacific Ocean basin, where arcs lie offshore from the Asian continent. Possibly, when all the continents became amalgamated into the supercontinent Pangea, the continent retreated from the trench and the old Devonian arc on its western boundary, leaving the new arc separated from the continental margin by a back-arc basin.

The Cordillera Emerges
—185 to 60 Million Years Ago—

Between early Jurassic and late Cretaceous time, from about 185 to 90 million years ago, the western margin of the newly formed North American Plate changed from one featuring island arcs to one carrying a continental arc. The change coincided with the breakup of Pangea into the present continents and their subsequent dispersal. As the Atlantic Ocean basin opened on the eastern side of the North American Plate, the North American continent may have advanced so rapidly toward the trench on its western margin that the oceanic crust to the west was unable to subduct fast enough, and there was a collision between continental and oceanic lithosphere. The collision scraped up and squeezed the mush of hot, weak arc rocks caught between the North American Plate and the oceanic plate to the west, thickening the crust and eventually building a mountain belt that featured continental arcs on the site of the former island arcs, back-arc basins, and passive continental margin.

The change seems to have occurred in two stages. First, terranes of the Intermontane Belt were plastered onto the passive margin rocks to the east between about 185 and 170 million years ago. The region where the two came together eventually became the Omineca Belt. Second, rocks of the Insular Belt were finally accreted to those to the east by about 95 million years ago, with magma from the youngest accreted arc terranes and from the new continental arc combining to form the granitic core of the Coast Mountains. The change from marine to nonmarine deposits across the entire Canadian Cordillera between about 185 and 95 million years ago provides conclusive evidence of the emergence of the region above sea level. By about 65 million years ago, at the end of Cretaceous and beginning of Tertiary time, the Canadian Cordillera probably resembled the modern Andes, where the crust is up to 70 kilometres thick in places.

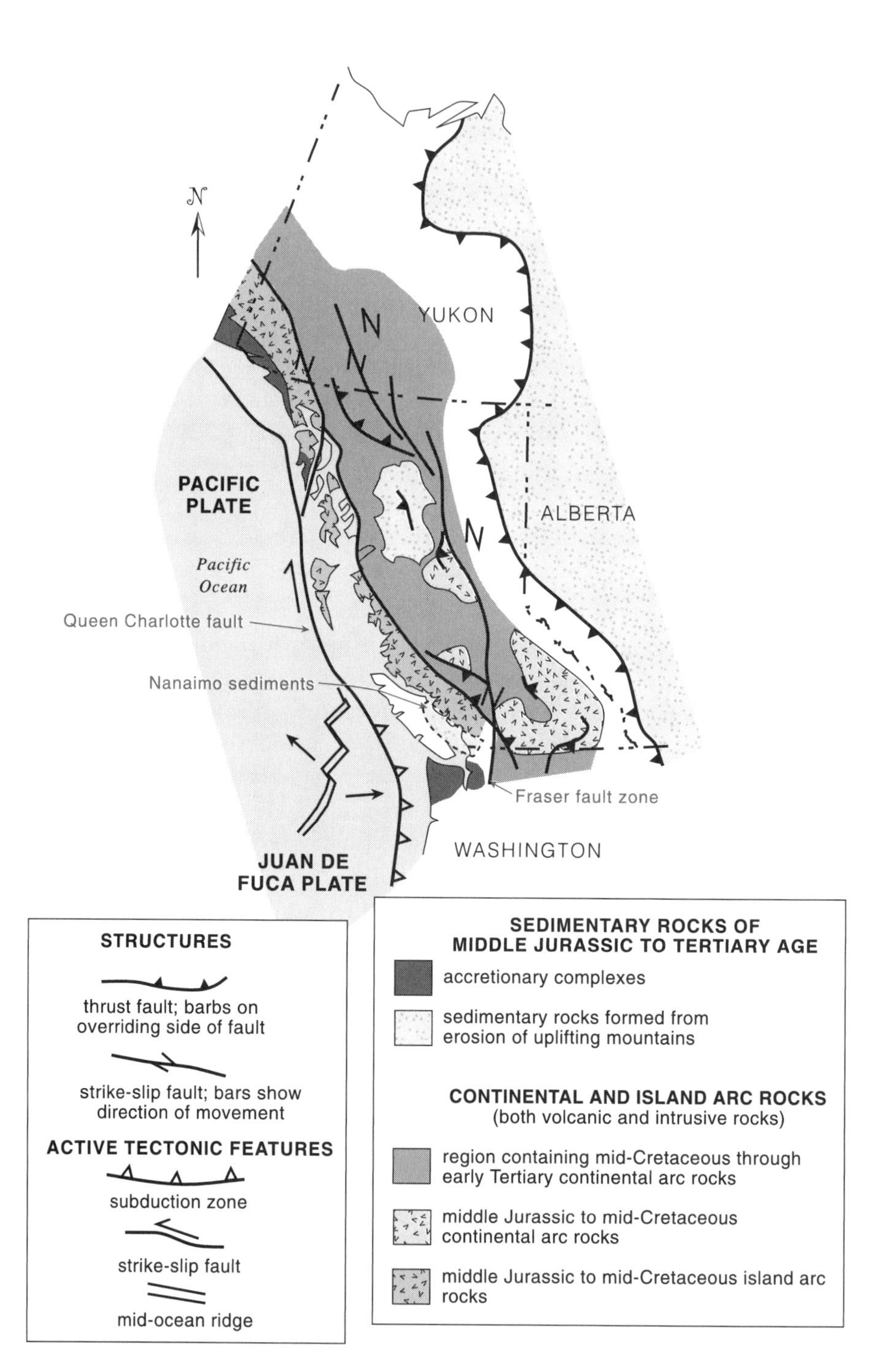

Faults, continental and island arcs, and extensive sedimentary deposits that formed after early Jurassic time record mountain building events when the terranes came together. Note the regions of continental arcs. The island arc in the westernmost Cordillera was not finally accreted until mid-Cretaceous time.

Stretching and Northward Sliding

—55 Million Years Ago to the Present—

About 55 million years ago, the Canadian Cordillera started to change to something like its present configuration, with a crust that today is mostly between 30 and 35 kilometres thick. The crust was stretched and thinned in many places, rather than squeezed and thickened as it had been previously. In addition, parts slid northward on great *strike-slip faults*, along which one side of the fault moved past the other. The change may be due to a shift in the direction of relative motions of oceanic plates to the west, from one that was more or less at right angles to the North American Plate margin in mid-Cretaceous time, to one with the strong component of northward relative motion that we see today. The change may have relieved the lateral compression that built the mountains, allowing them to partly collapse. In addition, parts of the western Cordillera may have been temporarily linked to the northward-moving oceanic plates and moved north with them.

Fire and Ice

—The Past 2.5 Million Years—

Today, the Canadian Cordillera is bounded west of Vancouver Island by a subduction zone, west of which is the small, young oceanic Juan de Fuca Plate. The plate boundary north of Vancouver Island is the active Queen Charlotte strike-slip fault, west of which the enormous Pacific Plate moves northward to dive beneath Alaska. Dormant volcanoes, such as Meager Mountain and Mount Garibaldi in the southwestern Coast Belt, are at the northern end of the Cascade magmatic arc, which continues southward into northern California above the small, subducting oceanic Juan de Fuca and Gorda Plates.

Ice covered most or parts of British Columbia several times in the last 2.5 million years, sculpting and molding the land surface. At times, such as today, the ice has been restricted to high mountains, but at other times the mountain glaciers coalesced to form a single Cordilleran ice sheet. Many low-lying areas were plastered with *till*, rock detritus ground up beneath the moving glaciers. Higher mountains rose above the ice, but lower mountains were covered and rounded off, and glaciers carved deep U-shaped valleys. The ice rearranged some stream drainages so that rivers that once flowed in one direction now flow in another. In places, volcanoes erupted through the ice.

Other landscape features formed as the ice retreated. In the interior of British Columbia, many large valleys contained temporary lakes that formed behind ice dams; finely laminated sediments of silt and mud deposited in the lakes form terraces that are a distinctive feature of many interior valleys.

GEOLOGIC TIME SCALE

Age	*Period*	*mya*	*Geologic Events in British Columbia*
CENOZOIC	Holocene Epoch		Juan de Fuca Plate continues to subduct under the West Coast.
	Pleistocene Epoch Quaternary	*.01*	Last ice age ends about 12,000 years ago. Recently active volcanoes in southwestern BC, such as Mount Garibaldi, belong to the Cascade magmatic arc.
	Pliocene Epoch	*1.8*	First of several ice ages begins about 2.5 mya. Latest uplift of the Coast Belt begins about 10 mya. Plateau basalt flows across the Intermontane Belt about 10 mya. Cascade magmatic arc starts about 35 mya. Fraser fault is active 45 to 35 mya. Metchosin basalt of Eocene age is accreted to the western margin of Wrangellia (Vancouver Island) between 55 and 40 mya. Beginning 55 mya, the region undergoes stretching; magmatic activity and normal and strike-slip faulting are widespread; much of the Cordillera moves northward for several hundred kilometres along strike-slip faults. Sediment eroded from uplifted Foreland and Coast Belts is deposited, respectively, eastward across the plains and westward into the Insular Belt and Pacific Ocean.
	Miocene Epoch	*5*	
	Oligocene Epoch	*24*	
	Tertiary Eocene Epoch	*36*	
	Paleocene Epoch	*58*	
MESOZOIC	Cretaceous	*65*	Rocks of the Foreland Belt are thrust eastwards onto the North American continent. The Pacific Rim complex is accreted to the western edge of Wrangellia (Vancouver Island) between 86 and 65 mya. Sediments eroded from the rising Coast Mountains accumulate in the Georgia Depression between 90 and 65 mya, forming the Nanaimo group. Wrangellia, carrying a middle Jurassic to early Cretaceous island arc, collides with the continental margin 95 mya, generating the first uplift of the Coast Belt 95 to 85 mya. Granites of the continental arc that forms much of the Coast Belt intrude between 95 and 45 mya. Continental arc forms 105 mya in the Intermontane Belt, extruding the Spences Bridge volcanics. Uplift and erosion of the Omineca Belt and parts of the Intermontane Belt result in deposition of sandstone, shale, and conglomerate to the east (some preserved in the Fernie basin) and west.
	Jurassic	*145*	Arc forms on eastern Wrangellia 170 to 110 mya. Cordilleran mountain building begins 180 mya as the continent collides with the first of several island arcs; folding, thrust faulting, and metamorphism in the Omineca Belt is associated with accretion of an island arc (Quesnel terrane), the intervening ocean basin (Slide Mountain terrane), and the arc's accretionary complex (Cache Creek terrane).

mya=millions of years ago

Geologic Events in British Columbia	*mya*	*Period*	*Age*
About 170 mya the Atlantic Ocean begins to open on the eastern side of the North American Plate. As the supercontinent Pangea breaks up, starting about 200 mya, the new North American Plate advances toward the trench on its western side. The Bonanza island arc begins to form on Wrangellia in latest Triassic time. The Quesnel island arc continues to form offshore of Pangea, separated from the supercontinent by a deepwater basin.	208	Jurassic	**MESOZOIC**
	248	Triassic	
The Quesnel island arc begins to form, possibly because the Laurentian continent retreated from the subduction zone on its western margin as Pangea was assembled beginning 300 mya. Devonian granites in the Omineca Belt are the roots of a continental margin arc that formed between 390 and 360 mya, indicating a change from a passive continental margin to a converging margin with a subduction zone. During early Paleozoic time, limestone, mud, and some sandstone are deposited on the Laurentian continental shelf, now in the Foreland Belt; mud, sand, minor limestone, and some basalt are deposited further out to sea on the continental slope, now in the Omineca Belt.	286	Permian	**PALEOZOIC**
	360	Carboniferous	
	417	Devonian	
	443	Silurian	
	495	Ordovician	
	545	Cambrian	
About 550 mya, deposition of marine sediment begins on the margin of the Laurentian continent, the predecessor of North America, as seawater fills the rift. Sand, shale, and local conglomerate and basalt of the Windermere supergroup accumulate in the continental rift basin between 750 and 570 mya. About 750 mya, the supercontinent Rodinia begins to slowly rift. The Rodinia supercontinent is assembled about 1,000 mya. Between 1.5 and 1.4 billion years ago, the sand, mud, and some basalt of the Belt-Purcell supergroup accumulate in an enormous basin within a continent. Two-billion-year-old rocks in the Monashee Mountains of the Omineca Belt may be the westernmost exposure of rocks from the Canadian Shield. The Canadian Shield, the bedrock nucleus of the North American continent, forms mostly between 3 and 2 billion years ago.	2,500	Proterozoic Eon	**PRECAMBRIAN**
		Archean Eon	

mya=millions of years ago

Geology and People in British Columbia

Geology has had an enormous influence on the development of British Columbia. Early colonists on Vancouver Island exploited its coal in the mid-1800s, and at about the same time, the search for placer gold was responsible for much of the early settlement of the interior of British Columbia by people of European and Chinese background. The coming of the railways in the 1880s, with their great transportation potential, meant that coal and other bulky mineral deposits could be exploited in remote regions and carried to markets. Today, coal is mined from enormous deposits in the Foreland Belt and transported by rail across almost the entire Canadian Cordillera to ports on the west coast. By the late 1800s, deposits of lead, zinc, and silver ore were discovered and exploited in the eastern Cordillera, and gold and copper in the western Cordillera. Most early mines were developed by tunnelling, but starting in the late 1950s, when heavy-duty machinery became more widely available (and miners more costly), open-pit mining became the preferred method for extraction of metal ores and coal. Some of the open-pit mines are so large that they are visible from space and have completely changed the local landscape. Since the 1940s, oil and gas have been of importance, with recovery mainly from sedimentary rocks in the easternmost Cordillera and the continental platform, but the offshore region also shows potential for these commodities.

Today, and particularly in the heavily settled regions of southwestern British Columbia, there is a need for geological knowledge that goes beyond the traditional skills required for commodity discovery and extraction. Safety and environmental considerations are major concerns. The west coast of North America is an active plate margin with associated earthquakes and volcanoes. Highways, railway lines, and oil and gas pipelines traverse steep mountainous terrain with high precipitation in places. Rockslides, debris flows, and avalanches interrupt the transportation routes from time to time. Dams built for hydroelectric generation require safe footings and vast amounts of material for their construction. The enormous and rapid growth in recent years of cities like Vancouver creates great demand for sand and gravel for buildings and roads, safe land for building, safe water for drinking, and sites for safe disposal of waste.

For resource and environmental reasons, geology will continue to have an enormous impact on this region.

Insular Belt

Vancouver Island and the Strait of Georgia

As its name suggests, the Insular Belt contains the islands along the coast of British Columbia and southeastern Alaska, but it also includes the seafloor surrounding those islands out to the western edge of the North American Plate, which is roughly 200 kilometres west of the mainland coast. In the vicinity of Vancouver Island, the eastern boundary of the belt runs near the eastern side of the Strait of Georgia, so many of the adjacent smaller islands and most of the seafloor east of Vancouver Island are part of the belt. The western boundary, in a water depth of about 2,000 metres at the toe of the continental slope west of Vancouver Island, is the active Cascadia subduction zone, west of which is the small oceanic Juan de Fuca Plate. North of Vancouver Island, the plate boundary is the Queen Charlotte fault.

The general topography of Vancouver Island is like an overturned canoe. The highest mountains, which form the Vancouver Island Ranges, lie along the keel approximately midway between either side and midway along its length. There, the mountains are high enough, in places over 2,000 metres, and the climate cool and wet enough to support small ice fields and glaciers. Scattered lowlands occur on the flanks of Vancouver Island, at both ends, and in the Cowichan and Alberni Valleys, two major valleys in the southern third of the island. The lowland topography and geology on the eastern side of Vancouver Island descends below sea level to underlie most of the Strait of Georgia.

Most of Vancouver Island is made of volcanic and sedimentary rocks, together with granitic intrusions and some metamorphic rocks. The rocks range in age from Devonian to Jurassic, about 380 to 170 million years old. These rocks belong to the suspect terrane called Wrangellia. Unconformably lying on top of the Wrangellian rocks in the lowlands and many islands on the east side of Vancouver Island are late Cretaceous coal-bearing sedimentary rocks about 90 to 65 million years old, mostly formed from detritus eroded from the Coast Belt to the east. Forming the south end of

Vancouver Island and a narrow strip along the coast near Pacific Rim National Park on the west side of Vancouver Island, and underlying the continental shelf and slope, are late Jurassic to Holocene deposits formed on the ocean floor that have been carried beneath Wrangellia during late Mesozoic, Tertiary, and Quaternary subduction.

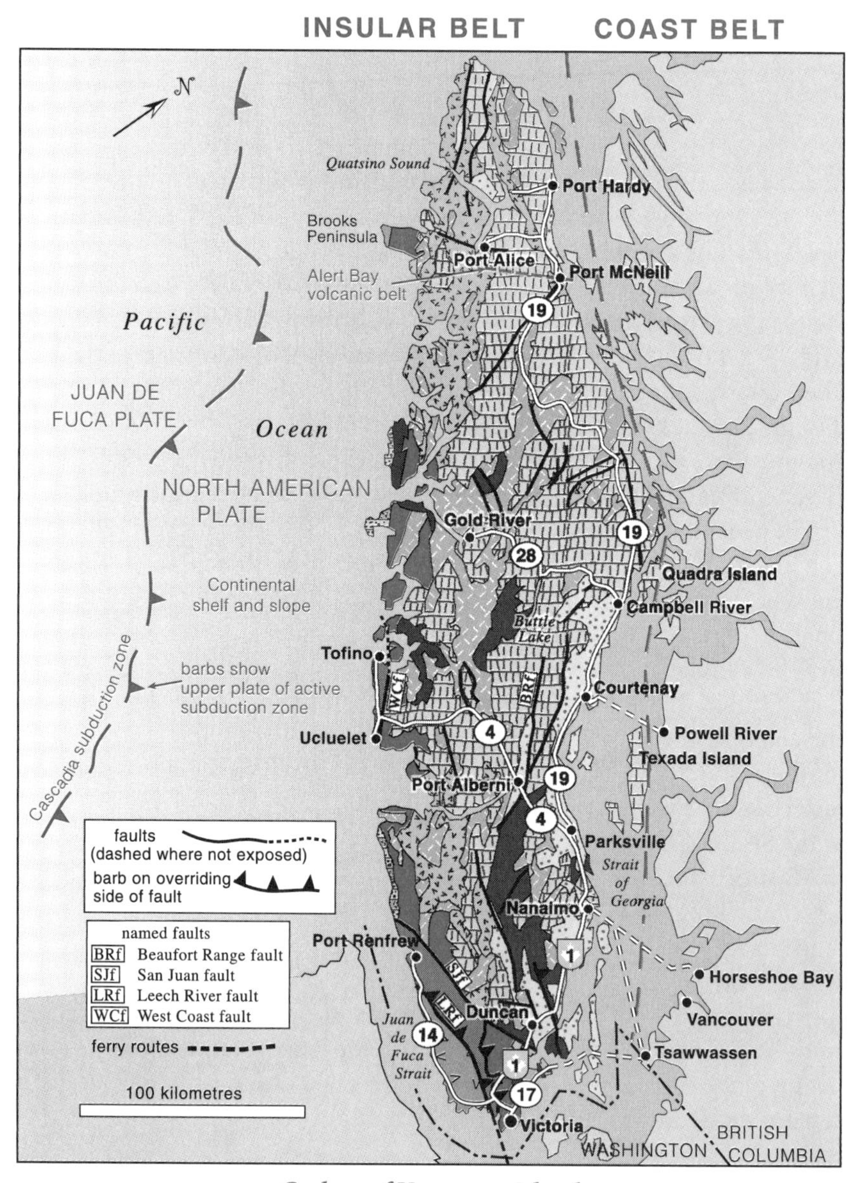

Geology of Vancouver Island

Wrangellia: A Suspect Terrane

Wrangellia occurs on Vancouver Island, the Queen Charlotte Islands, and in southern Alaska, a distance of more than 2,000 kilometres, and draws its name from the Wrangell Mountains in southern Alaska. On Vancouver Island, the rocks of Wrangellia can be divided into five distinct units that collectively span over 200 million years of Earth's history. Most of the rocks

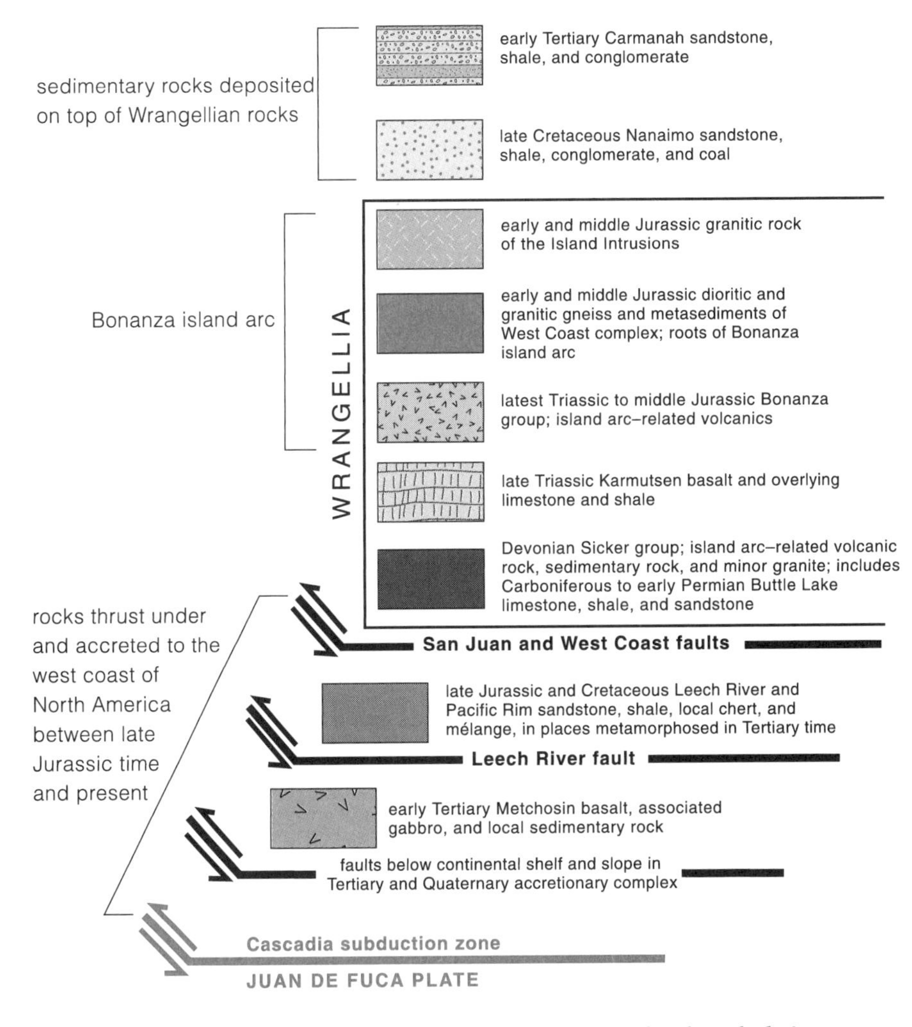

Rocks on the geological map of Vancouver Island and their stratigraphic and structural relationships to one another.

are gently tilted and warped, and only slightly altered by low-grade metamorphism, although metamorphic rocks derived from Wrangellia form a narrow, discontinuous belt in western and southern Vancouver Island.

Sicker group. The oldest-known part of Wrangellia on Vancouver Island is made of volcanic and sedimentary rocks of the Sicker group, exposed from the southeastern to the west-central parts of the island. Although few, if any, fossils are known from these rocks, we know they are middle Devonian or older because small granitic bodies that intrude the group near Duncan and on Salt Spring Island are about 380 million years old, so the rocks they intrude must be older. Sicker group volcanic rocks near Buttle Lake in central Vancouver Island yield similar isotopic ages. The volcanic rocks include lava with lesser amounts of fragmental volcanic rock and ash, together with some bedded chert and argillite, and their nature suggests they formed in an island arc.

Buttle Lake strata. Overlying the Sicker group are Carboniferous to Permian Buttle Lake strata that contain a much higher percentage of sedimentary rock—sandstone, shale, and limestone—than the older rocks. The sandstone contains grains eroded from volcanic-rich source areas. The limestone ranges from thin lenses to cliff-forming sections up to 300 metres thick that in places contain fossils, including brachiopods (marine invertebrates with bivalve shells), corals, and bryozoans (marine invertebrates that form branched or mossy colonies).

Karmutsen basalt. Triassic basalt, in places probably more than 6 kilometres thick, overlies the Paleozoic strata. The lava, called the Karmutsen basalt on Vancouver Island, is the most distinctive rock unit in all parts of Wrangellia from here to southern Alaska and is by far the most widespread unit exposed on the island. The basalt passes upward into limestone containing ammonite, clam, and coral fossils of late Triassic age. The origin of the enormous outpouring of Karmutsen basalt is not clear, but it may have taken place above a *mantle plume*, a localized upwelling of very hot rocks in the Earth's mantle.

Rocks of the Bonanza island arc. The volcanic rocks of the Bonanza group, the Island Intrusions, and the West Coast complex represent, respectively, surface, intermediate, and deep parts of the Bonanza arc. This was an island arc built on and through the older Wrangellian rocks in latest Triassic to middle Jurassic time, about 203 to 172 million years ago. Limy to mud-rich marine sediments above the Karmutsen basalt grade upward into volcanic flows and tuffs of the Bonanza group, which range in composition from basalt to rhyolite. The Island Intrusions, extensive granodiorite and quartz diorite bodies of early to middle Jurassic age, have a similar age range to most of the Bonanza volcanics. The West Coast complex, which extends from Victoria in the south to Brooks Peninsula on the northwest coast of Vancouver Island, consists of granitic and dioritic gneiss associated with amphibolite, a metamorphic sedimentary rock. The granitic rocks are of the same age as the Island Intrusions but their association with metamorphic rocks shows they cooled much deeper in the crust.

Where in the World Has Wrangellia Been?

Tracking the geographic position of Wrangellia over time has been, and remains, a challenging task. Fossils and paleomagnetic data, plus the presence of an accretionary complex to the east in the Intermontane Belt, suggest that most of the rocks on Vancouver Island formed a long way from rocks to the east in the continental interior and in most other parts of the Cordillera, but just how far away and in what direction remain uncertain. As in today's world, the position of the continents and variations in climate with latitude controlled the distribution of ancient plants and animals. Early Permian fossils in the Buttle Lake limestone on Vancouver Island are similar to fossils of the same age in the Ural Mountains in Russia, and early Permian fossils in Wrangellian rocks in southern Alaska are similar to fossils of the same age in the Canadian Arctic. In early Permian time, both the Urals and the Canadian Arctic probably were at latitudes between 30 and 60 degrees north, whereas the ancient continental margin of southwestern Canada, today in the southern Foreland Belt, was closer to 15 degrees north. By contrast, the Triassic and early Jurassic fossils on Vancouver Island are more similar to those of similar age in the southwestern United States than they are to those in southwestern Canada. This contrast suggests that Wrangellia moved well over 1,000 kilometres south between early Permian and late Triassic time.

The paleomagnetic record preserved in the Karmutsen basalt on Vancouver Island potentially can give us the latitude of the basalt in Triassic time. In today's Earth, the lines of force of the magnetic field in polar regions are steep to vertical with respect to Earth's horizontal surface, and the dip (or inclination) of the magnetic field decreases systematically from the pole toward the equator, where the field approaches horizontal. When lava cools below about 500 degrees Celsius, it freezes in a remnant magnetism that records the direction of the Earth's magnetic field at the time of cooling. Thus, lava flows laid down horizontally near the poles have a large angle between the plane of the flow and the magnetic inclination, whereas lava flows that cool near the equator have a small angle. Measurement of the *paleomagnetic inclination* in old lava potentially can give us some idea of the latitude at which the lava cooled long ago.

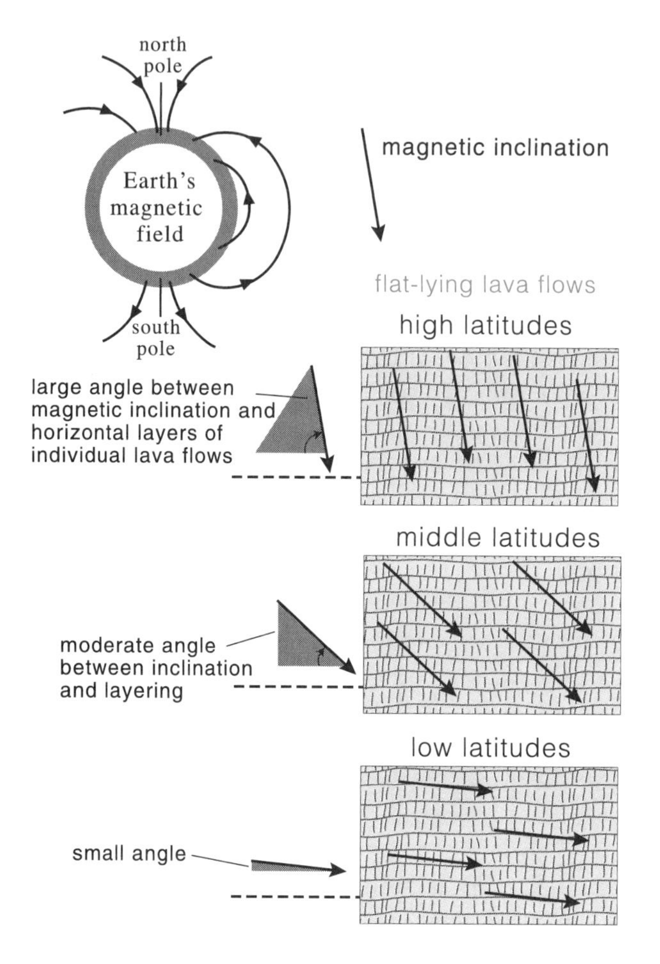

Magnetic inclinations preserved in horizontal lava flows at high, middle, and low latitudes dip at steep, moderate, and shallow angles with respect to the flows.

The ancient magnetic record preserved in the late Triassic basalt of Wrangellia on Vancouver Island shows it cooled at a latitude not too dissimilar from that of rocks of the same age on the old continent. The Triassic and early Jurassic fossils on Vancouver Island do not support the paleomagnetic record of the Karmutsen basalt as they are more similar to those that lived farther to the south on the North American continent.

Other paleomagnetic studies of Cretaceous sedimentary rocks deposited both on top of Wrangellia and on older mainland rocks compound the problem and have led to a big scientific controversy surrounding what is known as the Baja B.C. hypothesis. The controversy centres on the vastly different amounts of displacement suggested by paleomagnetic studies on the one hand, and by the geological and fossil record on the other. We know that Vancouver Island was near the Coast Belt by late Cretaceous time because most sedimentary rocks of the island's Nanaimo group were derived from newly uplifted eroding mountains in the Coast Belt. However, several studies on the paleomagnetism of the 90- to 65-million-year-old Nanaimo sedimentary rocks (as well as rocks within the Coast Belt) suggest that these rocks were deposited well to the south, and after about 70 million years ago these rocks moved northward about 2,500 kilometres relative to rocks in the continental interior. Other paleomagnetic studies in the southern Coast Belt and farther north in the Canadian Cordillera also support the hypothesis that in latest Cretaceous to earliest Tertiary time large parts of western British Columbia moved northward from latitudes that correspond to those of late Cretaceous Baja California. This displaced region has been called Baja British Columbia. It is argued that in latest Cretaceous to earliest Tertiary time, western British Columbia was temporarily attached to northward-moving offshore oceanic plates, much as those parts of present-day California southwest of the San Andreas fault move with the northward-moving Pacific Plate.

Geological and fossil evidence, however, does not support displacement of this magnitude. Several big strike-slip faults are recognized in the interior of British Columbia, such as the Fraser fault in the Coast Belt, and were active in latest Cretaceous to early Tertiary time. Rocks on the western sides of the faults moved northward relative to those on the eastern sides, in agreement with the *sense* of displacement required by the paleomagnetic studies, but the amount of displacement, measured by the offset of rock units that straddle the faults, is far less, with cumulative movement on all these faults less than 1,000 kilometres and perhaps closer to 500 kilometres. This would place the Nanaimo rocks north, instead of south, of the latitude of California in late Cretaceous time. In addition, late Cretaceous fossils in the Nanaimo group appear to be cool-water forms, and there appears to be a gradual change of Cretaceous faunas from cool-water to tropical fossils southward along the west coast of North America between British Columbia and Mexico, much as we see today. Finally, a few detrital zircon sand grains from the Nanaimo group give isotopic dates of about 2 billion years. These ages are common in the western Canadian Shield, but not in the continental interior near Baja California, where the nearest Precambrian rocks are about 1 billion years old.

Regardless of how far northward Wrangellia and parts of the Coast Belt travelled between late Cretaceous and early Tertiary time, they evidently arrived near their present latitude by about 50 million years ago. Amounts of offset across Tertiary faults and paleomagnetic studies on Tertiary rocks in the interior of British Columbia that can be traced into the Coast Belt finally agree; there is no significant Tertiary latitudinal displacement. The difficulty geologists have had reconstructing the geographic position of the geologically young Nanaimo rocks at the time they were deposited is sobering: if we are uncertain where some of the rocks in this region were 70 million years ago, how certain can we be about where they were 200 million years ago, or even further back in time?

More Suspect Terranes

Two fault-bound slices of bedrock in southern Vancouver Island formed offshore at an unknown location and were accreted to Wrangellia in latest Mesozoic and Tertiary time. These narrow strips are nowhere more than 15 kilometres wide. The inner of the two strips reappears along the coast at a few places, such as between Ucluelet and Tofino on the west side of central Vancouver Island, and on the tip of Brooks Peninsula to the north.

The inner slice is composed of the Leech River schist on southern Vancouver Island and the Pacific Rim complex on western Vancouver Island. The San Juan fault and the West Coast fault are large reverse faults that separate the Leech River schist and Pacific Rim rocks, respectively, from Wrangellian rocks. The rocks originated as mudstone, sandstone, basaltic lava, and minor conglomerate and chert and probably were deposited on the late Jurassic and Cretaceous continental slope and the adjacent ocean floor. Some were probably scraped off the ocean floor during late Mesozoic subduction and accreted to Wrangellia. The Leech River schist was originally sedimentary rock that was metamorphosed about 40 million years ago.

In their nature and age, the Leech River and Pacific Rim rocks are like some rocks in the San Juan Islands of Washington, and in the western foothills of the North Cascades on the Washington mainland. Elsewhere along the Pacific Coast of North America, similar but far more extensive rocks of mainly late Mesozoic age that are identified as accretionary complexes are the Franciscan complex in California and the Chugach terrane in southern Alaska.

The outermost and youngest fault slice is exposed on southernmost Vancouver Island, where it is separated from the Leech River schist by the prominent Leech River fault. The eastern end of the fault passes near Victoria, is submerged beneath the eastern Juan de Fuca Strait, and may link up with the still-active Devils Mountain fault system in northwestern Washington. The slice consists of Metchosin basaltic lava and associated gabbro of Eocene age and minor sedimentary rocks. The lava contains pillow structures, indicating it formed underwater, probably on the ocean floor. Rocks of similar age and origin, called the Crescent basalt, rim the north, east, and south sides of the Olympic Mountains in Washington on the south side of the Juan de Fuca Strait. There, they were thrust-faulted over still younger rocks exposed in the uplifted core of the Olympics.

Exploratory drilling for oil and gas in the 1960s and several more-recent offshore seismic reflection and refraction studies carried out for scientific research show that young sedimentary rocks, of similar age and origin to those exposed in the uplifted core of the Olympic Mountains, underlie much of the continental shelf and slope southwest of Vancouver Island. Seismic images of these rocks indicate that they have been folded and thrust to the southwest, and the outermost and lowermost rocks belong to the active accretionary complex above the subducting Juan de Fuca Plate. Sand and mud of Tertiary and Quaternary age have been deposited in basins on top of older, inactive parts of the accretionary complex.

The faults separating the different terranes on southern Vancouver Island are relatively straight and continuous at Earth's surface, and their surface traces are marked by big, east-west-oriented valleys. Seismic reflection studies across Vancouver Island indicate that the faults dip at angles of 30 to 40 degrees to the northeast beneath Wrangellian rocks and penetrate the entire crust of the westernmost North American Plate. Thus, Wrangellian rocks on Vancouver Island appear to be the upper part of an enormous stack of thrust- and reverse-fault slices that makes up the entire crust in the region. Stacked in descending order and decreasing age below Wrangellia are Leech River and Pacific Rim rocks, Eocene Metchosin basalt, rocks in the offshore Tertiary and Quaternary accretionary complex (partly exposed on land in the core of the Olympic Mountains), and lowermost, the presently subducting Juan de Fuca oceanic plate.

Structure of the crust and upper mantle of Vancouver Island and adjoining regions.

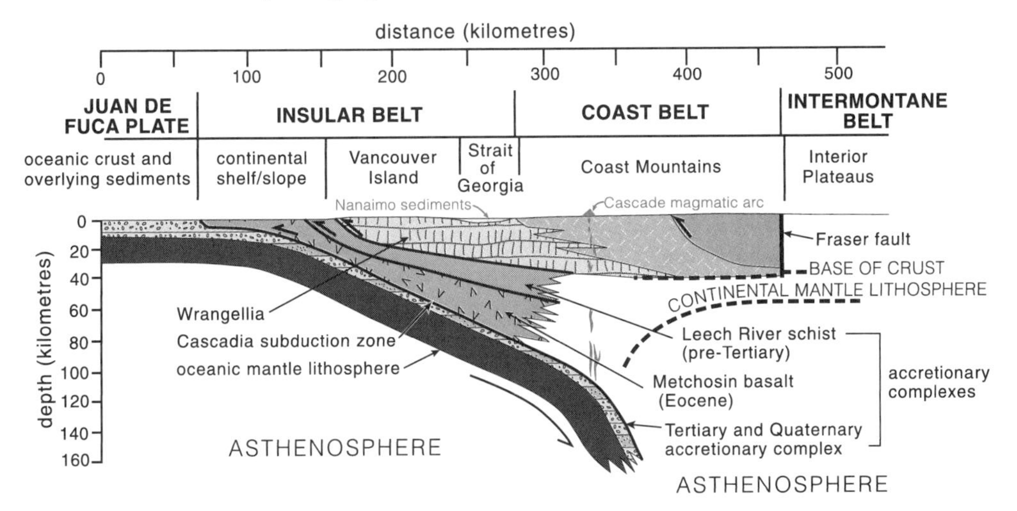

Cascadia Subduction Zone and the "Big One"

Two thousand metres below the ocean's surface about 100 to 200 kilometres west of Vancouver Island, the Juan de Fuca Plate is subducting below the North American Plate at the Cascadia subduction zone as the two plates continue to converge at a rate of 4.0 to 4.6 centimetres per year. Active deformation related to the subduction zone is of direct concern to all living in the coastal regions of southwestern British Columbia, Washington, and Oregon. The small Juan de Fuca Plate is nowhere more than 6 million years old, so its basalt is still warm and relatively buoyant, making it reluctant to sink back into the mantle. The nature of this subduction zone is more like a giant thrust fault—a megathrust—than a buoyancy-controlled structure.

Major earthquakes in the past evidently occurred when the Cascadia subduction zone periodically locked, building up strain in the crust, causing shortening, and bulging up the upper plate. The eventual catastrophic rupture of the locked subduction zone triggered very large earthquakes of

A megathrust earthquake occurs when a locked section of a subducting plate ruptures.

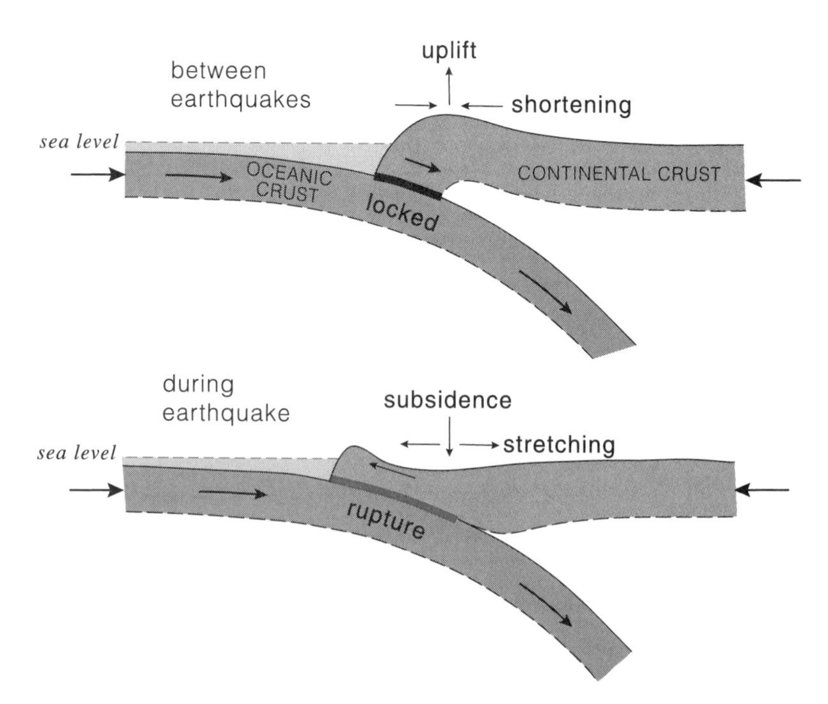

a type called a *megathrust earthquake*. The site of the ruptures was under the sea between the toe of the continental slope and the shore of Vancouver Island.

No megathrust earthquake has occurred in the short time that people of European origins have lived in the region. However, aboriginal oral histories and geological evidence of old earthquakes indicate that megathrust earthquakes occur periodically, at intervals of a few centuries. The last was about three hundred years ago. Studies of current and past high-water levels in tidal marshes in coastal regions show evidence of *tsunamis*, large earthquake-generated waves, and "drowned forests" along the coast formed when the land dropped suddenly and trees were killed by an incursion of seawater that immersed their roots. Ongoing tide-gauge measurements and repeated precise leveling suggest that current horizontal shortening, mostly on southern Vancouver Island, approaches 10 millimetres per year, and vertical uplift is at a rate of about 1 millimetre per year. Such rates cannot be sustained for more than a few hundred years without a major change. Our problem is predicting just when the next change—the so-called Big One—will happen!

Georgia Depression and Nanaimo Sediments

The Georgia Depression, a northwest-southeast trough whose axis lies along the Strait of Georgia, encompasses not only the strait but also the adjacent lowlands between the mountains of Vancouver Island and the Coast Mountains, and eastward links up with the Fraser Lowland south and southeast of the city of Vancouver. It is but one segment of the roughly 2,000-kilometre-long topographic depression that extends northward from the Willamette Valley in western Oregon to the straits and lowlands separating the islands of southeastern Alaska from the mainland. The maximum seawater depth in the Strait of Georgia is just over 400 metres and is located about 20 kilometres northeast of Nanaimo. Bedrock in this area is another 300 metres below the unconsolidated sediment on the seafloor.

Plate tectonic activity has been a major control on the origin of the trough since late Cretaceous time. Between 85 and 65 million years ago, southwest-directed thrust faults and reverse faults in the Coast Belt that were associated with mountain building probably loaded and pushed down the crust. Erosion accompanied the uplift in the Coast Belt, and sand, gravel, and mud accumulated in the trough in front of the rising mountains, forming much of the Nanaimo group. Although conglomerate at the base of the Nanaimo group was derived from underlying Wrangellian rocks, most of the other sediment came from the uplifted Coast Belt to

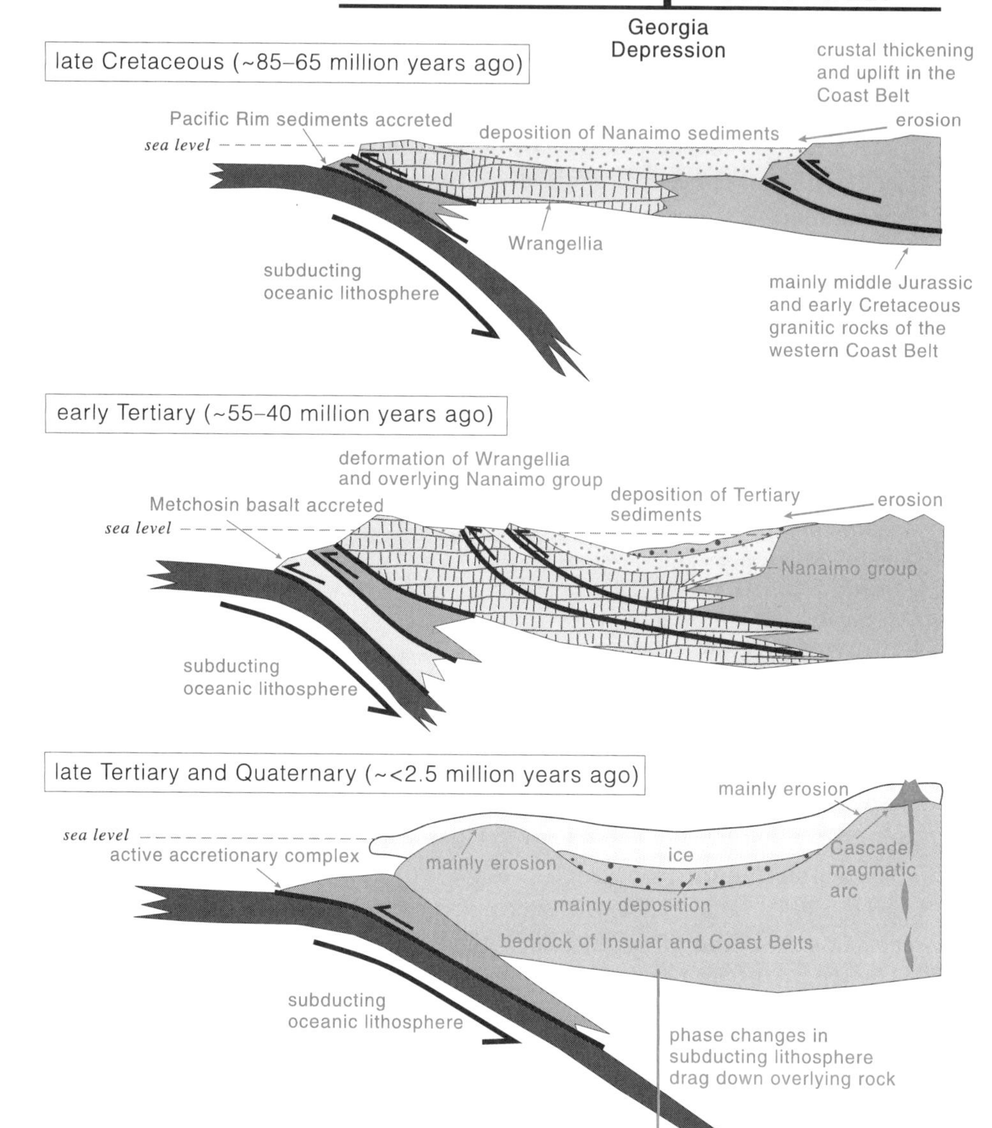

Sections across Vancouver Island and the Strait of Georgia in late Cretaceous, early Tertiary, and Quaternary times show factors that may have influenced the formation of the Georgia Depression.

the east and southeast: stream current directions measured in sandstone beds indicate the water flowed from the east, and the composition of most sand grains match those of rocks in the Coast Mountains. The sediments were deposited across Wrangellia and form many of the lowlands on the eastern side of Vancouver Island, which may have been mostly low-lying in late Cretaceous time. Most of these rocks are probably close to the elevations at which they were originally deposited. However, in the Vancouver Island Ranges, patches of Nanaimo rocks have been raised as much as 1,500 metres above their original level, and in the Georgia Depression, late Cretaceous and early Tertiary strata related to the Nanaimo sediments, measured in a borehole near Tsawwassen on the mainland, are as deep as 4 kilometres below sea level.

In early Tertiary time, oceanic rocks such as the Metchosin basalt were stuffed beneath western Vancouver Island and probably raised it relative to the eastern side and also tilted and faulted the rocks, breaking the sheet of Nanaimo rocks, which had once been more continuous, into discrete patches. The regional tilting is shown by the surface exposures of the West Coast complex—the most deeply buried rocks in Wrangellia—in western parts of Vancouver Island. The rocks of the Nanaimo group are exposed on land mainly on the eastern side of the island and Tertiary strata are found on the east side of the Georgia Depression.

Current earthquake activity along the depression, mostly at depths of less than 20 kilometres, testifies to continuing tectonic activity. In a few places faults offset sediments of Quaternary age in submerged parts of the depression. One hypothesis proposes that density changes in the subducting Juan de Fuca Plate, whose top is about 70 kilometres below the Strait of Georgia, may contribute to the sinking of the Georgia Depression. Some minerals under great pressure undergo *phase changes*, in which their crystal structures become more compact and dense. The increased density of minerals in the subducting plate below the Georgia Depression may drag down the overlying rocks.

Nontectonic factors have also contributed to the formation of the depression. The sedimentary rocks on its floor are soft and were far more readily eroded by glacial ice during the ice ages than the hard, resistant granitic, volcanic, and metamorphic rocks around the rim of the depression. In addition to erosion by ice in the last 2 million years or so, considerable sediment has accumulated in the depression, adding to the load. Quaternary sedimentation in and around the depression may be only as much as 200 to 300 metres thick, but the Tertiary and late Cretaceous sedimentary rocks within it are over 4 kilometres thick.

Quaternary Ice Ages

Erosion dominated the geologic record of the Quaternary ice ages in the mountainous areas of Vancouver Island; deposition dominated the record in the lowlands. Within the mountains, narrow valley walls confined the glaciers, or "streams of ice," and they moved relatively fast. Once the ice spread out over the open lowlands, it assumed a more leisurely pace. Erosion did not cease entirely in the lowlands, but deposits laid down directly by ice, or by streams issuing from it, had a better chance of survival. In lowland areas, even the less voluminous deposits of streams, lakes, swamps, seashores, and marine embayments that formed during relatively warm interglacial periods had a reasonable chance of being preserved.

Preservation does not guarantee exposure. The deposits of both the ice ages and the nonglacial intervals between are uncemented sand, gravel, and till that readily support the growth of vegetation. Only deposits in sites that have been actively eroded and thus exposed within the past decade or two reveal the sequence of beds and the history that they record. Sea cliffs, river cutbanks, and recent gravel pits are our principal sources of information, and many sites that revealed the record a few years ago are now covered.

Naturally, the best record is that of the last ice age, which reached its climax almost 15,000 years ago. Earlier ice ages are known, going back in time to very latest Tertiary time—about 2.5 million years ago—but are difficult to date, and older deposits are few and far between because they were destroyed or covered by younger events. Just how many ice ages occurred in this region is unknown. Possibly there were as many as twenty, but the current "best bet" is eight, and a good record exists of at least two.

The second-to-last glaciation is recorded in places by scattered occurrences of till, that mess of stones, sand, silt, and clay that was deposited directly from glacial ice, and at one time was called "boulder clay." Elsewhere, an erosion surface marks this penultimate glaciation.

The record of the last interglacial period is moderately well-known and is preserved in the Cowichan Head formation, a till-like deposit near Victoria that contains fossil shells and casts of marine organisms, notably clams. This deposit probably was deposited in salt water close to a receding glacier. Bedding within it becomes more obvious higher in the formation where marine fossils are missing. Instead, there are the remains of land plants and iron-stained sand is common, recording either the withdrawal of the sea or the rise of the land. Pollen preserved in these sediments is dominated by that of scrub pine and spruce, suggesting a forested land and a nonglacial climate somewhat like that of the Queen Charlotte Islands today. Isotopic dates from carbon in preserved organic material place this interglacial period from more than 40,000 to about 25,000 years ago.

Sediments of the Last Glaciation

Pale, nearly barren crossbedded sands, which may form striking white bluffs where nonvegetated, overlie the Cowichan Head sediments near Victoria and are found in many places around the Strait of Georgia. The sands are at elevations ranging mostly from about 100 metres above sea level down to about 10 metres above the present high-tide mark, although in places they are below mean sea level. A good example forms the sea cliffs below the University of British Columbia on Point Grey in westernmost Vancouver. The deposit was first named the Quadra formation after this sea cliff in the Vancouver-Quadra electoral district, which in turn was named in honor of the historic 1772 meeting near this cliff of the British and Spanish explorers Captains Vancouver and Quadra. Later, a better example about 175 kilometres to the northwest at the other end of the Strait of Georgia was chosen as the *type locality*—the location where a rock unit

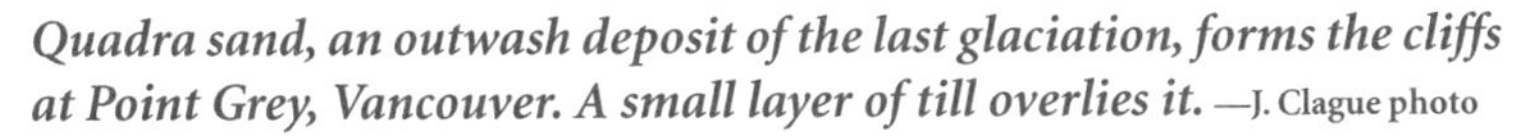

Quadra sand, an outwash deposit of the last glaciation, forms the cliffs at Point Grey, Vancouver. A small layer of till overlies it. —J. Clague photo

is described in detail and provides the rock's formal name. Fortuitously the new type locality just happened to be on Quadra Island, so the formation kept its original name.

The Quadra sand, over nearly all its lateral extent and for its full thickness, is composed dominantly of white feldspar, clear quartz, and lesser metamorphic minerals derived from erosion of the Coast Mountains. Only in the westernmost occurrences of the Quadra sand is there much material derived from the bedrock of Vancouver Island. Streams carried the sand to its present resting places. Pollen from a few peat layers indicates that the climate during deposition was cooler than that of the present climate, and cooler than that of the underlying Cowichan Head sediments. Isotopic dates of carbon in rare wood fragments range from 28,800 to 20,000 years. The oldest measured dates are from the northwest end of the Strait of Georgia and the youngest from the southeast end. Moreover, dates from any one locality show a range of only a few hundred years.

The Quadra sand occurs as numerous isolated patches on both sides of the entire length of the Strait of Georgia, on some islands within the strait, and also in the Puget Lowland of northwestern Washington. These outcrops seem to suggest there was once a continuous deposit more than 250 kilometres long, up to 40 kilometres wide and 100 metres thick, with a nearly level top surface that was close to the present 100-metre contour. How could this tremendous volume of sand be deposited in a mere 9,000 years, and why do radiocarbon dates from near the base of the Quadra sand become progressively younger from northwest to southeast?

It is possible that the Quadra sand never existed as a single continuous deposit. One hypothesis is that streams laid down the sand about 29,000 years ago in the northern part of the strait in front of the slowly advancing glaciers, but then ice plowed much of it up and streams issuing from the ice deposited it farther to the southeast. Like a giant bulldozer, the advancing ice attacked the new deposit and moved it still farther to the southeast over a period of several thousand years. This bulldozer concept reduces the problem of sediment volume and explains why the sand gets progressively younger to the southeast. The streams that carried the sand originated from the advancing glaciers instead of coming directly from the Coast Mountains. The glaciers were fed for the most part by snowfall on the high Coast Mountains, rather than on the lower mountains of Vancouver Island, and the debris they brought to the lowland was mostly eroded from the granitic heart of the mountains.

The ice advancing down the Strait of Georgia reached the latitude of Vancouver about 21,000 years ago, where it paused and briefly melted back. Forests regenerated on the newly exposed ground and record a cool midmountain to subalpine environment with an inland rather than coastal character. No longer was there a Strait of Georgia nearby to modify the harsh climate; because sea level had fallen, the nearest large body of open water was the Pacific Ocean, west of Vancouver Island. This condition persisted for 2,000 to 3,000 years. Then, from 17,000 to 15,000 years ago, the ice advanced along Puget Sound, well into Washington. An equally rapid retreat followed, and the southern Strait of Georgia was exposed again about 13,000 years ago. In places, a layer of till as much as 25 metres thick records this last advance and retreat of ice; elsewhere an erosion surface on the Quadra sand or, at higher levels, on earlier glacial sediments or bedrock records the ice's advance and retreat. Beach and delta deposits of sand and gravel, with occasional marine shells or casts, commonly overlie the till. As with the earlier penultimate glaciation, an incursion of the sea immediately followed the retreat of the ice. However, a marine incursion did not precede the ice invasion. To better understand this disparity, we must consider the ups and downs of the seashore caused by the growth and decay of glaciers.

Rising and Falling Sea Levels

Great changes in sea level during ice ages are both real and apparent. Such changes are of particular concern today, with global warming threatening to melt the polar ice caps and with so many major population centres located near sea level. Vast quantities of fresh water in the form of glaciers or ice caps are stored on land during an ice age. During the last ice age, nearly all of Canada and parts of the northern United States were buried by ice as much as 3 kilometres thick. A somewhat smaller volume of ice covered land in northwestern Europe. The volume of frozen water during the ice age was large enough to account for a 50- to 75-metre drop in sea level. Enough ice remains in Antarctica and Greenland to produce a rise in sea level of about 60 metres if it all melted. The rise and fall of sea level as continental ice sheets decay and grow is referred to as *eustatic*.

Land sinks under the weight of a large ice cap, producing an apparent rise in sea level. As the ice cap melts, the land rebounds and sea level appears to fall. The regional subsidence of the land under a load and its rebound on unloading is referred to as *isostatic*. The phenomenon provides good evidence that the nonrigid, underlying asthenospheric mantle allows material that was displaced laterally by the load of ice to flow back when the load is removed.

The amount of depression of the land is related to the size of the load of ice. Imagine a floating barge loaded with gravel. The greater the load is, the lower the barge settles in the water. A more realistic analogy has a central barge tied tightly to surrounding barges, fore, aft, and at the sides. When the central barge is loaded and settles lower in the water, it is partly supported by the adjacent barges, which tilt toward the central barge. The effect of the load is thus felt well beyond the limits of the load. In isostatic loading of glacial ice, the depression extends some scores of kilometres beyond the edges of the ice cap and forms a dish-shaped hollow surrounded by an almost imperceptible bulge. A still more realistic model would have the barges floating on a layer of warm tar rather than on water. The settling of the barge under its load would then occur slowly as the tar is gradually squeezed out sideways, and in this case a full adjustment to loading would require a substantial time interval.

This delayed response to loading and unloading is particularly noticeable in eastern Canada and in Scandinavia, where the melting of the ice caps took place mostly between 18,000 and 8,000 years ago, but where rebound is still going on, albeit at a reduced rate. Hudson Bay is a remnant of the depression. For these two regions it is estimated that the first half of the total rebound required 5,000 years, half of the remaining amount of rebound required another 5,000 years, and so on, until rebound becomes vanishingly small.

In British Columbia the response was much faster. Half the recovery was achieved in a few hundred years, and isostatic rebound was essentially completed some 9,000 to 8,000 years ago. The explanation for this fast rate of recovery is not known for certain but probably reflects the state of the asthenosphere that underlies the relatively rigid lithosphere. It seems likely that here in the active plate margin, the asthenosphere is warmer and flows more readily, or else is closer to the surface, than under eastern Canada or Scandinavia, which are underlain by cold Precambrian shields with thick lithospheres.

Different combinations of eustatic and isostatic shifts in sea level are recorded along the southern British Columbia coast. On the western side of Vancouver Island, where the ice load was slight and only lasted a short time, eustatic drop in sea level dominated during glacial times. In the Georgia Depression, where the ice load was much greater and persisted for thousands of years, isostatic effects dominated. During the advancing stage of the last ice sheet, 28,000 to 15,000 years ago, depression of the land was less than it was at the end of the ice advance because of the delay in the isostatic response. Hence, the nonmarine Quadra sand and the overlying till were laid down before flooding by the sea took place. In addition, the accumulation of ice reached a global maximum about 18,000 years ago, so the associated eustatic fall in sea level was at its greatest during the advance of ice in the Georgia Depression. At the time the ice withdrew, about 13,000 years ago, the land was depressed by about 300 metres. At the same time, the eustatic lowering of sea level was almost 100 metres, so the land stood about 200 metres lower with respect to the sea level of its time and seawater flooded the depression as the glacier withdrew. Combined, the two effects account for the lack of marine beds directly under the till of the last glaciation and their presence above it. The earlier glaciation probably followed a similar pattern.

Isostatic rebound was essentially complete on the British Columbian coast by about 8,000 years ago. At that time there was still a major ice cap over Hudson Bay. A 7,600-year-old buried peat bed, discovered during drilling of a bridge foundation along the lower Fraser River, records a sea level 13 metres below the present level. As the ice over Hudson Bay melted, eustatic sea level rose and reached its present position by about 5,500 years ago; it has shifted very little since.

HIGHWAY 1
Victoria—Nanaimo
112 kilometres

Glacially Sculptured Bedrock in Victoria

Knobs of hard bedrock crop out throughout Victoria and adjacent areas and consist for the most part of dark, intrusive rocks called *gabbro* and *diorite*, together with metamorphosed volcanic rock called *greenstone* and some pale granitic dikes. All belong to the West Coast complex, the deep root of the Jurassic Bonanza arc, uplifted and exposed by later erosion. All these rocks are hard and resistant to erosion, but some, notably the pale granitic dikes, withstood abrasion by overriding glaciers better than others.

Southward-moving ice, which reached a maximum thickness of about 1,000 metres in this area, passed over the site of Victoria about 14,000 years ago and retreated from the Puget Sound region about 13,500 years ago. Rock fragments embedded in the base of the ice produced scratches in the bedrock called *striae* and broader troughs called *flutes*, which record the direction of ice flow. The northern, upstream faces of the rocky knobs display the most abrasion and are distinctly rounded. The downstream faces, which were subjected to plucking by the ice, are more ragged.

A convenient place to see the sculpted bedrock is in Confederation Square Park, directly northwest of the Parliament Buildings. Fluting is well displayed there, and pale granitic dikes stand in relief over the weaker, enclosing rock and protect it in places on the downstream side. In addition, a

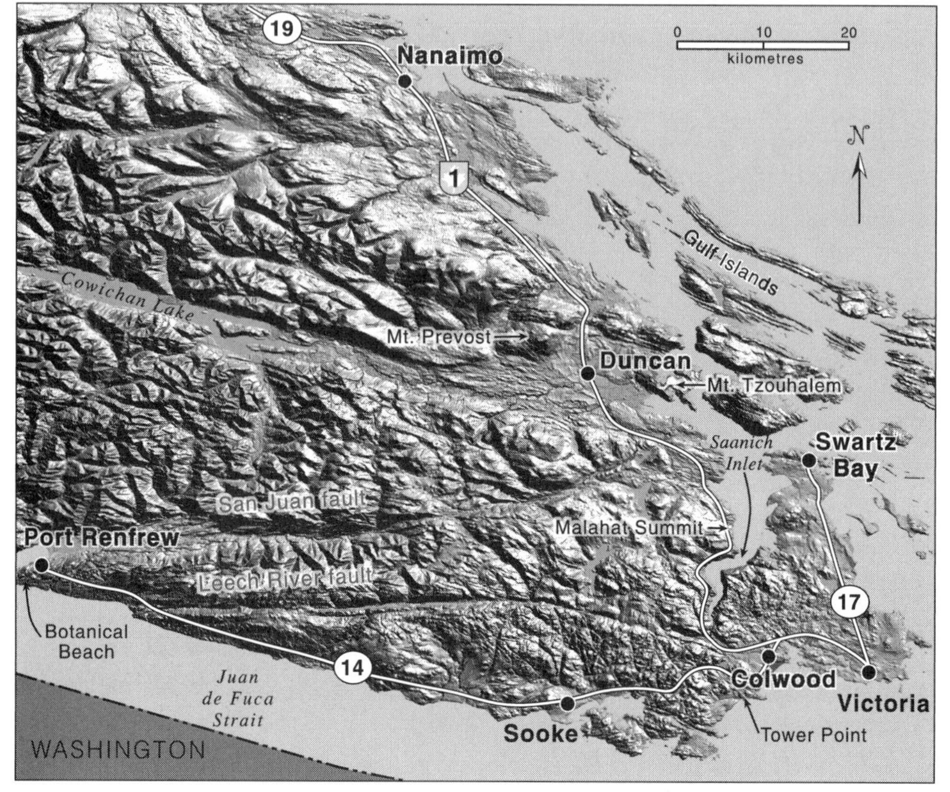

Shaded-relief image of topography of southern Vancouver Island. The two prominent east-west-trending valleys in the south follow the traces of the San Juan and Leech River faults. A third prominent valley containing Cowichan Lake also follows fault traces.
—Image constructed using Canadian Digital Elevation Data obtained from GeoBase

few of the sculptured forms here may have been engraved not by dirty ice but by sediment-bearing water flowing between the rock and the ice.

A second site illustrating the fine details of sculpture by ice is near the high-tide level on Finlayson Point, in the southwest corner of Beacon Hill Park. Here, delicate scratches are well preserved on blocks of fine-grained greenstone that are embedded in the granitic rock.

Marine Clay and the Empress Hotel

Much of the topography of urbanized Victoria consists of broad saucer-shaped basins with ever-decreasing slopes leading down from bedrock highs. Marine clay or mud blankets the smooth surfaces of these basins up to about the 60-metre elevation contour. During the time when the mud was deposited, at the end of the last ice age, the mud probably moved down

Glacially striated bedrock is directly northwest of the Parliament Buildings in Victoria.

the slopes to form deposits that are thicker in the centre of the basins than around the margins. At elevations between about 60 and 80 metres above present sea level, which was close to the sea surface at the end of the ice age, ancient wave erosion probably removed the clay, leaving only sand or gravel. The concentration of mud toward the centre of the basins makes undesirable building foundations, because it can be compressed under heavy loads. The marine clay contains scattered fossil shells of small cold-water clams, but near the surface and within the weathered zone of modern soils, soil acids and rainwater have leached away the calcium carbonate, leaving behind only shell casts.

The Empress Hotel shares with the nearby Parliament Buildings a commanding position facing the Inner Harbour, through which most visitors to Victoria had to pass before the rise of highway and air travel. "What a site for a major hotel," thought the president of the Canadian Pacific Railway. "What a headache!" thought his chief engineer. The north end of the hotel rests on shallow bedrock, but dredge spoil, bay mud, and marine clay underlies the south end of the structure. The hotel complex was built in stages, with the foundations laid from November 1904 to March 1905. The central part of the hotel opened in January 1908. By June of that year, the foundations of the south wing had sunk 9 centimetres below those of the north wing. By 1912, when construction of the south wing was begun, the settlement of the hotel was so obvious that engineers leveled it at monthly intervals until 1915 and subsequently at approximately one-year intervals until 1969. The north end of the hotel has settled little or not at all, but the southwest and southeast corners have sunk by as much as 73 centimetres in the fifty-seven-year interval. When this amount is combined with the unmeasured but estimated subsidence between 1904 and 1912, the total amount of settling by 1970 probably exceeded 1 metre. Remedial efforts have reduced the rate of settlement, and the building survives, albeit with one end lower than the other.

Colwood Plain

Colwood lies about 10 kilometres west of Victoria on a broad 12-square-kilometre plain about 80 metres above present sea level. *Kettles*, irregular to conical depressions as much as 30 metres deep, interrupt the plain's surface, especially at the foot of the hills on its southwestern margin. Most of the kettles hold no water, a testimony to the good drainage of the gravelly substratum. Many kettles are forested, and most are hidden from easy inspection or are modified by human activities. During the last glaciation, when ice blocks buried in the glacial outwash slowly melted, the gravel cover and its surface collapsed, forming the kettles. Geologists call such a surface a *pitted outwash plain.*

A dry meandering river channel heads at Langford Lake between the top of Saanich Inlet to the north and the Juan de Fuca Strait on the southeast, interrupting the plain. The channel extends 3 kilometres to the east as far as the Royal Colwood golf course and records drainage of glacial meltwater at a time when ice occupied Saanich Inlet and barred drainage to the north. The outflow followed the lowest available course, which was through the Colwood gap. The delta of this fossil river is at the eastern end of the golf course and formed where the river drained into the ocean; the former level of the sea was 75 metres above the present shoreline because the land was still depressed under the dwindling load of ice.

The southern part of the pitted outwash plain, crossed on the way to Tower Point, has been extensively quarried in recent years to provide construction materials, not only for Victoria and vicinity, but also to meet a substantial part of the demand from Vancouver. The artificial excavations reveal, in the alternation of gravel and sand, rapid variations in carrying capacity of the glacial streams. Beds inclined at 20 to 25 degrees mark the former site of the underwater delta front, down which the sand and gravel spilled as the front became oversteepened by material deposited on the delta top.

Metchosin Pillow Lava at Tower Point

Pillow lava is common throughout the Metchosin basalt of Eocene age that forms much of the fault slice south of the Leech River fault on southernmost Vancouver Island. The Metchosin basalt includes some of the youngest rock exposed on land in British Columbia that formed on the ocean floor and later was accreted to the continent. A good place to see the pillow lava is on the beach near Tower Point. From Highway 1A at Colwood drive south on Sooke Road (Highway 14), turn left on Metchosin Road, and continue for about 5 kilometres as far as Duke Road, and thence about 1 kilometre to the entrance to Witty's Lagoon Regional Park, Tower Point. Use the parking lot off Olympic View Drive and take the trail from the parking lot across the field and down to the shore of Juan de Fuca Strait. We suggest visiting at medium or low tide so you can walk along the shoreline and see the waterfront exposures.

In this area, lava accumulated into what looks like a stack of pillows or sandbags rather than as the extensive sheetlike forms typical of on-

Pillows in Metchosin basalt of Eocene age at Witty's Lagoon Regional Park, Tower Point. Note the convex upper surfaces of the pillows and some concave lower surfaces where the pillow has molded to underlying pillows; such features in ancient deformed lavas can be used to indicate the original top of the lava.

land basaltic flows. Closer inspection shows that some lava cooled and froze into tubes that intermittently pinch and swell like a string of giant sausages. These curious structures perplexed early geologists, although the common association of pillow lava with quiet-water sediments suggested to them that the lava was deposited in water. In 1910, people saw lava from a South Pacific volcano entering the ocean, where it formed bulbs with flexible frozen skins and still-molten cores that ballooned into sacks before solidifying. Some of the pillows shattered, others rolled intact into deeper water, and still others remained attached. Not all lava entering water necessarily forms pillows; some may explode into angular fragments to form *pillow breccia* and other lava cools as it flows without forming pillows.

Individual pillows tend to be elongate in the horizontal plane at the time of their formation and mold themselves to the shape of the underlying lava on which they have been laid. Thus, if a pile of pillow lavas is later tilted or overturned, it is possible to identify the direction and amount of tilting and also to recognize the original top and bottom. Upon cooling and contracting, pillows commonly develop radial fractures.

SIDE TRIP TO BOTANICAL BEACH

At Botanical Beach near Port Renfrew, deep tidal pools scooped out of soft Carmanah sandstone are exposed at low tide and form natural aquariums containing sea urchins, sea anemones, chitons, barnacles, small fish, and seaweed. The beach also provides the opportunity to see a well-exposed unconformity between the Tertiary Carmanah sandstone and the underlying late Mesozoic Leech River schist. The scenic trip to Port Renfrew from Victoria takes two to three hours one way.

Take Sooke Road (Highway 14) westward past the Sooke Basin and through Sooke and then drive parallel to the coast for about 75 kilometres to Port Renfrew. Botanical Beach lies just south of Port Renfrew; signs mark the way.

Scattered roadcuts along the highway for about the first 50 kilometres west of Sooke are in massive and locally fractured Metchosin basalt of Eocene age and also related intrusions of gabbro. These rocks are part of the lowest major fault slice exposed on Vancouver Island and continue underwater below the continental shelf to the west. At a major bend, the road crosses the valley of Loss Creek, which marks the trace of the east-west-trending Leech River fault. The Metchosin basalt was thrust, or subducted, beneath the Leech River schist along this fault. The Leech River schist is derived mainly from mud, silt, sand, and a minor mélange of probable Jurassic and Cretaceous age that was metamorphosed in places in Tertiary time. Roadcuts in

Outcrop of Leech River schist in the parking lot above Botanical Beach. The original rock was probably shale (darker matrix) *containing lighter-coloured fragments of sandstone and siltstone and initially was disrupted, folded, and faulted, possibly while the rocks were still wet, and later metamorphosed and further deformed.*

An undulating angular unconformity exposed near the foot of the trail leading to Botanical Beach. Dark grey Leech River schist with a strong foliation that crosses the middle of the photo was deformed and eroded before the nearly flat-lying Carmanah sandstone in the background and foreground was deposited on top of it. Foreground rocks are the lowermost Carmanah, directly above the unconformity, and contain angular fragments of Leech River schist. Erosion has, in essence, provided a window through the Carmanah sandstone, revealing the schist below.

the Leech River schist along Highway 14 as far as Port Renfrew are dark grey to black phyllite weathered to a rusty colour in places.

In Port Renfrew take the road to the west to the parking area for Botanical Beach. A good trail about 1 kilometre long leads down to the beach from the parking area. Try to visit the area at low tide.

An *angular unconformity,* in which sedimentary rock was deposited on the weathered surface of older deformed rock, is exposed in several places near high-tide level at the end of the trail down to the beach. Below the unconformity are dark grey rocks of the Leech River schist with shiny foliation. Above it are more massive, buff-coloured early Tertiary Carmanah sandstone and fine-grained conglomerate and breccia whose base contains numerous angular fragments of the underlying rock. The surface of the unconformity is irregular, so the schist appears to bulge up in places into the flat-lying sandstone. Northwest of Port Renfrew, Carmanah sediments extend across the San Juan fault, overlapping onto Wrangellian rocks, showing that major movement on the fault pre-dates deposition of the Carmanah sandstone.

Saanich Inlet from Malahat Lookout

West of Colwood, Highway 1 turns north and gradually climbs high above the western side of Saanich Inlet. Viewed from the lookout near the high point of Malahat Summit on Highway 1, the inlet appears at first to be just another of the many inlets on the British Columbia coast. It is some 25 kilometres long, as much as 7 kilometres wide near its north end, and tapers southward into a narrow valley with steep walls, a feature that suggests great water depth. (Note: The lookout is only legally accessible to those driving from south to north.)

Investigations of the underwater environment of Saanich Inlet show that it possesses a distinctive, if not unique, character. A shoal about 70 metres deep at its north end limits access of seawater from outside Saanich Inlet. The salinity of seawater in the southern Strait of Georgia is somewhat reduced by freshwater contributions from the Cowichan and Fraser Rivers. The diluted seawater that enters Saanich Inlet floats on the underlying salt water trapped behind the bar, mixing only with difficulty. This situation is comparable to an atmospheric thermal inversion, a phenomenon well-known in the Los Angeles basin and above the city of Vancouver. In Saanich Inlet, salinity rather than temperature controls the densities of the overlying confining layer and the underlying trapped layer.

The shallow, brackish layer of water in Saanich Inlet is oxygenated and supports a summer growth of floating organisms. These, together with

View northward from the lookout near Malahat Summit over the north end of Saanich Inlet and the lowlands and islands on the east side of southern Vancouver Island.

organic matter from the land, sink into the deep water, decay, and deplete the limited supply of dissolved oxygen at the bottom. Some 4 millimetres of black, organic-rich sediment accumulate each year on the bottom of Saanich Inlet, where it lies undisturbed by scavenging or burrowing animals. Some oxygen replenishment of bottom water occurs in the winter months and the character of the accumulating sediment changes accordingly, only to revert to black mud during the next summer. Two long cores obtained from the bottom of the inlet in 1996 span the record of the last 10,000 years and show the yearly changes in sediment. The inlet has proven to be a scientific gold mine for research into diverse topics ranging from late Quaternary stratigraphy, to sedimentary chemistry and paleomagnetism changes in bacteria and fish populations, climate, and the vegetation in the surrounding region.

Cowichan Valley

The Cowichan Valley lies west of Duncan on Highway 1. Near Duncan the valley is 10 to 12 kilometres wide, but about 30 kilometres to the northwest it narrows to a width of only 2 kilometres. Relatively soft and easily eroded Nanaimo shale of late Cretaceous age makes up the valley floor and lower slopes, but most of the high ground surrounding the valley is formed of more resistant metamorphosed volcanic rocks of the Paleozoic Sicker group. Nanaimo sandstone and hard, resistant conglomerate cap some summits, notably those of Mount Prevost, about 8 kilometres northwest of Duncan, and Mount Tzouhalem, about 6 kilometres to the east. The presence of Nanaimo sediments low in the valley and capping summits to the north is due to a combination of their tilts to the northeast and their repetition on a series of reverse faults that elevate and expose older rocks on their north-northeast sides.

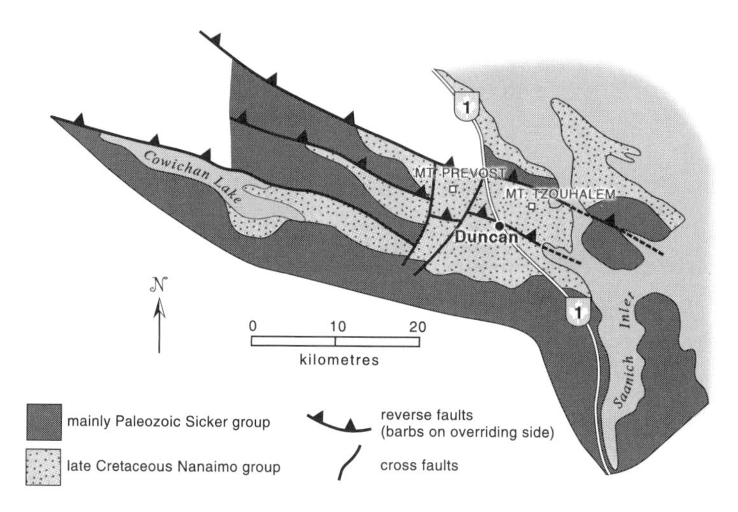

In the Duncan–Cowichan Valley area, Nanaimo rocks lie on top of the Sicker group of Wrangellia on their south-southwest sides and are bounded by reverse faults on their north-northeast sides.

Nanaimo: A Historic Coalfield

The Nanaimo group comprises marine and nonmarine sandstone, shale, conglomerate, and coal of late Cretaceous age, about 85 to 65 million years old, that was deposited near an ancient shoreline. Where deposition was above sea level, the stream-borne sediments were laid down as floodplain sand and alluvial fan gravel. On flat, poorly drained ground, on top of river deltas, in swamps between deltas, and in swampy lagoons behind beaches, organic material, peat, forest litter, and logs accumulated in substantial thicknesses to form the source beds for coal seams. Some coal seams lap up against low, pre-Nanaimo hills and pinch out. Elsewhere, marine fossils in beds interlayered with coal-bearing strata show that the organic material accumulated in nearshore settings. At the shore, sand was deposited as deltas and beaches, and in deeper marine waters, silt and mud as well as coarser-grained sediments accumulated at least in part by slumping and gravity flows down the marine topography to a depth of several hundred metres.

The Nanaimo coalfield played a major role in the early development of British Columbia. About 1859, First Nations people made the existence of coal known to early settlers of the Hudson's Bay Company. Exploration and exploitation of the coal quickly followed. Over the following century, the combined production of this field and its counterpart, 100 kilometres to the north near Comox, exceeded 70 million tons. Much of the coal was sold to the major market in California.

Production began on the Douglas seam in near-surface workings in what is now Nanaimo's city centre. The seam was named after James Douglas, chief factor of the Hudson's Bay Company and later governor of the British colony of Vancouver Island. Mining was soon extended to the Newcastle seam, about 20 metres below the Douglas seam. In 1869, an enterprising coal miner named Robert Dunsmuir discovered a third mineable seam, the Wellington, 180 to 300 metres below the Douglas seam. Dunsmuir raised funds for its development and managed the company, later buying out the other principals and becoming one of the richest men of the time in western Canada. By the end of the nineteenth century, coal production reached 1 million tons a year and utilized a workforce of two thousand miners of Chinese and European origins.

Coal mining has always been a dangerous business. In the Nanaimo area, a major explosion in 1887 killed 150 men in one mine, and an explosion in a different mine the next year killed 77 miners. In mines mainly deeper than 300 metres, there were periodic outbursts of methane gas, which formerly had been contained in the coal by pressure. In less violent outbursts, the coal merely became flaky and friable. In more violent outbursts, the coal and the mine timbers could be physically displaced. Catastrophic outbursts could instantly pulverize hundreds of tons of coal and blow up the tunnels. The workers faced not only the hazard of these violent outbursts but also the danger of ignition of the methane gas and coal dust by the open-flame lamps then in use.

Most of the coal was extracted by the room-and-pillar method, in which about half the coal was removed and the rest was left as pillars to support the roof of the mine. As mining progressed on the three seams, the workings were extended eastward, down the dip of the strata, beneath the shoreline, and under Nanaimo harbour. The miners could hear overhead the beat of the propellers of ships approaching the docks! As the workings advanced, the quality of the coal became poorer, and as the depth increased, there was increased risk of the explosive escape of methane. One by one the mines shut down, the last in 1968. By then, the Douglas, Newcastle, and Wellington seams had been mined continuously or intermittently for an outcrop length of 16, 3, and 20 kilometres, respectively, and for distances down the dip of the seam for up to 3, 2, and 1 kilometres. Although individual seams proved very small compared with those of the great coalfields of the world, the coal production came at a most opportune time and place for the development of British Columbia.

Nanaimo Parkway

Nanaimo Parkway is a new route that connects Highways 1 and 19, passing high around the western side of Nanaimo. Magnificent fresh (in 2003) roadcuts mostly expose generally flat-lying or gently tilted sandstone, shale, and conglomerate of the late Cretaceous Nanaimo group, which weathers to a buff colour. However, one roadcut on the western side of the highway as it climbs northward to the top of the hill exposes massive, dark green Karmutsen basalt of Triassic age and also the unconformity between it and the overlying Nanaimo group.

Stopping along the parkway is ill-advised, so leave the highway at exit 18 and head west on Jingle Pot Road; access is easier heading south on the

Gently tilted Nanaimo sandstone and conglomerate of late Cretaceous age unconformably overlies massive Karmutsen basalt of Triassic age; look for the contact in the road cut behind the lamp standard.

parkway. Just west of exit 18, turn left on Alder Road, then left on Salmon Road, and left again on King Road (all within 1 kilometre), which brings you to a convenient parking place near exit 18. From the parking area, walk a short distance to the highway and head south behind the barrier.

Beds of coarse Nanaimo marine sandstone (greenish when fresh and buff-coloured when weathered) with white shelly fragments pass downward into greenish conglomerate with well-rounded pebbles of volcanic rock, which become larger and more angular downward. (You'll also see holes drilled to obtain rock cores for paleomagnetic studies.) Just above the contact are angular boulders of the underlying dark green, massive Karmutsen basalt. Because both rock units are somewhat faulted and fractured near the contact between them, care is needed to precisely locate the unconformity. The basalt surface is sharp but irregular and probably had a local relief of up to several hundred metres in this area prior to deposition of the Nanaimo sediments.

HIGHWAY 4
Parksville—Tofino
174 kilometres

Near Parksville, about 35 kilometres north of Nanaimo, Highway 4 heads west from Highway 19 and, after a distance of about 150 kilometres, arrives at the western coast of Vancouver Island. The first 10 kilometres or so is across relatively gentle topography underlain by Nanaimo sandstone and shale of late Cretaceous age, with the rugged Vancouver Island Ranges to the west, which here are just over 1,800 metres high and made of late Triassic Karmutsen basalt and Jurassic Island Intrusions.

Beaufort Range Fault

Between Cameron Lake and the summit 5 kilometres to the west, Highway 4 climbs a hill across a structural high—a faulted arch—in which the oldest rocks along the route are exposed: Paleozoic metamorphosed volcanic rocks or greenstone of the Sicker group. At the top of the hill the roadcuts expose nearly flat-lying Cretaceous Nanaimo conglomerate and sandstone. The roadcut in a bend of the highway about 2 kilometres west of this reveals the faulted contact between the Cretaceous and Paleozoic rocks. The most conspicuous fault in the roadcuts dips steeply northeast into the hill and places Paleozoic greenstone on the northeast side against and above gently dipping Cretaceous sandstone and conglomerate on the southwest.

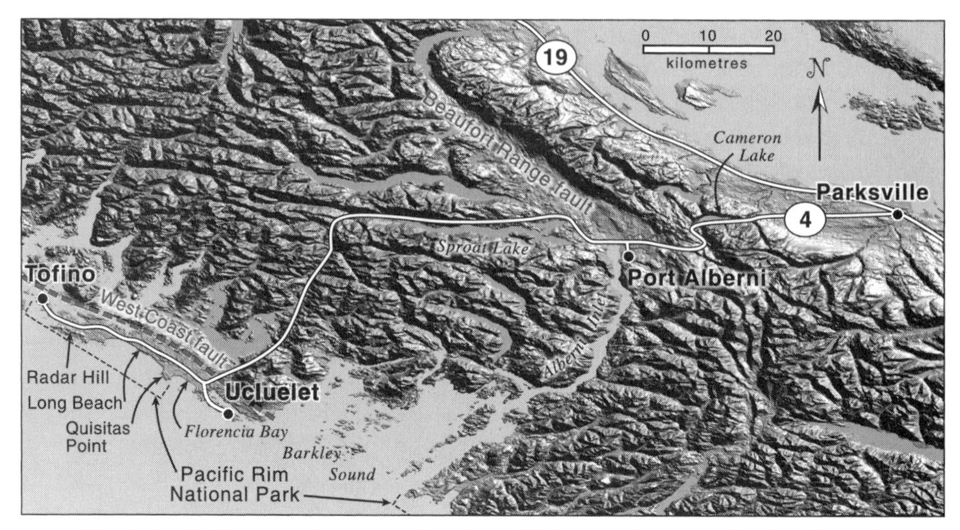

Shaded-relief image of topography along Highway 4 between Parksville and Tofino. —Image constructed using Canadian Digital Elevation Data obtained from GeoBase

Northeast-southwest-directed compression during Tertiary time produced this reverse fault.

The fault system exposed here, the Beaufort Range fault, can be traced northwestward from this exposure along the lower slope of the Beaufort Range for many kilometres, with resistant Paleozoic greenstone supporting the higher slopes and the readily eroded Cretaceous beds underlying the low ground. The topographic feature formed by the differential erosion of resistant and recessive rocks brought together by faulting is known as a *fault-line scarp*, and the front of the Beaufort Range is a prime example. Similar discontinuities in the rock types in the Cowichan Valley near Duncan, 90 kilometres to the southeast on Highway 1, may mark the southeastern extension of the Beaufort Range fault.

Port Alberni

The valley containing Port Alberni is linked to the Pacific Ocean by Alberni Inlet, a 40-kilometre-long, narrow, glacially scoured fjord, and Barkley Sound, an island-studded waterway that extends another 25 kilometres to the open ocean. With highly productive lowland forest in the Alberni area and a navigable connection to the ocean, it is small wonder that settlement began here early. The bustling town of Port Alberni contains a major sawmill, a large pulp and paper mill, associated service industries, and approximately 18,000 inhabitants. Most of the industrial facilities and some of the residential development are located close to the water's edge.

Activities in Port Alberni were rudely interrupted at 1:35 AM on March 28, 1964, by the arrival of a tsunami, or tidal wave, which raised the water surface of Alberni Inlet several metres above an already high tide. Movement of the seafloor during the Alaskan earthquake of 1964, which measured magnitude 8.5 on the Richter scale, generated the tsunami in the Gulf of Alaska. A series of very broad surface sea waves, or giant ripples, spaced perhaps 100 kilometres from crest to crest but only a few metres high, spread laterally from the Gulf of Alaska at rates in deep water of close to 600 kilometres per hour. By the time they reached the western coast of Vancouver Island, the height of the waves, from trough to crest, had declined to about 1.5 metres as measured by the tide gauge at Tofino. However, the shape and length of the Barkley Sound–Alberni Inlet waterway amplified the wave height by funnelling the waters. The rising waters swept inland across the head of Alberni Inlet and the river delta beyond, flooding the sawmill and pulp mill and demolishing homes. There was no loss of life, but homeowners suffered several million dollars of losses from damage or destruction, and damage and lost production at the sawmill and pulp mill added a few million more. Fortunately the lunar tide began to fall after the first wave arrived, because the second wave proved to be even greater than the first. Luckily, waves that followed were smaller than the first two.

Subsequent investigations showed that although the 1964 tsunami was of a scale that might be expected only once or so each century, recurrences could occur. Accordingly, the mills established formal safety practices, the municipality strengthened the dikes and introduced zoning and building restrictions on low-lying land, and the Canadian government rejoined the tsunami warning system for the Pacific Ocean, a service established by the United States Coast and Geodetic Survey.

The earthquake that caused the damage was centred about 2,000 kilometres away in the Gulf of Alaska. What if, as predicted, an enormous megathrust earthquake took place below the continental shelf and slope off the west coast of Vancouver Island? The effects are likely to be far more disastrous.

West of Port Alberni, Highway 4 runs along the north side of Sproat Lake. The somewhat monotonous black and dark green rocks exposed in roadcuts along this part of the highway are slightly metamorphosed Karmutsen basalt of Triassic age containing pillows in places. The highway eventually descends to the low coastal region and, at the T-junction, turns north, roughly parallel with the coastline, heading toward Pacific Rim National Park and Tofino.

Rocky Headlands and Scalloped Beaches in Florencia Bay

Conspicuous near Florencia Bay, west of the T-junction, and northward along the coast toward Tofino, rough rocky headlands separate broad bays. Between the headlands, gently sloping beaches of rippled sand extend landward to a smoothly curving high-water mark at the foot of relatively steep and generally timbered low cliffs or banks made mostly of unconsolidated glacial deposits. If observed on a typical day when the swell sweeps in from the open ocean, the migrating wave crests become nearly parallel with the shore as they approach it. In deep water the waves do not display this effect, but instead remain relatively straight. As the waves enter the shallow water, their rate of advance is slowed, and depending on the shape of the sea bottom, they bend, turning toward the headlands and spreading out on the concave beaches. The sweep of the surf moves loose sand away from the point where a particular wave crest first reaches the shore to the location where it last breaks. This mechanism is responsible for the smoothly curving shape of the beaches.

Depending on the wave characteristics, the surf may drive sand up the beach to the high-water mark, carry it back into deep water, or contribute only to the movement along the shore as described above. Over hundreds and thousands of years, loose sediment gradually shifts seaward, a shift maintained by transport of material from the eroding sea cliffs. In winter there is generally a loss of material from the intertidal zone to deeper water, particularly of finer-grained sediment: most of this material is returned in summer. Ripple marks on the sand surface can change form and orientation within hours, and rill marks, cut by water running down from the upper beach, can change within minutes.

The rocky headlands, made mostly of hard, slightly metamorphosed sedimentary and in places volcanic rock, are much more resistant to change in spite of the concentration of wave energy on them, but even they do not go unscathed, particularly in big storms. Most obvious is the removal of broken rock from fractures and shear zones in the rock to produce channels that the surf passes through. Continued erosion, particularly on the intersection of two or more shear zones, can lead to the development of sea caves, several of which are present on the headland northwest of Florencia Bay.

At one time, the headlands were covered by glacial deposits that survive in the sea cliffs behind the bays. In places, a residue of ice-transported boulders persists on the beach near the inner edge of the headlands.

Gold at Florencia Bay

Over the past century, Florencia Bay has yielded more than $30,000 in placer gold—not enough to make anyone rich, but sufficient to arouse the hopes of ever-optimistic prospectors. The gold has been found with the black sand along the beaches, made of the black, iron mineral magnetite. Waves probably concentrated the gold and magnetite in the beaches from lower-grade outwash deposits in the sea cliffs.

Bedrock between Ucluelet and Tofino

Rocks of the Pacific Rim complex underlie much of the coastal lowland between Ucluelet and Tofino. These rocks are very different from the Wrangellian rocks that form the mountains to the east of the lowland, from which they are separated by the West Coast fault, a large, near-vertical fault. Near Florencia Bay, much of the bedrock forming the headlands is slightly metamorphosed volcanic rock. The outer part of Quisitas Point at the northwest end of Florencia Bay, and much of the Pacific Rim complex to the north, is a dark grey-brown, poorly bedded muddy sandstone, called *greywacke*, containing lenses of very dark mudstone called *argillite*. The rocks were eroded from a mixed plutonic, volcanic, and metamorphic upland source. In a few places the rock contains blocks of chert or recrystallized limestone. Visible folding is rare, but what is visible suggests that at least some deformation took place when the rocks were still wet sand and mud and before the sediment was consolidated into rock. The nature of the rocks in the Pacific Rim complex and their style of deformation suggests to most geologists that the rocks formed in an accretionary complex, here a wedge of sediments eroded from the continent, deposited on the continental slope and adjoining parts of the ocean floor, and mashed up against the continent during subduction.

Rocks exposed at the south end of Long Beach, north of Florencia Bay, are mainly massive greywacke. However, directly below the building on the headland at the Pacific Rim National Park visitors centre, and also exposed a few hundred metres to the south, the rocks comprise a dark, muddy matrix surrounding blocks of different kinds of rock such as chert, volcanic rock, diorite, and sandstone. Elsewhere are blocks of pillow basalt and, rarely, ultramafic rock. This distinctive rock type is called *mélange*, from the French word for "mixture." The rocks were deposited in deep water, probably near the toe of the continental slope, as the result of slumping and sliding of a variety of boulders, sand, and mud down the continental slope, and possibly onto the ocean floor, from whence they eventually were incorporated in the accretionary complex.

Rocks in the Pacific Rim complex span most of Mesozoic time. Beds of limestone and chert in volcanic rocks along the coast west of Ucluelet contain microfossils and molluscs of late Triassic and early Jurassic age, but these rocks are relatively restricted in distribution, and they may be

Mélange and deformed bouldery mudstone along the beach just south of the Pacific Rim National Park visitors centre. Note that the layers and blocks are broken and discontinuous but have a through-going irregular foliation.

enormous blocks in the mélange. The far more widespread and abundant sandstone, mudstone, and chert, and most of the mélange, contain mostly early Cretaceous and in places late Jurassic clams and siliceous shells of *radiolarians*, minute single-celled organisms.

A small body of 52-million-year-old granitic rock intrudes Pacific Rim complex rocks just west of Tofino. Granitic bodies of similar age occur in Wrangellian rocks and also intrude the West Coast fault. From this, it seems that folding and faulting in the Pacific Rim complex must have ended by the time the early Tertiary granite intruded.

Radar Hill and the Strandflat

The top of Radar Hill, reached by a side road about 8 kilometres southeast of Tofino, provides a view of the coastal lowlands, 8 to 10 kilometres wide, and of the mountains that rise steeply to the northeast. The contrast in the landforms is striking. The mountains rise 700 to 1,250 metres above sea level and are separated from one another by *fjords*, valleys that in places descend below sea level. In the lowlands, scattered bedrock hills rise no more than about 200 metres above the water, and still less above a plain of unconsolidated sediments of glacial origin. Why is there such a great difference in relief and in the depth of the valleys between the coastal lowlands and the inland mountains?

In this area the topography might be at least partly explained by the different nature of the bedrock underlying the lowlands and mountains, but the question above has been raised for similar lowlands fringing other mountainous coasts at high latitudes where a big bedrock change is not present. Some of the best examples are in Norway, where the lowland is called the *strandflat.* Off the Norwegian strandflat there is commonly a rocky platform studded with small islands and rocky shoals called the *skerrygard.* This feature is not well developed here but is conspicuous on the mainland coast north of Vancouver Island.

The famous Arctic explorer Fridtjof Nansen proposed that strandflats develop close to sea level in cold climates. Sea fog and spray provide moisture, particularly on the poorly drained lowlands, which facilitates the breakdown of rocks by frost action. The steeper mountain slopes escape much of the frost action by virtue of their good drainage. The fact that strandflats often are better developed in the lee of high-standing islands than in areas more fully exposed to great ocean waves supports this theory.

This view from the headland south of Long Beach shows typical heavy wave action on the headland, and the coastal lowland, or strandflat, with the Vancouver Island Ranges made of Wrangellian rocks rising sharply beyond.

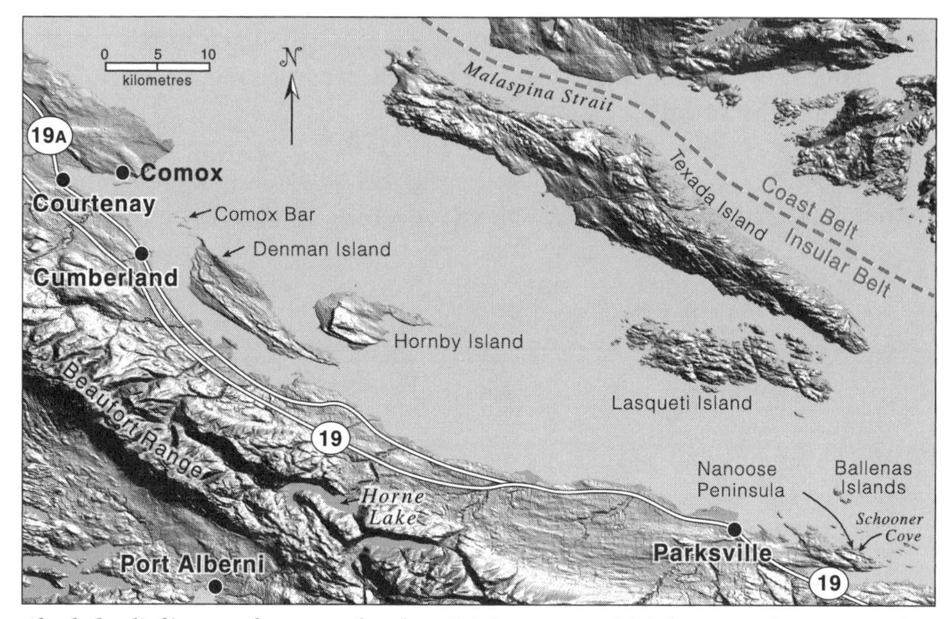

Shaded-relief image of topography along Highway 19 and Highway 19A between Parksville and Courtenay. —Image constructed using Canadian Digital Elevation Data obtained from GeoBase

HIGHWAY 19/19A

Nanaimo—Port Hardy

391 kilometres

Highway 19 is the continuation of the Nanaimo Parkway north of Nanaimo. You can see the Coast Mountains across the Strait of Georgia from several places on the highway between Nanaimo and Parksville. Texada Island, the long island on the east side of the strait, is underlain by Wrangellian strata, mainly Karmutsen lavas, and the granitic Coast Belt lies just east of it. North of the turnoff to Highway 4 near Parksville, the old Highway 19, now called Highway 19A, more or less follows the coastline through Courtenay to Campbell River. The new, broader, and straighter Highway 19 (Island Highway) runs inland, west of Courtenay, along the rolling lowland made of Nanaimo sedimentary rocks.

Nanoose Peninsula and Ballenas Islands

Paleozoic rocks are exposed on Nanoose Peninsula, on the offshore Ballenas Islands, and on the southeastern tip of Texada Island. The Nanoose Peninsula, about 20 kilometres north of Nanaimo, is by far the most accessible.

The excellent wave-washed exposures along the shore of the Strait of Georgia on the east side of the peninsula and their ready access by either land or sea justify a hike along the water's edge from Schooner Cove north to Dolphin Beach, less than 1 kilometre to the northwest. Private land partly bars access to the waterfront here, but you can reach it by short trails across extensions of road allowances. Note that in Canada the intertidal zone is publicly held, and the route can be traversed in all but higher tides.

Schooner Cove is cut into a granitic intrusion at least 1.7 kilometres in diameter. Heat from the intrusion affected the rocks along its edge. The first, very slender promontory northwest of Schooner Cove exposes thin-bedded, tightly folded, and slightly metamorphosed siliceous rock or chert, interbedded dark micaceous schist probably derived from mud, and a few lenses or pods of siliceous limestone. Faults separate these beds from fragmented volcanic rock, called *breccia*, and a variety of limy and clay-rich rocks, minor sandstone, tuff or volcanic ash, and tuffaceous chert. The limestone has been reworked into sand and pebbles, and in places disks derived from the breakdown of the segmented stems of fossil sea lilies, or *crinoids*, are common. Similar rocks on the Ballenas Islands, in the Strait of Georgia to the east, contain calcareous microfossils of late Carboniferous age, about 310 million years old. Also present on the peninsula are pillow lava and dark green altered volcanic rocks, some of which intrude the sedimentary rocks. In places, sea spray has eroded the fine-textured matrix of the lava, leaving the larger crystals standing in relief.

Horne Lake

About 20 kilometres north of Parksville, look for the turnoff to Horne Lake, which lies west of the highway. There, cliffs of massive but bedded Buttle Lake limestone, named from exposures along Buttle Lake to the northwest, flank the north side of Horne Lake. In places, the limestone contains late Carboniferous to earliest Permian fossils that are more like those in the Ural Mountains in Russia than any fossils of similar age presently at the same latitude on the old continental margin. This hints that these rocks were deposited at a higher latitude than their present position relative to the continental interior would suggest. The dark horizontal layers in the cliff are sills of basaltic rock that are probably related to the overlying Karmutsen basalt. At Horne Lake Caves Provincial Park, you can explore caves dissolved into the limestone and observe dripstone features made from calcium carbonate.

Cliffs of Buttle Lake limestone on the north side of Horne Lake. The dark layers are sills of basalt that probably are feeders to the Triassic Karmutsen basalt.

Coal near Courtenay

Courtenay, on Highway 19A, is at the head of Comox Harbour. The town of Comox is about 2 kilometres to the east, on the eastern side of the harbour. To meet a growing demand for Vancouver Island coal, the Dunsmuir interests in 1888 opened another coal mine in the Nanaimo group, the Comox Coalfield. The field produced 70 million tons of coal between 1888 and 1953, when the last mine, at Cumberland, closed down. Another 1.5 million tons were added before an outlying mine on the Tsable River, 15 kilometres southeast of Cumberland, ceased to operate in 1966. Peak production occurred in 1911, when almost 500,000 tons of coal were mined and shipped by boat from a loading terminal at Union Bay, about 12 kilometres south-southeast of Courtenay.

The coal came from strata lower in the Nanaimo group, and hence somewhat older, than strata in the mines at Nanaimo. The Comox coal was high-volatile bituminous, as at Nanaimo, but produced better *coke*, the high-carbon, hot-burning residue made by heating coal and expelling the volatiles as coal gas. The seams dip regularly to the northeast at about 5 degrees, and faults are rare compared with the Nanaimo coalfield. Three seams provided nearly all the coal. The most extensive, the No. 4 seam, was mined from outcrops overlooking Comox Lake down the dip of the beds for a distance of as much as 2 kilometres, and parallel to the surface trace of the seam for a maximum distance of 8 kilometres.

Comox Bar and the Postglacial Sea Level

In the past, steamships moved people from place to place along the British Columbia coast, and Comox was one of the busy ports. However, very few ships used the direct route between Comox Harbour and the Strait of Georgia. Instead, most ships coming from the north sailed past the southeastern tip of Hornby Island, past southern Denman Island, and north-northwestward up the narrow channel between Denman and Vancouver Islands, a detour of more than 50 kilometres. The ships were avoiding Comox Bar, a chain of sandy banks, shoals, and low islands that link the north end of Denman Island with the Comox Peninsula. Although the bar no longer receives sand carried by longshore currents from the sea cliff at the north end of Denman Island, it presents an obstacle to navigation with a depth of only 2.4 metres below lowest normal tide.

Following the last glaciation, sea level at Comox was at least 6 metres below its present level, and Vancouver Island was linked to Denman Island by the completely emergent Comox Bar. Waves washed along a continuous intertidal beach extending northwest from the sea cliff on Denman Island to the vicinity of Comox. Sand dune construction had already started by 8,700 years ago and a date from the base of an organic deposit built on a beach bar less than 1.5 metres above the present high-tide level suggests that dune construction ceased by 5,700 years ago, when the sea rose to its present level with the melting of the ice cap, submerging Comox Bar. The shore below the dunes at Comox is now well armored with coarse debris and no longer serves as a local source of windblown sand.

Courtenay Earthquake of 1946

At 10:15 on Sunday morning, June 23, 1946, the land shook. With a magnitude of 7.3 and an epicentre (the place on Earth's surface directly above the earthquake) near Courtenay, the ground motion was felt as far north as Vanderhoof, British Columbia, 500 kilometres to the north, and Portland, Oregon, 500 kilometres to the south. Notwithstanding the severity of the shock, damage was relatively slight, thanks to low population density and wood-frame construction of most dwellings and commercial buildings near the epicentre. Chimneys collapsed, with bricks plunging through roofs and floors around the northern Strait of Georgia, and if the quake had occurred on a school day, there could have been serious injuries to children. The only fatality was a man who drowned on Comox Lake after his boat was overturned by a wave initiated by a quake-triggered landslide.

The epicentre of this earthquake was initially judged, on the basis of the times required for the shock waves to reach the various seismic recording stations, to lie in the middle of the Strait of Georgia. By the 1980s, seismology had become more sophisticated, and seismic velocities under Vancouver Island now are calculated to be somewhat faster than they were thought to be in 1946. The recalculated velocities shifted the position of the epicentre northwest to a site on the Forbidden Plateau, about 20 kilometres northwest of Courtenay.

The rupture causing the earthquake was about 20 kilometres deep, which is within the crust of the upper North American Plate and not at the plate boundary, which is the potential site of a big megathrust earthquake. Such a shallow depth for an earthquake of this magnitude indicates the likelihood that the surface of the ground had moved, so surveyors remeasured sites in the mountains of eastern Vancouver Island. This provided a hint of a shift in the easternmost survey stations with respect to western ones, but not enough to be beyond the limit of surveying uncertainties. Air photos revealed some suspicious northwest-trending fault scarps, but the small scarps could not be ascribed with confidence to the 1946 quake and may have formed during some earlier seismic event.

Seymour Narrows and Ripple Rock

About 13 kilometres north of the town of Campbell River is a lookout where Highway 19 turns from its northward course along the west shore of Discovery Passage, between Vancouver and Quadra Island, to a more westerly course along the south shore of Menzies Bay. To the north is Seymour Narrows, an 800-metre-wide channel between the islands. This is the most constricted part of the entire Inside Passage along the coast of British Columbia, and the site of some of the most vigorous tidal currents—up to 7.7 metres per second—among the world's shipping lanes.

Complicating navigation of Seymour Narrows is a shoal of Triassic Karmutsen basalt called Ripple Rock, whose presence is indicated by a patch of extremely turbulent water during a rapidly falling or rising tide. Part of the rock once was as little as 3 metres deep at low tide. Its toll between 1875 and 1959 included 24 large ships, more than 100 fishing boats, yachts, and tugs, and at least 114 lives. Even a hydrographic survey ship in 1943, having just completed a detailed survey of the shoal and sailing northward for the next task, gave way to a ship passing in the other direction and snagged her bottom on Ripple Rock. She was beached in a bay on the east side of the

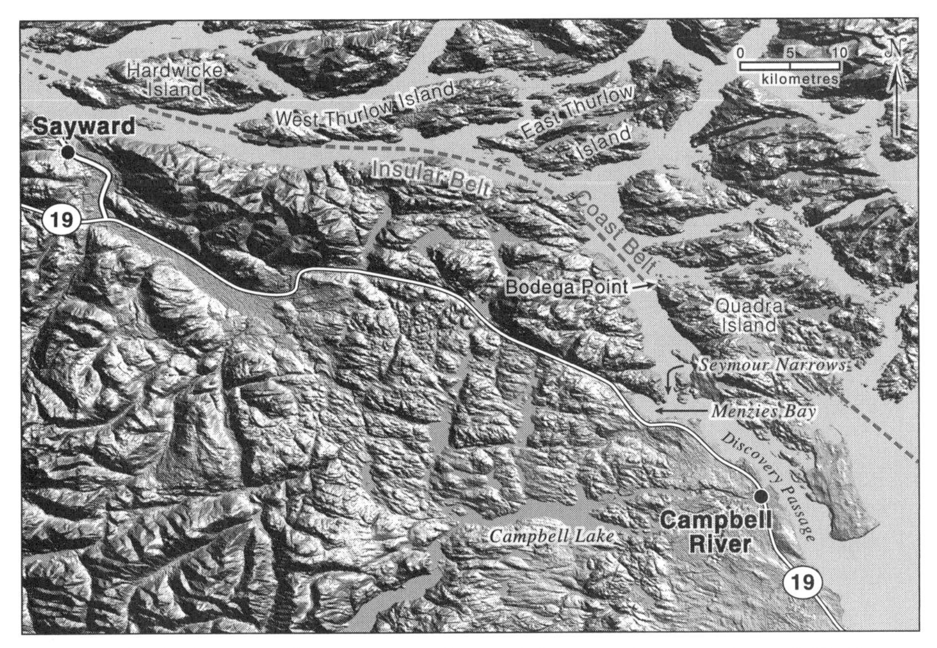

Shaded-relief image of topography along Highway 19 between Campbell River and Sayward. —Image constructed using Canadian Digital Elevation Data obtained from GeoBase

passage and all the crew made it to shore, but with the rising tide, the stern swung into the current and the ship slipped into deep water with only its masts showing.

The Canadian Government undertook the demolition of higher parts of Ripple Rock in the late 1950s. It was no easy task. The turbulent water hampered early attempts involving drilling from an anchored barge. Anchor cables fluttered in the strong currents and snapped, on average, only forty-eight hours after installation. Two overhead cables across the passage, high enough to permit passage of ships beneath them, were then used to hold the drilling barge in place, but they only kept it steady enough for efficient drilling for about an hour during slack tide.

The third attempt at demolition was subterranean. A shaft was sunk on the east shore to a depth of 175 metres and then driven 730 metres horizontally out to midchannel, where a vertical shaft was excavated upwards for 90 metres into Ripple Rock. Finally, two networks of small-diameter tunnels were dug to within about 10 to 15 metres of the seafloor and filled with explosives. At 9:31 AM on April 5, 1959, the ground shook from what was billed as the largest nonnuclear, man-made explosion up to that time. The blast removed the top 12 metres from the reefs and scattered debris

widely over the channel floor. The channel now has a minimum depth of 14.3 metres at low tide, enough to provide clearance for all but supertankers and large bulk carriers. Ship navigation through Seymour Narrows is still restricted at times by the strength of the tidal currents, and passage during periods of one-half to three hours from slack tide is recommended.

Quadra Island

Quadra Island, across Discovery Passage from Vancouver Island, is one of the few places where the boundary between the Insular and Coast Belts is exposed on land. You can also see the belt boundary on Hardwicke and West Thurlow Islands to the east of northern Vancouver Island. Most of the boundary is submerged beneath the eastern side of the Strait of Georgia. Texada and South Thormanby Islands on the eastern side of the Strait of Georgia are largely Triassic Karmutsen basalt, and the adjacent Coast Belt across the narrow channel is mostly middle Jurassic to Cretaceous granitic rock between about 167 and 95 million years old.

At Bodega Point on northwestern Quadra Island, granitic rocks dated at 164 million years old intrude and metamorphose Karmutsen basalt.

Discovery Passage from the southern part of Quadra Island looking west to Vancouver Island. The rocks across the channel are mainly Karmutsen basalt; at the northeast end of Discovery Passage, 164-million-year-old middle Jurassic granitic rock intrudes the basalt. The contact marks the geologically defined western limit of the Coast Belt in this area.

The youngest Jurassic granitic rocks in the Insular Belt, the Island Intrusions, which are part of the Bonanza arc, are about 170 million years old, just slightly older than the oldest intrusions known in the southern Coast Belt. In this region, the western boundary of the Coast Belt is the western limit, sometimes called a *magmatic front*, of the 167-million-year-old and younger granitic intrusions that form over 80 per cent of the rock in the Coast Belt. In middle Jurassic time, island arc magmatic activity evidently moved eastward from the location that it had occupied since latest Triassic time on what is now Vancouver Island, into what eventually became the western Coast Belt.

Alert Bay Volcanic Belt

The small Alert Bay volcanic belt lies along a southwest-trending line from near Port McNeill on the eastern side of northern Vancouver Island to Brooks Peninsula on the western side. The belt includes at least eight cones, vent fillings, and shallow intrusions that are geologically very young, between 8 and 3 million years old. The most northeastern body, on Haddington Island, 6 kilometres northeast of Port McNeill, is a volcanic rock called *dacite* and is the source of the building stone for the Parliament Buildings in Victoria. Another dacite body, visible dead ahead when driving west on Highway 19 at the Port McNeill turnoff, forms a steep-sided hill that rises 200 metres above its surroundings. Other volcanic rocks in the belt include rhyolite and basalt.

The position and trend of the Alert Bay volcanic belt correspond closely to the northern end of the offshore Juan de Fuca Plate in a region where that plate breaks up into a number of microplates. Why the volcanic belt has such a geologically limited duration and such a great range in composition remains unexplained.

North Island Coal

The earliest commercial coal mines on Vancouver Island were at Suquash on the south shore of Queen Charlotte Strait, directly north of the lookout located 16 kilometres west of the Port McNeill turnoff on Highway 19, and 5 kilometres east of the turnoff to Port Alice. There, late Cretaceous strata, equivalent to the lower part of the Nanaimo group, contain at least one coal seam up to 2 metres thick with numerous interbeds of shale and sandstone.

With the aim of supplying their ships and trading posts with coal, the Hudson's Bay Company brought in groups of Scottish coal miners in the 1840s to work the Suquash deposit. An estimated 10,000 tons were mined, but poor living conditions, labor disputes, the hostility of the natives, the lure

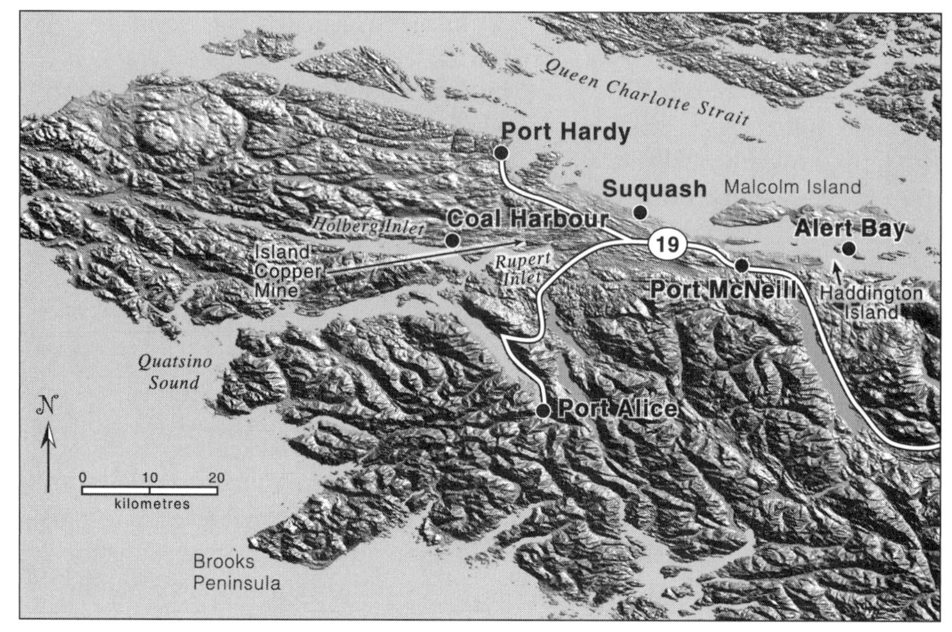

Shaded-relief image of the Port Hardy region. Note the relatively low-lying tip of Vancouver Island north of Holberg Inlet. It may be low lying because the crust has been stretched and thinned east of the north end of the Juan de Fuca Plate.
—Image constructed using Canadian Digital Elevation Data obtained from GeoBase

of the 1849 California gold rush, and better opportunities at the newly discovered coalfield at Nanaimo led to abandonment of the mining operation in 1852.

A limited coal-bearing formation of early Cretaceous age is responsible for the name of a small settlement, Coal Harbour, on Holberg Inlet about 15 kilometres south-southwest of Port Hardy. Nowhere are the beds known to contain coal seams thick enough to invite mining. The widespread occurrence of marine fossils in associated strata indicates that a substantial part of the unit accumulated in salty or brackish water, which is not conducive to forming coal.

Port Alice: A Modelled Community

The bedroom community called Port Alice, or occasionally Rumble Beach, is 32 kilometres by paved road south of Highway 19. The original town site adjoined and served the existing pulp mill. It was relocated in 1965 to a new site 5 kilometres to the north that was considered more desirable from an environmental standpoint. The new planned community sits on an apron of alluvial fans at the foot of a steeper hillside, which slopes mostly at 20 to 25 degrees and rises as much as 650 metres above the fan heads. Glacial sediment a few metres thick mantles the hillside of Triassic bedrock. Brushy, partly deciduous vegetation has replaced the coniferous forest that was logged in the 1940s and 1950s.

During a stormy night in 1973, and again in the autumn of 1975, runoff from heavy rain, and perhaps from snowmelt, spilled from a logging road, initiating a small soil slip that moved down the slope into a gully. There, it incorporated a large volume of loose debris and moved down the gully, growing into a *debris flow*, which is a slurry of wet mud, sand, and boulders, plus roots and branches. The debris flow entered the town, sweeping parts of buildings off their foundations and damaging others before coming to a stop.

The provincial government, justly alarmed by two such events, authorized a study of ways to protect this new community and its several hundred inhabitants. The study concluded that a warning system based on meteorological forecasts of heavy, prolonged precipitation would lead to too many false alarms, which might result in a lack of response to a real emergency. In view of the limited supply of loose debris within each of the gullies, it was felt that a system of dikes could provide protection against individual debris flows. But how high must the dikes be built to be effective?

To solve this last problem, a scale model was made of the hillside, the fans, and the existing buildings. Then a series of simulated debris flows were generated to duplicate the volumes and limits reached by the actual debris flows. Model dikes of various heights and patterns helped engineers select the most effective design. The real dikes were built at a cost of $250,000, a figure significantly less than the damage claims of $700,000 for the 1973 and 1975 debris flows.

To date there have been no debris flows above the town that could challenge the dikes. Others have occurred between the town and the mill, one of which swept an observer to his death, but these were in an area where no defences had been built.

Island Copper

The Island Copper Mine was a major source of income for northern Vancouver Island between 1971 and 1996. It was located 13 kilometres south of Port Hardy, on the north shore of Rupert Inlet, which is the easternmost branch of Quatsino Sound and is 24 kilometres from the Pacific Ocean. Oceangoing ships could dock at the mine property, but they had to navigate a constriction in a 5-kilometre section of the channel, which was only about 225 metres wide at each end. The constriction, with a depth of only 18 metres in a channel where tidal currents are up to 3.1 metres per second, creates extreme turbulence. A right-angle turn midway through the channel and strong winds add to the challenge faced by captains of oceangoing ships.

The Island Copper orebody was 1,250 metres long, up to 450 metres wide, and 300 metres deep. Originally it was not exposed, but in places where only 9 metres of overburden covered the orebody, geochemical prospecting showed that the surface soil and hemlock bark and needles were anomalously rich in copper. Following intense surface prospecting, geochemistry and accompanying geophysical surveys guided the drill program that outlined the orebody. More than 35,000 metres of drilling between 1966 and 1969 indicated that enough ore was present to justify the cost of developing an open-pit mine, the mill for processing the ore, and the associated facilities.

The orebody was in fragmental volcanic rocks of early to middle Jurassic age that formed in the Bonanza island arc of Wrangellia. The ore lay

View looking southeast at the open pit of the Island Copper Mine in 1992.
—George Onuska photo, courtesy of BHP-Utah Mines Ltd.

within these rocks on either side of a large igneous dike that contained large crystals of quartz and feldspar in a finer-grained matrix. The dike may have been the source of the mineralizing solutions, and almost certainly was a focus for the severe fracturing that concentrated the ore minerals. The valuable ore minerals are chalcopyrite (a copper-iron sulfide), associated gold, and molybdenite (a molybdenum sulfide). Also present are bornite (another copper-iron sulfide) and pyrite and pyrrhotite (both iron sulfides). A variety of minerals are associated with the ore, including lilac-coloured dumortierite, of interest to mineral collectors, but unfortunately the area is no longer accessible.

Production at the mine began in 1971, and 250 million tons of ore were processed by 1991. In the same period, about 600 million tons of rock covering and containing the ore were dug out and dumped in the shallower part of Rupert Inlet. Mining ceased in the big open pit in 1996, and it now is flooded with seawater.

The mine deserves comment on a matter of geographical trivia. For a geological instant, from the time the pit floor was dug below the minus-86-metre elevation until the pit was flooded, the floor was the lowest point on dry land in the western hemisphere. Moreover, after mining was completed, the floor temporarily became the second-lowest point on dry land on the whole globe, outdone only by the shore of the Dead Sea.

HIGHWAY 28
Campbell River—Gold River
88 kilometres

For about 3 kilometres west of the town of Campbell River, Highway 28 crosses the modern floodplain of the Campbell and Quinsam Rivers. For the next 2 kilometres, it climbs the steep front of the postglacial delta of Campbell River, which had been built into the sea before the land rose by postglacial rebound to its present elevation. A branch road turns north into Elk Falls Provincial Park and follows the break in slope at the top of the east-facing delta front. The branch road crosses a row of large pipes, called penstocks, for the John Hart hydroelectric power plant, which was built to take advantage of the sharp drop in the Campbell River, the largest river on Vancouver Island. The road then turns west to follow the shore of an artificial head pond and the crest of a broad earth-filled dam. About 600 metres from Highway 28, the road reaches the concrete spillway that controls the water level in the head pond.

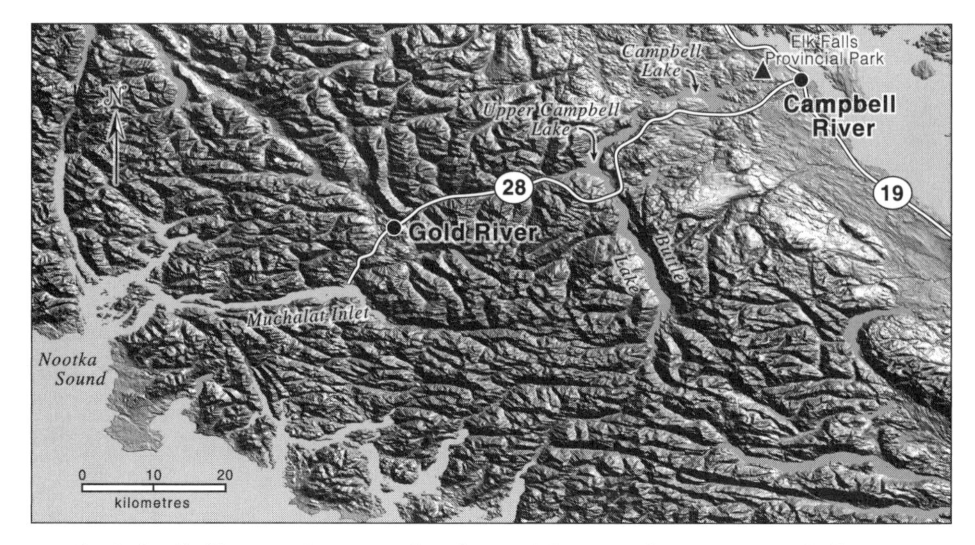

Shaded-relief image of topography along Highway 28 between Campbell River and Gold River. —Image constructed using Canadian Digital Elevation Data obtained from GeoBase

In earliest postglacial time, the Campbell River established its high delta at a place that coincides with the top of a buried bedrock hill of Triassic Karmutsen basalt. The hard rock formed a natural spillway that was sufficiently resistant to prevent any deep erosion of the channel floor and at the same time precluded any lateral shifting of the Campbell River. Downstream, the river, aided by postglacial uplift, cut a rock-walled canyon holding Elk Falls, a chain of waterfalls. Today, Elk Falls is a sad remnant of its former glory because most of the stream flow now passes through the man-made penstocks.

When the John Hart power project was being built in the mid-1940s, the engineers seemed to be oblivious of potential geological hazards. Had they been aware of them, they surely would not have built a structure that ponded water at the outer edge of a leaky former delta top. The power project was still under construction when the Courtenay earthquake shook the region in 1946.

Leaky foundations to dams and head ponds present a variety of problems. One is the loss of water that otherwise could be used to generate power. In addition, leaking water may cause problems where it emerges from the dam. At Lake McIvor, the head pond to a second dam on the Campbell River southwest of the John Hart Dam, leakage from the dam raised the water table in the area below it and increased the discharge of streams flowing through a freshly planted experimental forest. Then nature took over: beavers moved in and soon converted the newly wetted site into a swamp to the detriment of the recently planted trees. A third problem with water leaking from dams happens when the water velocity between the grains of the fill of the dam increases to the point at which it can flush away silt and sand, rendering the remaining coarser sediment even more leaky. This phenomenon, called "piping" by engineers, creates voids ranging from miniature pockets to substantial caves. Perhaps the most serious of the problems caused by raising the water level in a head pond is the increase in the pore water pressure in the banks of the pond, which increases the chance of landslides. If a landslide leads to breaching of a dam, then there is real trouble!

Steps to increase earthquake resistance at the John Hart power project were undertaken in the late 1980s. These required an extensive drilling program to define the lower limit of the leaky delta sediments. Much of the sediment was excavated and replaced by a broader embankment of till, sand, and crushed rock. Excavation in the head pond was a special problem because the power plant was required to keep operating to meet electricity demand. Moreover, since the water was also used to supply the community of Campbell River, the local paper mill, and hatching and rearing beds for trout and salmon, great care was needed to minimize muddying the head pond, especially when tree stumps were removed from its bottom. Finally, for the length of the rebuilt earth-filled dam, a trench was dug, mostly down to impermeable glacial till or bedrock, and filled with a slurry of bentonite clay and Portland cement. This belated application of geological engineering was expensive, but the land downstream is now much safer.

Ghost of Robert Dunsmuir

Coal mining on Vancouver Island died in 1966 at Nanaimo and Comox because of a lack of markets, a shortage of good quality coal, and rising labor costs. In the 1980s, new markets appeared in eastern Asia, mechanization eased some of the labor costs, and a coal seam hitherto undeveloped in the hilly country west of Campbell River was available for mining. Production began in 1988. By a twist of fate, the seam is on land once held by Robert Dunsmuir, the nineteenth-century Scottish miner who discovered a coal seam near Nanaimo and became one of the richest men in Canada.

The seam of high-volatile bituminous coal is 3 to 4 metres thick, dips gently to the east, and is cut by few faults. That part of the seam located near the ground surface was removed by *strip-mining*, in which the cover is removed to expose a strip of coal. After the coal is mined, the cover of the next strip is placed on the site of the first, and the second strip of coal extracted. This procedure is repeated, and nearly all of the coal extracted, until the depth of cover is so great that the cost of its removal exceeds the return from the newly exposed coal. In 1991, this stage was reached along much of the length of the seam, and additional mining proceeded by underground methods.

Reclamation involves flattening the steeper slopes on the piles of waste rock, replacing the original topsoil, and seeding to provide a vegetative cover designed to resist erosion, to discourage oxidation of sulfide minerals such as pyrite, and to aid evaporation of groundwater. Infiltration of rainwater is minimized, and water emerging from the piles of waste rock is intercepted and pumped back to be recycled.

Basalt along Upper Campbell Lake

Highway 28 bypasses the second and third dams on Campbell River but descends to the level of Upper Campbell Lake and follows the lakeshore fairly closely for about 16 kilometres. The road passes through dark green to black basaltic lava flows, some pillow lava, and fragmented volcanic rock, part of the enormous pile of Triassic Karmutsen volcanic rocks that underlies so much of Vancouver Island. Low-grade metamorphism has altered the fragmental rocks and pillow lava so that they mostly break across fragments and pillows rather than around them, and close-up it may be difficult to recognize any original layering.

If you look across Upper Campbell Lake to the basalt on the far side, you can see gently tilted layering in the enormous pile of lava. Some of the original flow contacts and the rocks deposited between flows are less resistant to erosion and are marked by recessed ledges. These may be the best and in some places the only indication of the original layering of the basalt flows. Close-up, particularly where the forest limits visibility, the continuity of the ledges may not be apparent. Only at a distance, as across the lake, can the layering be seen.

In this area, geologists can estimate the thickness of the Triassic Karmutsen rocks because they are sandwiched between limestone. The volcanic rocks are under late Triassic limestone near Upper Campbell Lake and overlie late Paleozoic Buttle Lake limestone exposed along Buttle Lake, which is in the southward continuation of the valley containing Upper Campbell Lake. The basalt flows tilt an average of 19 degrees from horizontal toward the northeast. The horizontal distance between the base of the volcanic rocks on the west shore of Buttle Lake and the top of the volcanic rocks near the west shore of Upper Campbell Lake is about 30 kilometres. If

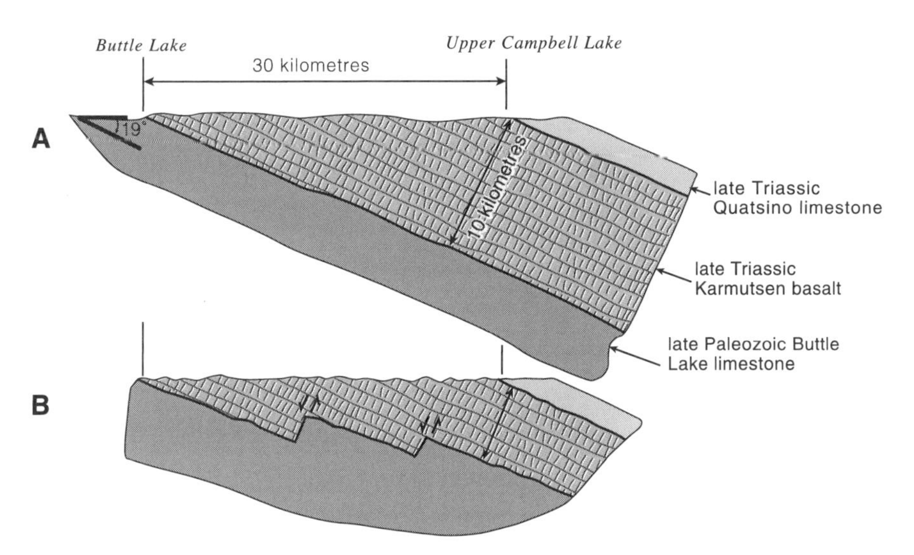

A. *The thickness of the Karmutsen basalt can be estimated from the average tilt and assuming no faults.* B. *Concealed normal faults will reduce the thickness estimate.*

the average tilt is used, the thickness of the pile of lava in this area can be calculated to be about 10 kilometres. Displacements of the strata along hidden faults could change this figure to the current "best bet" of about 6 kilometres, for although geologists do not recognize any big faults along the direct line between the two points near the lakeshores, they easily could be overlooked in the rather homogeneous pile of rocks.

The origin of the thick carapace of Triassic basalt on Vancouver Island is something of a mystery as it does not fit neatly into any known modern plate tectonic settings. Chemically it resembles basalt formed at mid-ocean ridges, but here the basalt overlies mainly island arc–related strata of Devonian to Permian age and not ocean floor rocks. Some of the underlying strata, such as the limestone on Buttle Lake, were deposited in shallow water. Although much of the Karmutsen basalt was deposited below sea level, terrestrial flows occur in places in the upper part, and at least some of the overlying Triassic limestone was deposited in shallow water. The basalt may have formed as a mostly submarine flood basalt above an upwelling plume of magma from deep within the mantle. A present-day analogue might be Hawaii, which lies entirely within the Pacific Plate, but the chemistry of the Karmutsen basalt differs from basalt extruded in modern within-plate settings such as Hawaii.

View across Buttle Lake. The mountains are capped by Triassic Karmutsen basalt overlying white cliffs of well-bedded, gently tilted Buttle Lake limestone of latest Carboniferous to earliest Permian age, which is intruded by thick, dark sills that probably fed the Karmutsen flows.

Boatside Geology

Four ferry routes cross the Strait of Georgia: two from Highway 1 near Nanaimo; one from Swartz Bay on Highway 17, 20 kilometres north of Victoria; and a northernmost route, not described here, from near Courtenay to Powell River. Even though the seafloor is hidden from view, the routes pass several islands, and although it is "boatside" rather than "roadside" geology, to omit any description of the routes leaves a big gap. The sections in the chapter introduction that discuss the Strait of Georgia and Nanaimo sediments, Quaternary ice ages, sediments of the last glaciation, and rising and falling sea levels provide a good primer for the ferry route guides.

FERRY FOR HIGHWAY 1
Nanaimo—Horseshoe Bay

Two ferries leave from the Nanaimo area. The one from Duke Point, just south of Nanaimo, arrives at Tsawwassen, near the international boundary about 35 kilometres south of downtown Vancouver. The other, from Departure Bay in Nanaimo, arrives at Horseshoe Bay in West Vancouver to reconnect with Highway 1. We present boatside geology for the Departure

Bay ferry, but the first leg of the trip follows the same route as the Duke Point ferry.

The ferry terminal on the south shore of Departure Bay is 2 kilometres north of the centre of Nanaimo and directly west of Newcastle Island. The terminal was constructed on the site of the Brechin Mine, from which the Newcastle coal seam was worked seaward under the northern half of Newcastle Island during the early twentieth century. Access to the workings was by means of a short shaft and an inclined passageway, called a "slope," which extended eastward down the dip of the seam for 1.5 kilometres to a depth of about 100 metres below sea level. The base of the Nanaimo beds

Shaded-relief image of the Gulf Islands, the western San Juan Islands, and the Fraser River delta. Ferry routes between Vancouver Island and the mainland are shown. —Image constructed using Canadian Elevation Data obtained from GeoBase

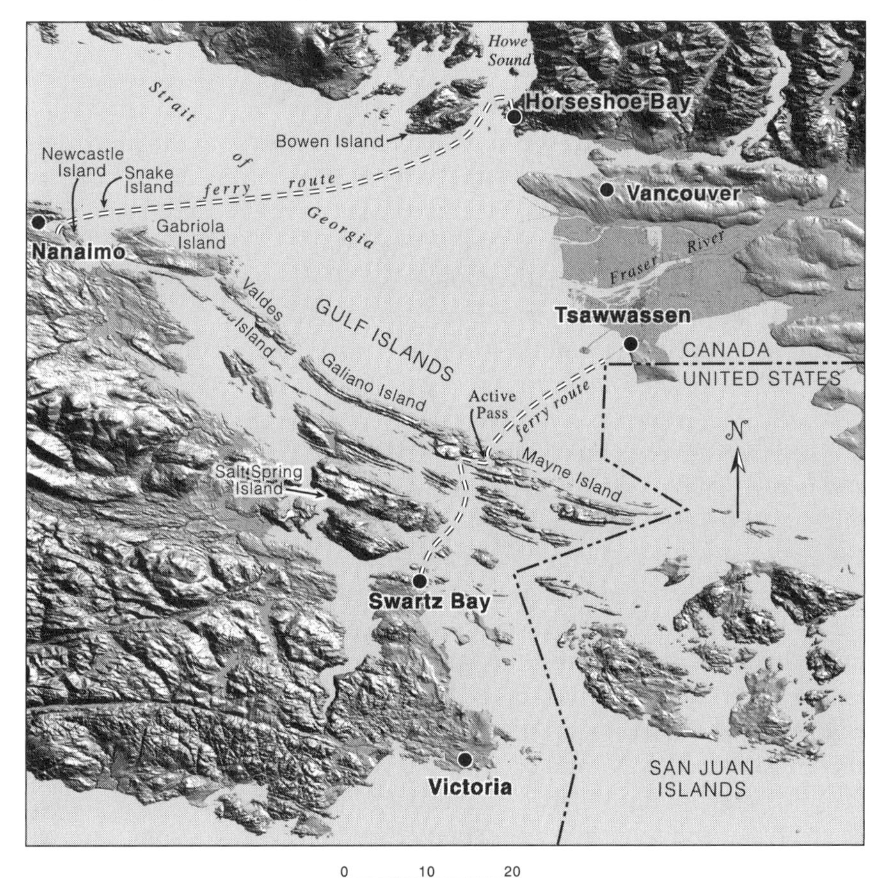

is about 200 metres below the terminal and rises to the west and north. You can see it in places above Triassic volcanic rocks on the north shore of Departure Bay.

About ten to fifteen minutes after leaving the Departure Bay terminal and nearly 4 kilometres from the harbour mouth, the ferry passes Snake Island, where resistant sandstone and conglomerate beds about 1,500 metres above the base of the Nanaimo group are exposed. These rocks are some of the youngest Nanaimo beds and probably belong to the same unit that forms the cliffs on the west shore of Gabriola Island, 6 kilometres to the south, and on the southwest sides of Valdes, Galiano, and Mayne Islands, for another 75 kilometres south-southeastward, nearly reaching the international boundary.

Turbidites from the Fraser River Delta

East of Snake Island the sea bottom drops off in what on land would be considered a gentle slope but which by marine standards is steep. About thirty minutes travel time from the dock at Departure Bay, the ferry crosses the deepest part of the Strait of Georgia on this route. The sea bottom here is a flat-floored trough about 400 metres deep that averages about 5 kilometres across and extends for more than 30 kilometres to the northwest and 20 kilometres to the southeast. The floor slopes continuously but ever more gently to the northwest, then becomes horizontal, and finally rises almost imperceptibly to the far end of the trough. A tongue of sediment that originates from the mouth of the Fraser River crosses the floor of the trough. High-powered echo sounders, instruments that measure water depth by bouncing sound waves off the sea bottom, reveal not only the shape of the upper surface of the tongue but also record layers within the tongue, which have different textures than neighboring layers of the same shape.

The Fraser River carries a mixture of sand, silt, and clay to the sea. The clay may remain suspended in the near-surface brackish water of the Strait of Georgia for days or even weeks and can be carried tens of kilometres from the river mouth before settling to the sea bottom. The sand moves down the riverbed and is deposited on the delta along with some of the silt and a little of the clay. Close to the river mouth, sedimentation may build the delta forward 8 to 10 metres per year. After an interval probably measured in decades or centuries, higher and steeper parts at the front edge of the delta become overloaded and unstable and collapse in a submarine landslide. Sand and mud in the landslide, stirred by its motion, form a soupy mixture that is significantly denser than the surrounding clearer water and flows down the submarine slope. On occasions it moves fast enough to erode and incorporate the finer sediment at greater depth.

Where the bottom slope becomes nearly horizontal, the murky mixture slows and deposits first the coarser particles and later finer materials. These coarser-grained beds that become finer-grained upward are called *turbidites*. The turbidites alternate with clay-rich deposits formed in the quiescent times, when clay-sized particles settle out from near-surface waters. Many sandstone beds at Pacific Rim National Park and some of those in the Nanaimo group were deposited in this way.

Texada Island and Malaspina Strait

From the ferry, the view to the north shows the high ridge of Texada Island, formed of Wrangellian rocks as are other, lower islands in this part of the Strait of Georgia. Most of Texada Island is Triassic Karmutsen basalt, with a small area of late Carboniferous rocks at the southeastern end of the island and late Triassic limestone and small Jurassic granitic intrusions at the northern end. The limestone is quarried for construction purposes and for making Portland cement. Iron and copper-gold deposits are associated with the hydrothermal alteration near the intrusions but are no longer mined.

By contrast, the Coast Belt mainland to the east of Texada Island, across the 5- to 10-kilometre-wide Malaspina Strait, is mainly late Jurassic and early Cretaceous granitic rock containing minor amounts of metamorphic rock derived mainly from Karmutsen and Bonanza volcanic rocks. The boundary between the Insular and Coast Belts lies beneath the water of Malaspina Strait.

More Seafloor

The eastern part of the ferry route passes over rough seafloor topography located at the southeastern end of two ridges that rise to within 22 metres and 38 metres of the water surface. Seismic sounding that penetrates the surface of these ridges reveals layering, or stratification, within them that is not as smooth and repetitious as the layering in the turbidite strata but is still nearly horizontal. The ridges are probably deposits laid down in a glacial environment.

On the last leg of the trip, southwest of Bowen Island in the entrance to Howe Sound, echo sounding reveals a stratified deposit dipping persistently at 10 degrees southwestward. This deposit is probably the same as the latest Cretaceous and early Tertiary sediments that underlie Vancouver and are exposed in places at the foot of the mountains north of the city, in sea cliffs and on beaches in the city, and periodically in downtown building excavations. The rough sea bottom southeast of Bowen Island is probably eroded in the underlying dark metamorphosed volcanic rock and pale granitic rock visible on the adjacent shores.

FERRY FOR HIGHWAY 17
Swartz Bay—Tsawwassen
See map of ferry routes on page 79

In roadcuts along Highway 17 between downtown Victoria and the ferry terminal at Swartz Bay are green, somewhat-metamorphosed Jurassic granitic and dioritic rocks that belong to the West Coast complex. The rocks visible at the ferry terminal are the overlying Nanaimo sandstone and shale.

Gulf Islands
The ferry route from Swartz Bay passes first through the Gulf Islands, named from the Gulf of Georgia, the original name given to the strait before its northern entrance was charted. Most of these are made of gently folded and reverse-faulted sandstone, shale, and conglomerate strata of the late Cretaceous Nanaimo group. The southern part of Salt Spring Island, which forms the mountain north of the ferry terminal, consists of Paleozoic granitic, volcanic, and sedimentary rocks that are some of the oldest rocks in Wrangellia. They are unconformably overlain by the Nanaimo strata, which form the northern two-thirds of Salt Spring Island.

The Gulf Islands comprise the island chain between Nanaimo in the northwest, Salt Spring Island in the southwest, and from there southeastward to the international boundary. A quick glance at a map of the outlines of the Gulf Islands is sufficient to show their distinctly linear character, which is found nowhere else along the coast of British Columbia. Nearly all the Gulf Islands display a strong northwest-southeast elongation as do, on a smaller scale, their promontories and bays. The variable resistance to erosion of the Nanaimo rocks, which are reverse-faulted with fault planes dipping steeply at the surface, folded in places, and tilted mainly to the northeast, controls the landforms. Where a resistant layer, such as sandstone or conglomerate, is at the surface, the land tends to stand high and form cliffs along the shoreline and ridges inland. Softer beds, such as shale, particularly where thick or accompanied by few resistant interbeds, are worn low, forming valleys and lying below sea level in channels or bays.

Geological Differences across the International Boundary
To the southeast of the Gulf Islands, the San Juan Islands in Washington separate the Strait of Georgia from Puget Sound. With the exception of a few islands of Nanaimo strata at the northwest end of the San Juan Islands, most of the rocks are very different from the Wrangellian rocks on Vancouver

Island. Paleozoic and early Mesozoic rocks are mainly chert, limestone, and basalt, and younger Mesozoic rocks are mainly sandstone and argillite. The latter are not very different in nature and age from some of the distinctive rocks of the Leech River schist and Pacific Rim complex on southern and western Vancouver Island. The older rocks are like those found to the east on the mainland in Washington in the North Cascades, and some of these can be traced north of the international boundary, about 80 kilometres southeast of the Greater Vancouver area. Not only are Wrangellian rocks not known in northwestern Washington, but so are the mainly late Jurassic and early Cretaceous granitic rocks of the southwestern Coast Belt.

Active Pass

Midway along the ferry route, about forty-five minutes after departure from Swartz Bay, the ferry passes through Active Pass, a narrow, Z-shaped channel between Galiano Island on the north and Mayne Island on the south. The pass is 5 kilometres long, 0.6 to 0.9 kilometres wide, and mostly 27 to 55 metres deep. Its origin is debatable; one suggestion is that it marks the site of a river that cut through the easternmost resistant ridges of Nanaimo sandstone and conglomerate in interglacial time.

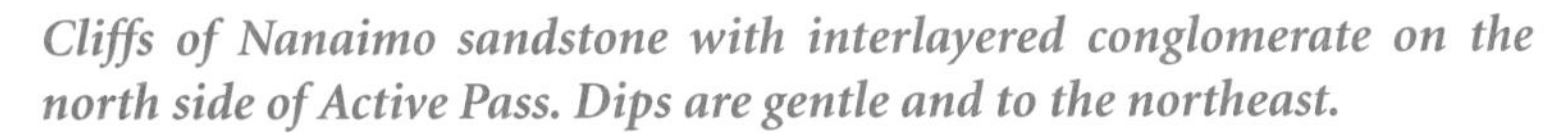

Cliffs of Nanaimo sandstone with interlayered conglomerate on the north side of Active Pass. Dips are gentle and to the northeast.

Between Active Pass and Tsawwassen

The ferry crosses the open and deep water of the Strait of Georgia, here just under 400 metres maximum depth. Seismic data and an exploratory borehole near Tsawwassen show that about 4.5 kilometres of Cretaceous and Tertiary strata underlie the region. At a deep drill hole just south of the mainland ferry terminus at Tsawwassen, more than 1,500 metres of late Cretaceous beds are overlain by 1,600 metres of Eocene beds and 1,200 metres of Miocene beds and capped by 250 metres of Pleistocene sediment.

The sea bottom drops off northeastward from Active Pass in a series of northwest-trending ridges and troughs. About 4 kilometres northeast of the eastern end of Active Pass, the ferry crosses a large fault, whose northeast side has dropped down and which may still be active. Bottom topography and echo-sounding profiles indicate the presence of northeast-dipping strata that can be traced 15 kilometres southeastward to an exposure of Paleocene sediments on Sucia Island in Washington State.

From the axis of the strait the sea bottom rises, abruptly at first, then very gradually, to a flat area at about 240 metres deep. The flat area terminates 10 kilometres west of the terminal, 13 kilometres south of it, and 16 kilometres to its southeast at an abrupt slope which forms a nearly perfect arc centred on Point Roberts, the headland just southeast of the ferry terminal. The submerged topography suggests that it formed as a classical sea level delta whose surface is now about 240 metres below sea level and whose sediment source was located at Point Roberts. Alas, there is no sign of a former river at Point Roberts or of a glacial outlet channel to the open ocean at a time when it was 240 metres below its present level. The origin of the deltalike area and the source of the sediments within it remain unexplained.

Fraser River Water

Commonly, during late spring and early summer the surface water of this part of the Strait of Georgia is turbid and brownish. Clear water is present only a metre or two below the surface and can be seen where churned up in the wakes of ferries or fishing boats. At times, tidally induced currents create sharp boundaries between clouds of murky water and less-murky water. At other times the surface exhibits a series of bands with sharp boundaries marked by concentrations of driftwood.

The murky water comes from the Fraser River during its flood stages following rapid snowmelt in the mountains. The fresh river water, being less dense than the salty ocean water, tends to float and spread over the water of the strait, mixing in slowly when stirred by winds, rapid tidal currents, or passing ships. A cloud of muddy water carried one direction from

This photo, taken in late July, shows a sharp boundary in the Strait of Georgia between clear seawater (left) *and turbid water from the Fraser River, which floats on the seawater. The Coast Mountains are visible in the distance.*

the river mouth during the flood tide will be carried in a different direction during the ebb tide, and even a shift in wind may produce a discontinuity in the muddiness of the surface water. In addition, the nearly horizontal boundary between less-dense water above and more-dense water below may have waves just as the sea surface does. These internal waves produce a regular banding on the surface of the water. Internal waves near the surface can displace surface waters, and driftwood becomes concentrated at the surface along a line where the surface water sinks. Curiously, Captain George Vancouver, who explored the Strait of Georgia in June 1792, failed to recognize that a large river emptied into the sea in this vicinity, and it was not until 1808 that Simon Fraser established its presence by descending the river from its upper reaches to tidewater.

Fraser Delta

The Tsawwassen ferry terminal is at the south side of the Fraser River delta and at the outermost limit of the broad tidal flats, with the high-water mark about 4 kilometres to the northeast. The tidal platform slopes almost imperceptibly seaward until it reaches the brink of the delta, where it drops sharply into deep water. The break in slope at the outer edge of the Fraser delta is a good site for docking facilities. Offshore there is adequate water depth for ships, inshore the shallow water permits easy construction of artificial islands for storage and loading facilities, and nearby is much loose sediment that was used to build the island on which the ferry terminal is located.

A second man-made island, located 2.5 kilometres northwest of the ferry terminal, holds a storage area and loading facilities for shipping containers and for coal mined in southeastern British Columbia and southern Alberta. Each day, an average of six trains arrive from the east, each carrying 10,000 tons of coal. The coal is either conveyed directly from the wagons to a waiting ship or else stockpiled.

The causeway from the ferry landing to the mainland crosses three belts of sediments on the tidal flats. The outermost belt extends along the outer delta between the low-water mark and midtide level, is composed of coarse to medium sand, and generally is well stirred by waves except in sheltered areas. The middle belt, extending from midtide level almost to high-tide level, is made of sand and silt that generally are well covered with vegetation. This belt forms an important rearing ground for fish. The innermost belt, a rather narrow strip of salt marsh near the high-water mark, is made up of organic silt.

Coast Belt

Coast and Cascade Mountains

The Coast Mountains are those mountains north and west of the Fraser River in southwestern British Columbia that extend from the city of Vancouver northwestward along the mainland coast of British Columbia and southeastern Alaska. South and east of the Fraser River the mountains are called the Cascade Mountains in British Columbia and the North Cascades in Washington. They continue southward to where the older rocks disappear beneath the extensive cover of volcanic rocks of the late Eocene to Holocene Cascade magmatic arc. Together, the Coast and Cascade Mountains make up the Coast Belt, a rugged mountainous region nearly 2,000 kilometres long and 100 to 200 kilometres wide, whose most distinctive geologic feature is that it contains abundant granitic rock.

In southwestern British Columbia, the Coast Mountains rise steeply from the Strait of Georgia and reach heights of about 1,000 to 1,500 metres near the coast, becoming progressively higher inland for a distance of about 100 kilometres. In the eastern Coast Mountains, many summits are nearly 3,000 metres high. Farther north, the mountains are even higher in places; Mount Waddington, about 300 kilometres north-northwest of Vancouver, is the highest point in British Columbia at 4,019 metres. About 200 kilometres east of the Strait of Georgia, the rugged topography, high elevations, and relief decrease, and the Coast Mountains gradually merge with the more subdued topography of the Interior Plateaus. Here, the belt boundary, delineated by bedrock geology and early Tertiary and older faults, differs from the topographic boundary, which is a bit farther east. They differ because the major geological boundaries were established by early Tertiary time, whereas the topographic boundary reflects late Tertiary and Quaternary uplift of the Coast and Cascade Mountains relative to flanking regions.

The topography of the Coast Belt reflects the combination of hard rock, high precipitation near the coast resulting in rapid erosion by both ice and

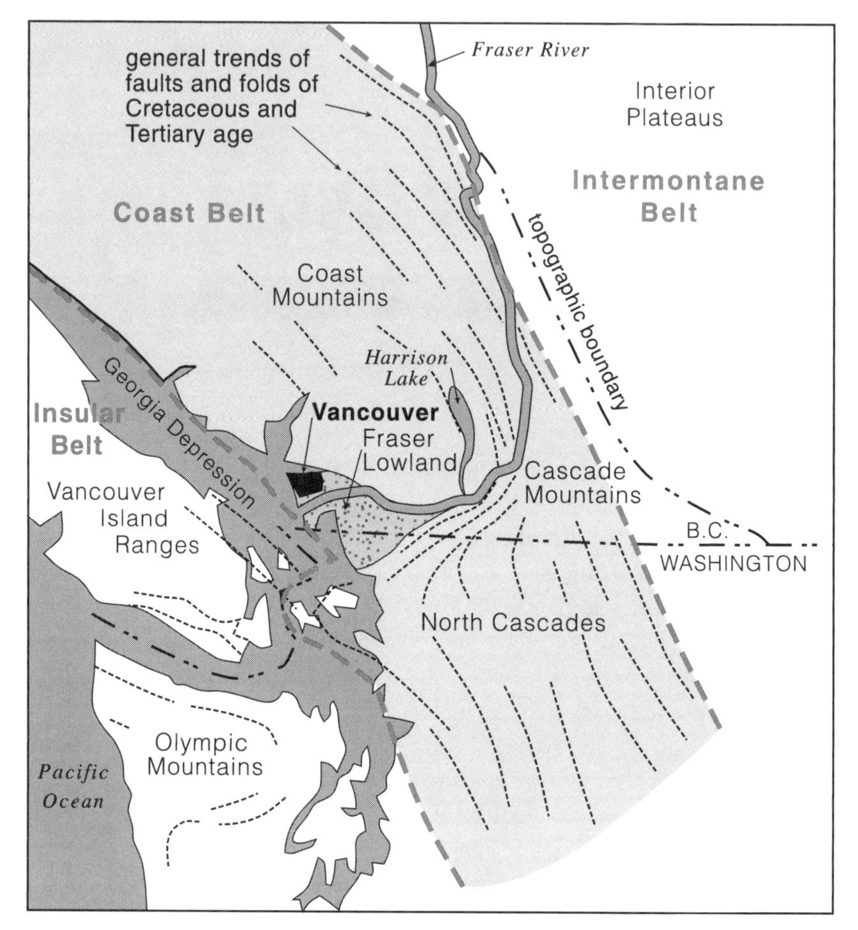

Major topographic divisions of the southern Coast Belt. Note that the eastern boundary of the topographic Cascades Mountains lies slightly east of the geological boundary of the Coast Belt. Faults and folds in the northwestern Cascade Mountains seem to wrap around the southwestern Coast Mountains.

running water, and geologically young uplift. The lower mountains nearest the coast are rounded because the Pliocene through Quaternary Cordilleran ice sheets covered and abraded them. The mountains higher than about 2,000 metres, farther inland, protruded through the ice sheet and have sharp summits and jagged upper ridges. Today, the higher summits, especially those on the wet western side of the mountains, carry permanent snowfields and small glaciers.

The mainland coast of British Columbia is deeply indented by fjords, long, steep-walled saltwater inlets that penetrate the Coast Mountains, and whose present forms owe much to sculpting by the glaciers that at one time descended the mountains to the sea. The coastline more closely resembles the coast of Norway than the comparatively smooth coastline of the conterminous United States, which was little affected by glaciation.

Granitic Rock of the Coast Mountains

Granitic rock makes up about 80 per cent of the Coast Belt. The term *granitic rock*, used throughout this book, is the general name for a family of rocks that cooled from molten magma and crystallized slowly in the crust, so that the entire rock is made of interlocking crystals. The most common variety of granitic rock in the Coast Belt is *quartz diorite*, which contains the white or grey mineral plagioclase feldspar as well as glassy quartz, and lesser amounts of the dark-coloured, iron- and magnesium-rich minerals biotite and hornblende. Another common variety is *granodiorite*, with increased but minor amounts of potassium-rich feldspar. *Diorite*, which is typically a black-and-white rock with plagioclase feldspar but no visible quartz, is less abundant, and true *granite*, defined by its high content of pink or white potassium feldspar, is very rare.

Typical granodiorite from the Coast Belt north of Vancouver; note roughly equidimensional, unoriented crystals of grey quartz, white feldspar, and black hornblende and biotite.

Bodies of granitic rocks are called *plutons*, named for Pluto, Greek god of the underworld, and granitic rocks are so abundant in the Coast Belt that it has been called the Coast plutonic complex. Plutons have an enormous range of size and shape. Some plutons are more than 50 kilometres long and perhaps about half that in width; the larger ones are called *batholiths*.

Different bodies of granitic rock may be distinguished from one another by differences in mineralogy, proportions of minerals, rock texture, degree of alteration and shearing, intrusive relationships, and age. In the Coast Belt, where plutons intrude plutons that intrude plutons, separating the different bodies from one another requires extreme care.

Boundaries between a pluton and the surrounding rock, called *country rock*, may be sharp in places but elsewhere are indistinct across a width of a kilometre or more. Clear signs of the invasion of granitic magma into the adjacent rock are cracks filled with veins of granitic rock, and fragments of the country rock frozen within the enclosing granitic rock. Boundaries may also be narrow faults or broad zones of sheared granitic rock, and a

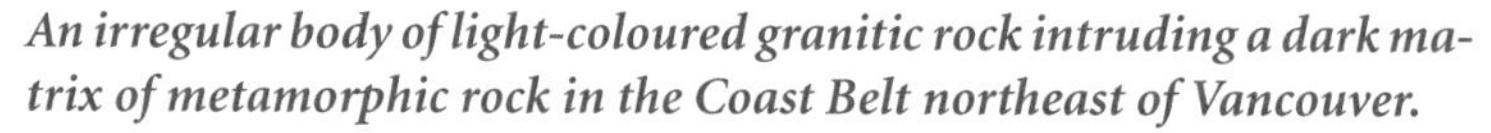

An irregular body of light-coloured granitic rock intruding a dark matrix of metamorphic rock in the Coast Belt northeast of Vancouver.

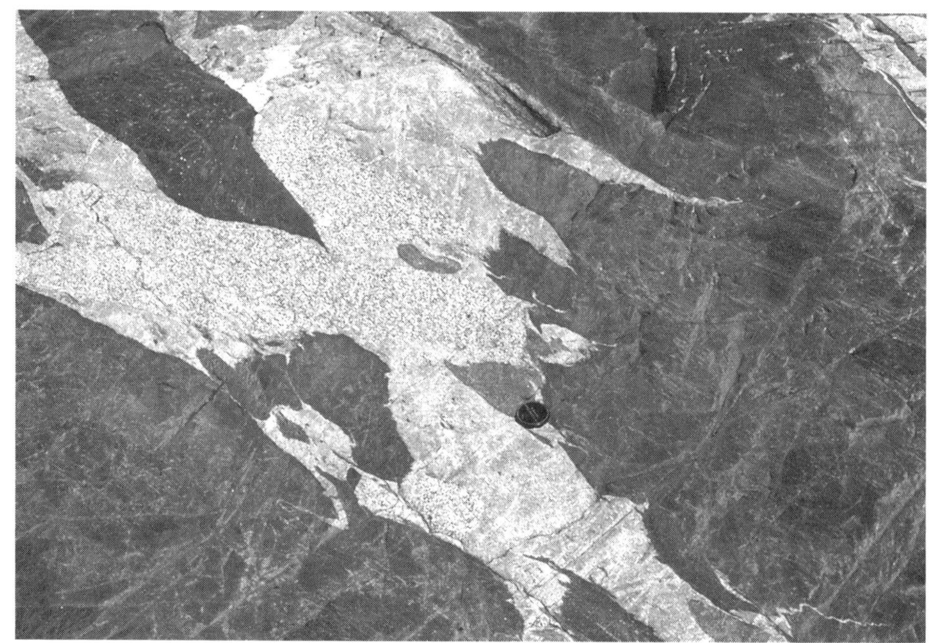

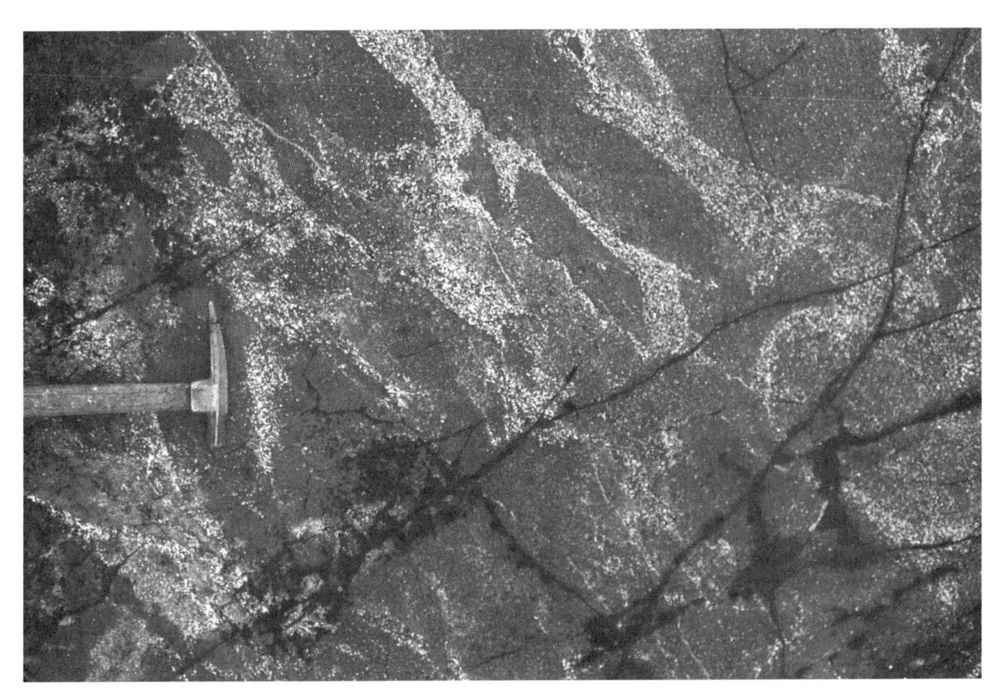

When granitic rock pervasively invades and alters blocks of dark country rock, it forms what superficially resembles a sedimentary breccia.

few boundaries are depositional, where volcanic rock or sediment was laid down on top of an older pluton that had been exposed by erosion.

Geologists determine the relative age of granitic rock by looking at its relationship to other rocks. Granitic rocks must be younger than the rocks they intrude and older than rocks deposited on top of them. As noted in the introduction, geologists determine the absolute age, the number of years before present that the rock crystallized, by using rates of radioactive decay of isotopes. Isotopic dating of the granitic rocks in the southwestern Coast Belt, west of Harrison Lake, shows that most of them range in age from about 170 to 95 million years old. Granitic rocks in the southeastern Coast Belt, east of Harrison Lake, and in the Cascade Mountains south and east of the Fraser River are mostly younger and range in age from 95 to 45 million years old and are associated with Cordilleran mountain building. Some even younger granitic rocks, about 26 to 18 million years old, straddle the Coast Belt, although they are concentrated in the eastern part, and represent older, deeper parts of the still-active Cascade magmatic arc.

Metamorphic Rocks between the Granites

Many of the remaining rocks in the Coast Belt are metamorphosed, made by subjecting pre-existing volcanic, sedimentary, and in some cases granitic rock to high temperatures and pressures, although in a few places the rocks are little metamorphosed. The increased temperature and pressure cause metamorphic minerals to grow at the expense of the former minerals in the rock may and may completely change its nature. The metamorphic rocks may form bands called *septa*, which can be many kilometres long and a few kilometres wide, or small, isolated bodies called *inclusions*, within the dominant granitic rocks.

Metamorphic rocks derived from volcanic rocks are common in the Coast Belt. Low-grade metamorphism of basaltic or andesitic volcanic rock produces a green rock called *greenstone*, and where deformation accompanied metamorphism and caused the mineral grains to grow in alignment, the rock is called *greenschist*. The alignment of the minerals creates a planar texture called *foliation*. If the same volcanic rocks had been exposed to higher temperatures and pressures they might consist in large part of the black mineral hornblende (one of the amphibole class of minerals);

Banded gneiss from within the southeastern Coast Belt. Light bands are quartz and feldspar. Dark bands are mainly amphibolite.

this rock is called *amphibolite*. Some coarsely crystalline amphibolites resemble diorites, and some diorite may have originated in this way. With still greater heating, and especially in the presence of water, the rocks may be partly melted. Any quartz and feldspar in the rock separate out first because they melt at lower temperatures than amphibole, and small bodies of melt may move away from their original site and coalesce to form granitic plutons, leaving behind a rock enriched in amphibole. This may result in a banded black-and-white rock called *gneiss* or *amphibolite gneiss*, with dark layers rich in minerals such as amphibole and biotite, and light layers rich in quartz and feldspar.

In the Coast Belt, metamorphic rocks derived from sedimentary rocks are generally less abundant than those derived from volcanic rocks. In the field, they often show up as bands, rusty-coloured when weathered, within the granitic rock, a characteristic that reflects their high content of the iron sulfide mineral pyrite. At low metamorphic grades, shale or mudstone form dark *argillite*, where no grains are preferentially orientated, or *phyllite*, which is a dark slaty rock with shiny surfaces. At higher grades, the rock may form glistening *mica schist*.

The temperatures and pressures indicated by metamorphic minerals vary considerably from place to place in the Coast Belt. In the southwestern Coast Belt, the former volcanic and sedimentary rocks appear to have been heated to high temperatures by the abundant granitic intrusions but probably never were buried very deeply. In places, such as on the western side of Harrison Lake and the east and west sides of the Cascade Mountains, the rocks are barely metamorphosed. However, about 10 kilometres east of Harrison Lake, the metamorphic grade is very high, and the metamorphic minerals there, typically either *staurolite* or *kyanite* and *sillimanite*, which are aluminum silicates derived from clay-rich mud and muddy sandstone, probably formed at temperatures of up to 700 degrees Celsius and at pressures corresponding to burial depths between 15 and 30 kilometres.

Components of the Southern Coast Belt

Terranes. In places such as the flanks of the Cascades and immediately east of Harrison Lake, the rocks are little metamorphosed and we can see that several different terranes are present in the Coast Belt, all brought together during mountain building. Wrangellia, the terrane that underlies much of Vancouver Island, can be traced eastward into the Coast Belt, where it occurs as disconnected shreds within the dominant granitic rock. Wrangellia and rocks linked to it in Jurassic time may extend from the western margin of the Coast Belt as far east as the valley containing Harrison Lake and Pemberton. East of this, and also south of the Fraser River and across the

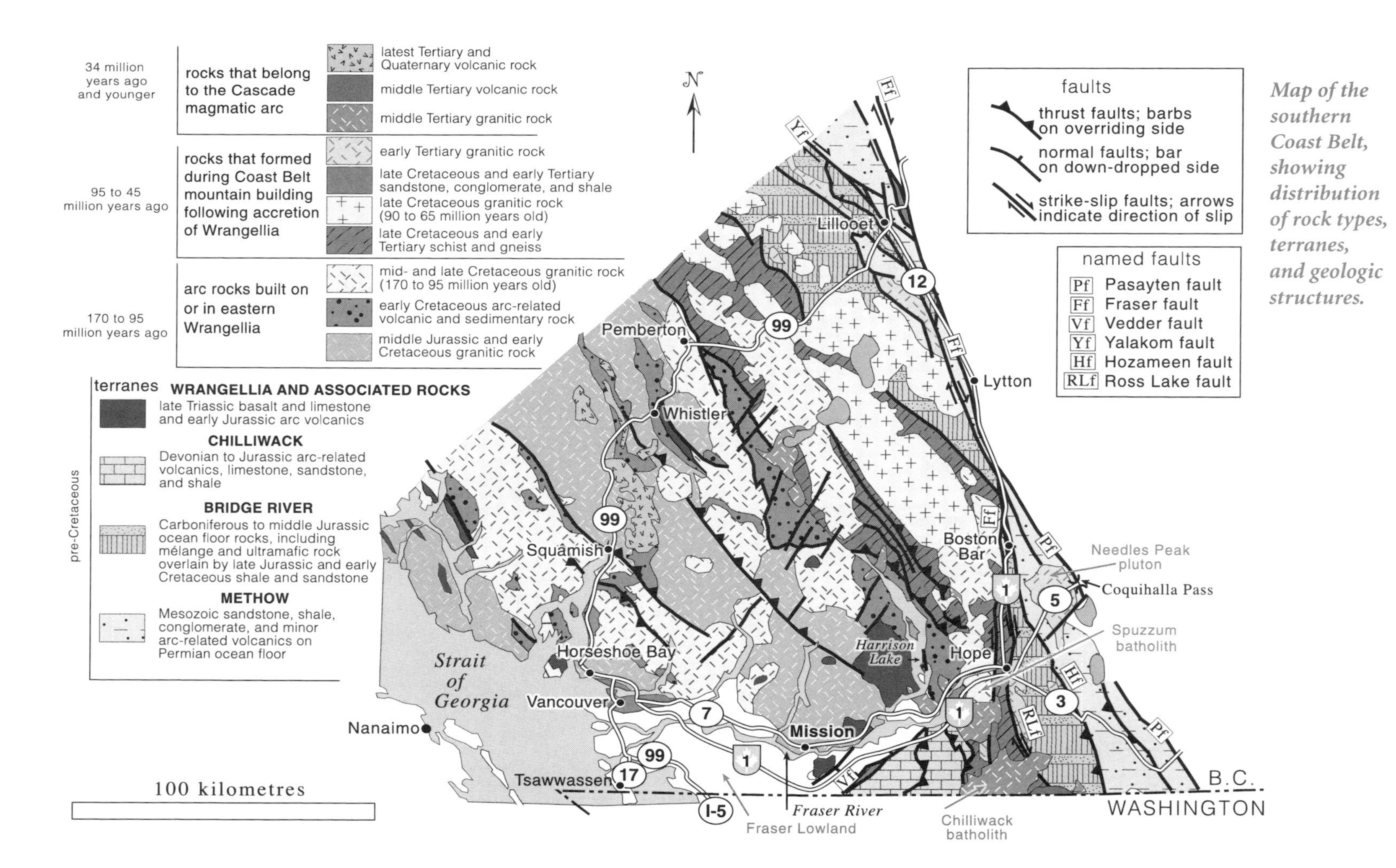

Map of the southern Coast Belt, showing distribution of rock types, terranes, and geologic structures.

international border, are three other terranes. The Devonian to Jurassic Chilliwack terrane contains mainly arc-related rocks that are widespread in the western Cascades but only form a sliver in the Coast Mountains at the southeast end of Harrison Lake. The Carboniferous to middle Jurassic Bridge River terrane appears to be remnants of an ocean basin floor incorporated in an accretionary complex in Mesozoic time and is overlain by late Jurassic and early Cretaceous marine sandstone and shale. The Permian to mid-Cretaceous Methow terrane consists of Permian oceanic crust, on which were deposited marine and mainly arc-derived shale, sandstone, and conglomerate of Mesozoic age. Its youngest sedimentary rocks contain detritus eroded from the Intermontane Belt to the east.

170- to 95-million-year-old granitic rocks. Middle Jurassic to early Cretaceous granitic rocks and marine volcanic and sedimentary rocks, all between about 170 and 95 million years old, are deeper parts of an island arc that formed on the eastern side of Wrangellia; higher parts of the arc are represented by variably metamorphosed volcanic rocks. The granitic rocks, exposed on Quadra and other islands at the north end of the Strait

Wrangellian rocks in the Coast Belt east of Texada Island include metamorphosed pillow basalt correlated with the Triassic Karmutsen formation of Vancouver Island. They retain their original pillow shapes but are now metamorphosed to mainly fine-grained amphibole with small red crystals of the mineral garnet.

of Georgia and underlying the mainland Coast Mountains, define the western boundary of the Coast Belt. These rocks are rare in the southeastern Coast Belt or in the Cascade Mountains.

90- to 45-million-year-old granitic and metamorphic rocks. Granitic rocks intruded during the folding, faulting, metamorphism, and initial uplift of the Coast Belt are about 90 to 45 million years old. In places, the granitic rocks are so intimately associated with metamorphic rocks derived mainly from the Bridge River and Methow terranes that the time of metamorphism can be determined by isotopic dating of the granitic rocks. The presence of granitic rocks about 95 million years old in both southwestern and southeastern parts of the Coast Belt provides the oldest firm evidence that all components of the belt were together by that time. Granitic rocks between 90 and 45 million years old are known only in the southeastern Coast Belt.

Cascade magmatic arc. Beginning about 40 million years ago, volcanic and granitic rocks of the Cascade magmatic arc were emplaced across all of the older rocks. The arc extends from southwestern British Columbia to northern California above the subducting Juan de Fuca and Gorda Plates. The older part of the arc in Canada is represented by volcanic rocks ranging from about 34 to 14 million years old, and by granitic rocks ranging from about 26 to 18 million years old. The younger part of the arc is more restricted in extent and is represented by volcanoes, lava flows, and a few small granitic intrusions less than 8 million years old. The young volcanic centres in Canada are aligned in a roughly north-to-south direction and lie between 50 and 200 kilometres north of Vancouver.

The volcanic cone of Mount Baker, about 100 kilometres east-southeast of Vancouver, last erupted in 1872. It lies just south of the international boundary and is the northernmost Cascade magmatic arc volcano in the United States.

Origins and Uplift of the Coast Belt

More than seventy years ago, Canadian geologist Colin Crickmay recognized that the highly folded and faulted rocks of the southeastern Coast Mountains east of Harrison Lake could be traced southward across the Fraser River into the Cascade Mountains. By contrast, older rocks of the southwestern Coast Mountains, between Harrison Lake and the coast, are barely represented in the Cascade Mountains south of the Fraser River. Crickmay also noticed that structures in the southeastern Coast Mountains bent around what he called the "buttress" of the southwestern Coast Belt. His observation provided the first major insight into the tectonic evolution of the southern Coast Belt.

It appears that in mid-Cretaceous time, about 95 million years ago, Wrangellia and a middle Jurassic to early Cretaceous arc built on its eastern part converged and collided with the western edge of the North American continent. Rocks of the intervening marine basin, represented by the Methow and Bridge River terranes in the southeastern Coast Mountains and the Cascades, were caught and squeezed in the tectonic vise between Wrangellia and the edge of the Cretaceous continent, folded and thrust upon one another, deeply buried in part and metamorphosed, and uplifted in part and eroded. It is these rocks, now in the southeastern Coast Mountains, that wrapped around the southeast side of Wrangellia and its associated arc—Crickmay's buttress—and continue southward across the Fraser River into the Cascade Mountains.

The first, and probably greatest, episode of uplift of the Coast Belt occurred during and shortly after the initial collision about 95 to 85 million years ago. East of Harrison Lake, granitic rocks that formed 95 million years ago are associated with metamorphic rocks formed at depths of nearly 30 kilometres. In the same area, granitic rock just 10 million years younger is associated with metamorphic rocks that formed at depths of about 15 kilometres. Geologists believe that the erosion that accompanied and followed crustal thickening and uplift thus removed a thickness of nearly 15 kilometres of rock from the area between 95 and 85 million years ago. The products of the erosion—sand, gravel, and mud—were carried westward by running water and deposited and consolidated as the sandstone, conglomerate, and shale of the Nanaimo group on Vancouver Island, in which the oldest rock is about 90 million years old.

Erosion continued through late Cretaceous and early Tertiary time and reduced the first-formed big mountains to rolling hills by Miocene time. We deduce this from three lines of evidence. First, Miocene fossil plants and pollen characteristic of a wet climate are found in sedimentary beds beneath 8-million-year-old lava flows on the presently very dry east side of

Three stages in the evolution of the southern Coast Belt.

middle Jurassic to early Cretaceous
Wrangellia Converges with North America

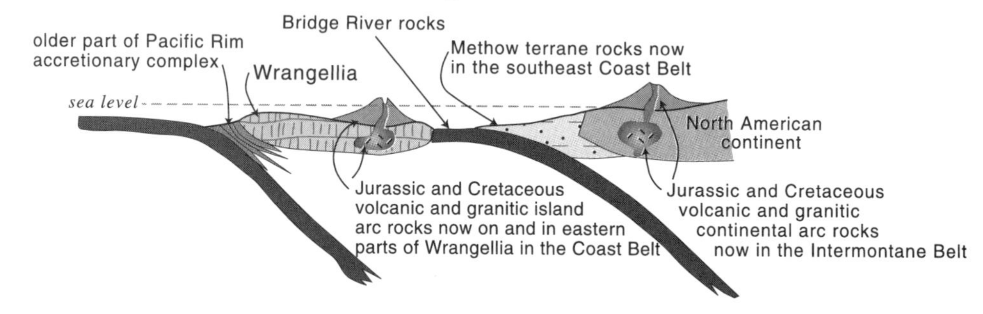

late Cretaceous to early Tertiary
Coast Belt Mountain Building

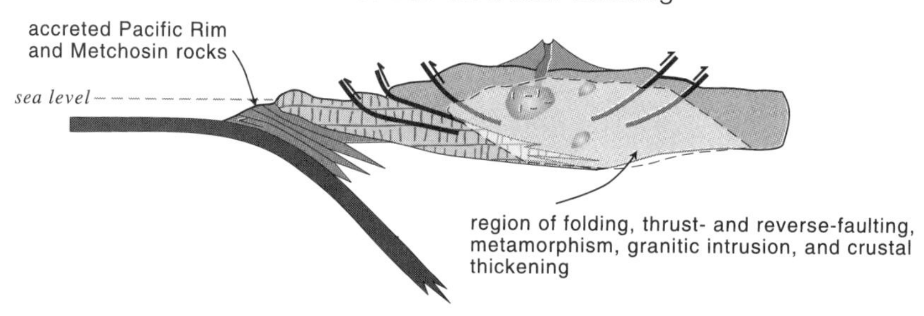

early Tertiary to Holocene
Cascade Arc Magmatism

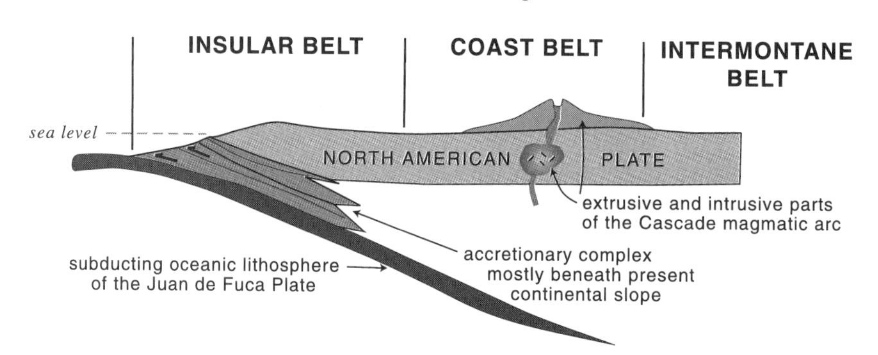

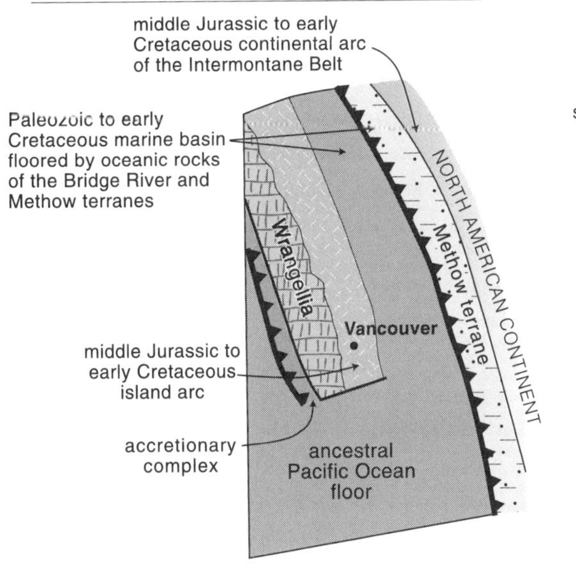

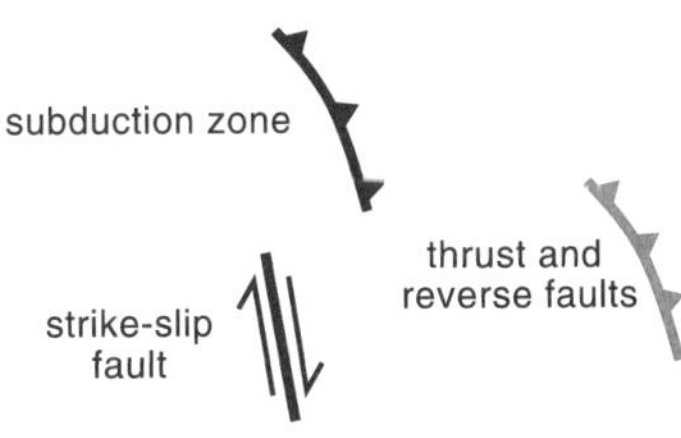

late Cretaceous to early Tertiary

now southwest COAST BELT

now southeast COAST BELT

Vancouver

Cretaceous to early Tertiary accretionary complex

accreted ocean basin sediments

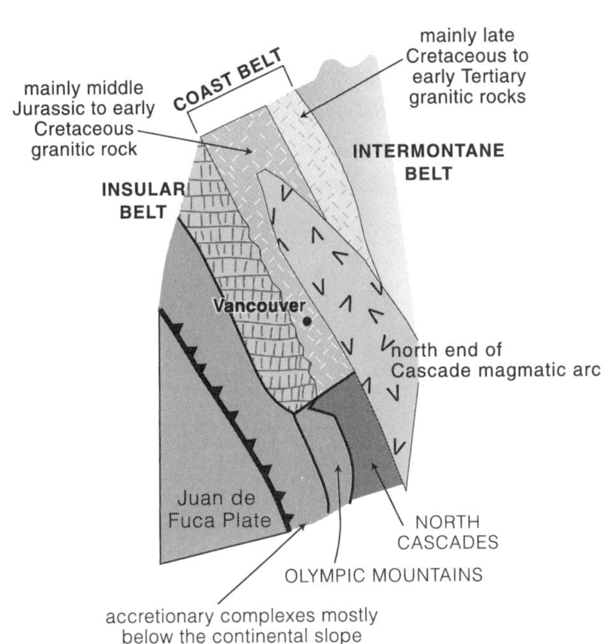

the Coast Mountains. If there were big coastal mountains in Miocene time, there would have been a rain shadow in the region as there is today, and the plant fossils would reflect a dry climate. The fact that big mountains exist today indicates that the region must have been uplifted again, this time geologically quite recently.

Second, on the eastern side of the Coast Mountains, 8-million-year-old lava flows have been gently tilted in places and are raised by as much as 1,800 metres above their flat-lying counterparts farther east in the Intermontane Belt. Signs of tilting are even seen in younger lava flows between 3 and 2 million years old.

More evidence of uplift comes from the tilts of bedding layers in sedimentary rocks that occur in patches at the foot of the mountains along the mainland shore of the Strait of Georgia and on the north side of the Fraser Lowland east of Vancouver. There, sandstone, shale, and conglomerate, in part equivalent to the late Cretaceous Nanaimo strata on Vancouver Island, and in part of early Tertiary age, were deposited as nearly horizontal beds by southward- or southwestward-flowing streams that brought gravel, sand, and finer debris eroded from the original Coast Mountains. The beds now tilt about 10 to 15 degrees to the southwest. The tilting resulted

View from Stanley Park in Vancouver looking northwest toward West Vancouver, which is built on the lower slopes of the southwestern Coast Mountains. The general dip of the slope approximates the tilt of the late Cretaceous and Tertiary strata and reflects uplift of the Coast Mountains in the last 10 million years.

from uplift of the mountains and subsidence under the Strait of Georgia sometime after they were deposited.

Finally, geologists think that erosion has removed between 2 and 4 kilometres of rock from parts of the southern Coast Mountains in the past 10 million years. The evidence for this is provided by examination of *fission tracks* in crystals of apatite, a mineral that is present in small amounts in granitic rock. The tracks, microscopic scars produced by the radioactive decay of minor amounts of uranium, tell us that some rocks of the Coast Belt cooled below 100 degrees Celsius about 10 million years ago. From boreholes and mines, we know this temperature generally occurs today at depths of 2 to 4 kilometres below the ground surface, depending on the local heat flow in the Earth.

It might be surmised that the divide separating streams flowing more or less directly to the coast from those draining to the interior was likely to develop at the site of greatest uplift, somewhere within the eastern part of the Coast Mountains. However, several streams now head near the eastern side of the mountains or on the Interior Plateaus east of it and flow westward right across the mountains in spectacular canyons. Examples include the Homathko River, its tributary Mosley Creek, and the Klinaklini River, all near Mount Waddington. These rivers may have originally crossed the site of the present mountains when they were little more than low hills and kept to their original courses as the mountains rose again at the end of Tertiary time. Such incised drainage patterns are called *antecedent drainage.*

HIGHWAY 1 AND HIGHWAY 7

Vancouver—Hope

150 kilometres

Two highways follow the Fraser River between Vancouver and Hope, Highway 1 on the river's south side and Highway 7 on the north side. Highway 1, also called the Trans-Canada Highway, is a divided highway with exits numbered with the distance, measured to the nearest kilometre, from the ferry terminal at Horseshoe Bay. Highway 7 is a slower road that follows the Fraser River more closely, winding its way to Hope. We've combined the roadlogs because many of the geologic stories along these roads involve the Fraser River valley as a whole. Look for the highway icon in the book margin to find sections pertinent to the road along which you are travelling.

Highway 1 and Highway 7 are built mostly on glacial and postglacial deposits of the Fraser Lowland, a roughly triangular area that extends from

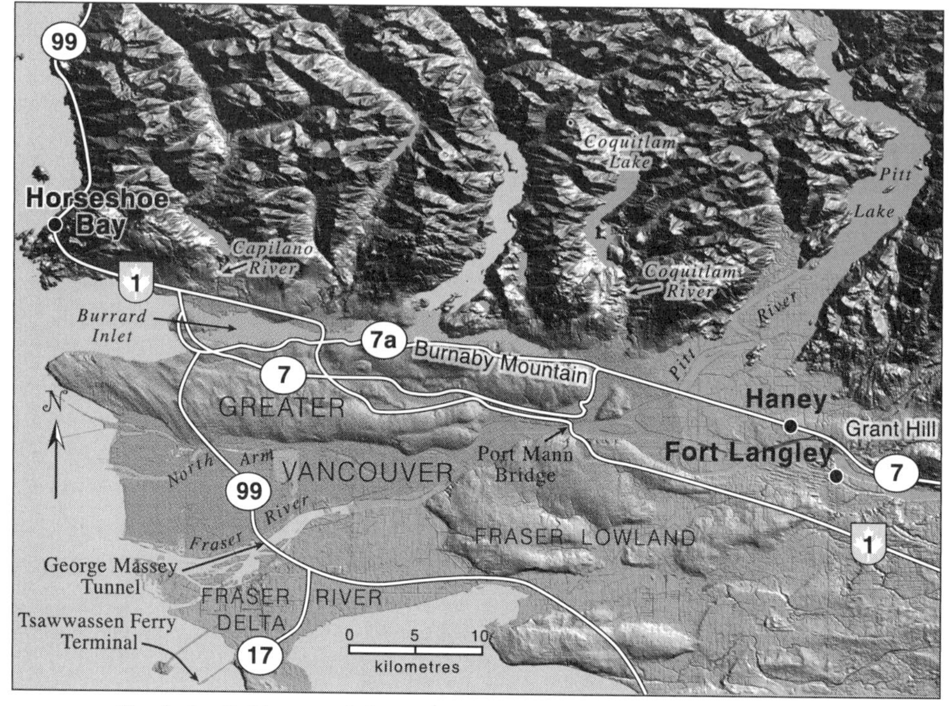

Shaded-relief image of the Fraser River valley near Vancouver.
—Image constructed using Canadian Digital Elevation Data obtained from GeoBase

Hope at its eastern point to the Strait of Georgia in the west and separates the Coast Mountains to the north from the Cascade Mountains to the southeast. However, between Chilliwack and Hope the valley narrows and is confined between steep mountain walls, and its easternmost tip is little more than the width of the Fraser River.

Hills, ridges, and plateaus within the Fraser Lowland fall into two categories: those that lie closer to the Coast Mountains, like Burnaby Mountain, Grant Hill, and Sumas Mountain are cored by bedrock; others lying more toward the centre of the Fraser Lowland are made of layers of glacial debris mantling older unconsolidated gravel and sand. In both types of hill, the glacial cover tends to be thinner and less continuous on the slopes facing the direction the glaciers came from.

Horseshoe Bay to Port Mann Bridge

Eastward from the ferry terminal at Horseshoe Bay, Highway 1 passes roadcuts in early Cretaceous granitic rocks and metamorphic rocks belonging to the southwestern extremity of the Coast Mountains. Between exits 14 and 16, Highway 1 crosses the Capilano River, which cuts through raised delta gravels into underlying late Cretaceous sandstone that rests, in turn, on weathered granitic rocks. The contact is exposed about 100 metres upstream from the bridge. Highway 1 crosses Burrard Inlet on the Second Narrows Bridge, recently renamed the Iron Workers Memorial Bridge. Weakly cemented Eocene sandstone and conglomerate occur near the south end of the bridge and form Burnaby Mountain to the east. East of this as far as the cloverleaf at Abbotsford, the highway crosses Quaternary sediments of the Fraser Lowland.

Highway 1 crosses the Fraser River on the Port Mann Bridge. Drilling for the foundations of this bridge (between exits 44 and 48) revealed a persistent peat layer in Fraser River sediments at a depth of 13 metres below present sea level. The peat is about 7,600 years old and dates back to a time when the last remnants of the ice sheets of eastern Canada and Scandinavia still survived. The subsequent melting of the ice and corresponding sea level rise accounts for much or all of the submergence of the Port Mann peat.

⑦ *Coquitlam Valley*

You can reach Highway 7 by taking the Hastings Street exit off Highway 1 just south of the Iron Workers Memorial Bridge. Follow Hastings Street eastward to join the Barnet Highway (7A), which circles around the north side of Burnaby Mountain, where barely consolidated Tertiary gravel is exposed in a roadcut. Pass through Port Moody at the head of Burrard Inlet, which is Vancouver's harbour, and join Highway 7 near the bridge over the Coquitlam River.

The Coquitlam Valley continues to be a major supplier in the Greater Vancouver area of gravel and sand for concrete and fill. However, this use of resources is under severe pressures from ecologists objecting to periodic if inadvertent discharges of muddy water into the Coquitlam River, and from homeowners objecting to heavy truck traffic through bedroom communities.

A few kilometres upstream from the bridge, the gravel deposits fill the bottom of the valley to a depth of 150 metres and have been incised by the Coquitlam River, particularly on the west side. Gravel and sand, brought in and deposited by glacier-fed streams, are overlain by till, an assorted mixture of silt, sand, and gravel deposited directly from glacier ice. The till is of no commercial value, being too rich in fine-grained material for easy sorting by washing or sieving and susceptible to heaving and collapse during freeze-thaw cycles. The till is cast aside so that the underlying gravel can be excavated.

Freshly excavated pit walls reveal not only the sequence of sediments but also organic-rich beds that yield material for radiocarbon dating and pollen analysis, enabling geologists to reconstruct some of the glacial history in this valley. A flood of outwash sediments from the north or west was followed by a mantle of till as glaciers of the last ice age first reached Coquitlam Valley somewhat less than 22,000 years ago. The ice front withdrew slightly, exposing the floor of Coquitlam Valley about 19,000 years ago, and streams deposited another thin layer of outwash. About 17,000 years ago the ice began a rapid advance to its outer limits, some 275 kilometres to the south in west-central Washington. It covered this distance in a brief span of approximately 2,000 years and deposited another layer of till. The fossil pollen record indicates a cool and continental climate accompanied by subalpine vegetation at the start of this advance.

⑦ *Pitt River and Pitt Lake*

Highway 7 crosses the Pitt River, a tributary of the Fraser River that flows south from Garibaldi Provincial Park and through the long, narrow Pitt Lake. The lake sits in a valley bounded by mountains about 12 kilometres upstream of the bridge. Just after the last ice sheet withdrew and the land was still depressed, Pitt Lake was a fjord linked directly with the Strait of Georgia. Though now isolated from the sea by the rebound of the intervening land and by the growth of the Fraser River floodplain and delta, it is still influenced by it.

With the rising tide in the Strait of Georgia, water backs up the Fraser River and, a couple of hours later, the effect is felt at the mouth of the Pitt River, 40 kilometres upstream. At this time the Fraser River starts sending part of its discharge up the Pitt River, and eventually Pitt Lake begins to rise. A few hours after the tide begins to fall, the surplus water in the lake begins to drain as Pitt River again flows seaward. The lake rises and falls with the tide by as much as 1.2 metres during low water on the Fraser River in late winter, and by as little as 0.25 metres during high water in June. The lakeward flow in the Pitt River has created a major reversed delta at the lake outlet. The delta is bigger than that formed by the upper Pitt River that flows into the lake at its head.

Though upstream flows on the Pitt River are short-lived and confined to the winter months and the spring tides, they do much to control the size and shape of the river channel. Below the lake, the Pitt River occupies a channel almost as deep and wide as that of the Fraser River even though its drainage basin is but a fraction of the size. In a single tidal cycle, Pitt Lake, with its 50-square-kilometre surface area, can rise and fall as much as 1.2 metres, requiring the inflow and escape of 60 million cubic metres of water. The rise and fall is accomplished in little more than six hours each way. Thus the mean stream flow can reach 2,750 cubic metres per second, only slightly less than that of the Fraser River itself.

The riverbed and the reversed delta at the outlet of Pitt Lake are enriched in sand carried upstream from the Fraser River. In any one tidal cycle, the turbid water of the Fraser River can be carried up the Pitt River for no more than about half of the 12 kilometres to the lake, so why does the sand not get carried back when the tide reverses? Three factors play a part in the sediment transport. First, the rising tide runs for a significantly shorter period but at a significantly faster rate than does the ebb. On many ebbing stages, the seaward-flowing water near the riverbed fails to reach velocities high enough to whip sand particles into motion, whereas on the corresponding flooding tide, the critical velocities needed to start sand moving upstream are exceeded early in the flood. Second, during the ebbing tide the high velocities tend to occur late, leaving a relatively short period for downstream transport, whereas in the flooding tide the high velocities occur early. Third, the first two factors, which dominate the sand transport in the Pitt River, also affect the form of the sediment accumulations on the riverbed. Underwater dunes are streamlined with respect to upstream flow, and this effect in turn permits higher water velocities close to the bed, favoring initiation of sand transport upriver. The influence of these three factors does not apply to the silt and finer-sized sediments, which are swept back down the Pitt River.

Much of the bottom sediment in the southwestern part of Pitt Lake exhibits rhythmic layering, with regular bands of homogeneous silt alternating with uniform to banded clay. The layering, once considered to be a tidal effect, is now known to be annual. A vertical core sample from the bottom of the lake was sliced into individual layers, which were then analyzed for radioactive cesium, a product of nuclear bomb tests in the atmosphere. The year-by-year fluctuation of the cesium fallout levels, as measured for example in analyses of cows' milk, shows a remarkable correspondence with the cesium pattern from Pitt Lake.

⑦ *Collapse of a Riverbank in Haney*

To see a large, collapsed riverbank, eastbound travellers should turn off Highway 7 at 216 Street in Haney and drive south for two blocks to River Road and thence 0.7 kilometres east. Westbound travellers turn off Highway 7 at Kanaka Creek, angle left on Haney Bypass, angle left again at River Road, and continue for another 0.7 kilometres.

The brushy recess in the otherwise gently curving riverbank, which is 60 metres high at this point, marks the site of a riverbank collapse on January

30, 1880. One excited newspaper editor in Victoria provided the following lurid headline: "A ridge of land breaks from its moorings and becomes a floating island. Curious and alarming event on the lower Fraser." He added, "A sound resembling a discharge of heavy cannon was heard." Residents rushed out to witness a gigantic wave apparently about 20 metres in height, although later reports made this more like 6 metres, sweeping across the river from the toe of the slump, which now obstructed the northern two-thirds of the river. The wave continued up and down the river, damaging docks and buildings along its shores for many kilometres. The wave also swept into the forested southern shore, battering down trees and killing a local farmer.

A slump at a site such as this is to be expected. A relatively high bank composed of soft clay and on the outside of a bend in a big river is prone to this type of slope failure. Collapse during the wet winter months is also common.

Could similar destructive events recur again nearby? There are records of this happening in the prehistoric past, but the railway, built in the early 1880s, provides protection and eases undercutting of the slope. Urbanization north

Shaded-relief image of topography along Highway 1 and Highway 7 between Abbotsford and Hope. —Image constructed using Canadian Digital Elevation Data obtained from GeoBase

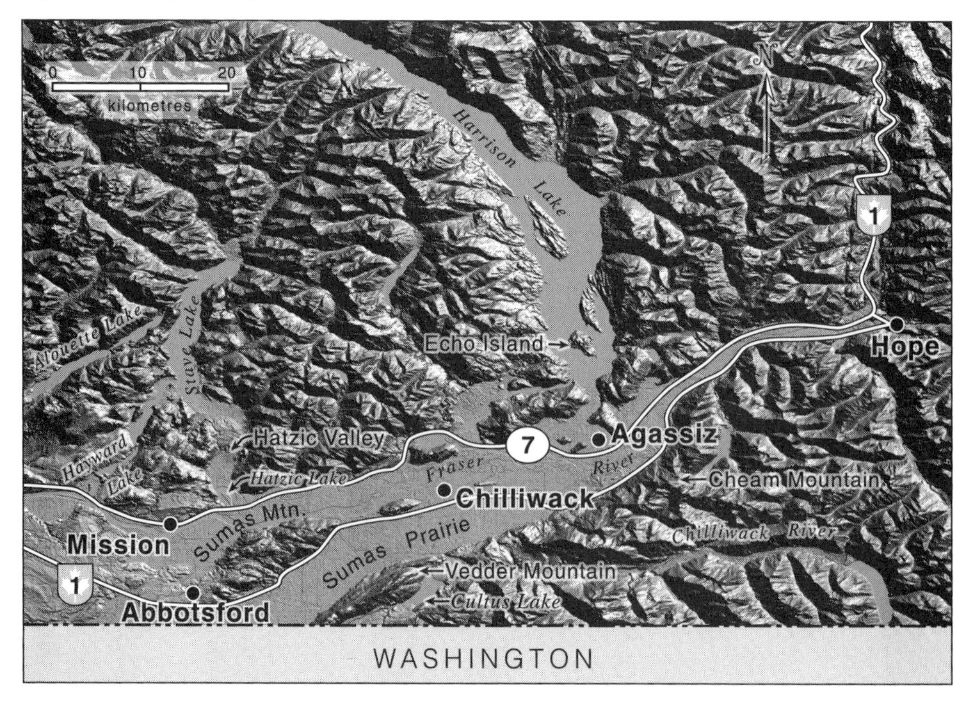

of the riverbank may have decreased the input of water into the clay and reduced its tendency to slump, but a leaky water main or an earthquake could still cause a repetition of the event.

Abbotsford Glacial Outwash Complex

Along Highway 1 just south and southwest of Abbotsford, the route crosses an area of glacial gravel and sand deposited by a major but short-lived river of glacial meltwater. It flowed from the south edge of an ice tongue that was about 3 kilometres north of the highway. A second ice tongue moving in from the east occupied the lower ground to the east and diverted the meltwater and its sediment load to the southwest. As the tip of the second ice tongue retreated eastward, the river flowed onto the freshly exposed low ground. Its sediment accumulated at successively lower levels to a final location a few kilometres to the east, where deposition abruptly ceased.

Shaded-relief image of Highways 1 and 7 near Abbotsford, showing the Abbotsford outwash, Hatzic Lake, and the very flat Sumas Prairie, the site of a former lake. —Image constructed using Canadian Digital Elevation Data obtained from GeoBase

A system like this forms a tier of river terraces with each bounded on the east by steep ice-contact faces and sloped gently from north to south or southwest. The pattern here is made more complicated by two additional processes. First, buried ice blocks melted to form kettles, some deep enough to extend below the present water table and hold standing water. Second, wind reworked the outwash sand, forming irregular dunes. The southern limit of the complex, where the river entered the sea, is only a few kilometres south of the international boundary, about 20 metres above the modern sea level.

⑦ *Channel Patterns of the Fraser River*

From a few kilometres above Hatzic Lake to about 80 kilometres downstream, near its mouth, the Fraser River flows mainly in a single channel about 750 metres wide in a series of straight reaches, each separated from the next by a relatively sharp bend. A few vegetation-covered islands may be present at these bends. The riverbed is mainly sand, and the river surface slopes very gently at an average of 5 centimetres per kilometre. At Mission, 75 kilometres upstream from the mouth of the Fraser River, the river surface is only 3.7 metres above mean sea level, and tidal fluctuations, although much less than at the Highway 7 bridge over the Pitt River, are still measurable.

Meandering river patterns are common in the Fraser River floodplain between Fort Langley and Chilliwack. Hatzic Lake, 6 kilometres east of Mission, provides an example. This small but attractive body of water is surrounded by the floodplain of the Fraser River on three sides and by a sharply rising bedrock hill on the fourth, western side. Called an *oxbow lake*, it is shaped like an inverted U. It was formerly a channel of the Fraser River but is now cut off from the main flow. The broad sweeping curve of its western shore is where the river formerly impinged on the high, resistant western bank made of bedrock.

About 3 kilometres east of Hatzic Lake, Highway 7 crosses another former meander channel, the semicircular loop of Nicomen Slough. Another example is the abandoned channel enclosing the settlement of Fort Langley, south of the Fraser River and 25 kilometres west of Hatzic Lake. Other examples occur as shallow looping depressions in the fields of Nicomen Island for a distance of 12 kilometres east of Hatzic Lake.

From the vicinity of Hatzic Lake, upstream for 75 to 100 kilometres as far as Hope, the Fraser River features a network of braided channels separated from one another by shifting gravel bars, many displaying their recent origin by the scarcity of vegetation on them. In its braided section, the river surface slopes at approximately 50 centimetres per kilometre,

The easternmost part of the Fraser Lowland, looking west: Cascade Mountains are left foreground; the Coast Mountains on the right; braided channels of the Fraser River in the middle foreground; the broader valley floor in the distance; and Sumas Mountain in the far distance on left.

ten times greater than the slope below Mission. The abrupt change in the channel pattern near Hatzic Lake coincides with an equally sharp change to coarser-grained riverbed sediments. Perhaps the supply of gravel from Fraser Canyon since the end of the last ice age has been trapped in the stretch between Hope and Hatzic, leaving none to be carried farther west. Or perhaps more recent landslides in this steeply walled part of the valley provided a new supply of gravel that has been swept downstream only this far. More data on the depth and composition of the gravel between Hatzic and Hope is needed to understand the reason for the change in texture.

⑦ *Hatzic Valley and Lingering Glacial Ice*

Hatzic Valley extends north from Hatzic Lake. Its south end is flat and close to sea level, but between 9 and 13 kilometres north of Highway 7 the valley is partly filled up to the 130-metre elevation with an unknown depth of unconsolidated sediment. This forms a drainage divide, north of which is Stave Lake, whose surface has been raised to 82.5 metres above sea level behind the hydroelectric Stave Falls Dam. Stave Lake continues 25 kilometres north of Hatzic Valley into rugged, mountainous terrain with glacier-clad summits.

On the west side of the unconsolidated fill that divides Hatzic Valley from Stave Lake, a dry, 10- to 20-metre-deep channel can be traced northward into Stave Lake. Surprisingly, the glacier-fed meltwater that cut this channel flowed northward. At the time the meltwater channel was being cut, the ice surface over southern Hatzic Valley must have been higher than the head of the channel, which now is over 100 metres above sea level. This tells us the ice must have lingered in the Fraser Lowland after it had withdrawn from at least the southern fringes of the Coast Mountains.

A similar picture holds for the southeastern side of the Fraser Lowland: ice-free conditions prevailed within the Cascade Mountains, while glacial conditions persisted in the Fraser Lowland. A major meltwater stream flowed west from Chilliwack Lake down the Chilliwack River for almost 40 kilometres into a lake dammed by glacial ice that covered the site of Cultus Lake and the Chilliwack River alluvial fan. Barred by this ice, the water escaped by a valley paralleling Sumas Valley on the east to join the Nooksack River in Washington.

The most obvious reason for the persistence of ice in lowland areas lies in the rapid rate of melting along the ice margin. There, bare rocks and soil soak up the warmth of the sunlight during the day and after dark radiate it back to the glacier, as well as to the night sky. Broad expanses of white to pale blue bare ice reflect much of the solar energy during the day and are too far from the ice margins to receive significant heat from them after dark. Thus, ice could melt away from confined mountain valleys before the broad glacial apron could melt in the Fraser Lowland.

Sumas Valley: Site of a Mid-Tertiary Graben

East of the Abbotsford cloverleaf, where Highway 1 crosses the north-to-south road from Abbotsford to Sumas, Highway 1 runs northeastward along the very flat floor of Sumas Valley, which is bounded by parallel northeast-trending ridges. Sumas Mountain is to the north and Vedder Mountain, which crosses the international boundary, is to the southeast. Sumas Mountain is mostly early Jurassic volcanic rock and middle Jurassic granitic rock, like rocks in the Coast Mountains on the north side of the Fraser River. The forested ridge of Vedder Mountain is made of amphibolitic gneiss, diorite, and Jurassic sedimentary rocks unlike those on Sumas Mountain but similar to others in the Cascade Mountains to the southeast.

The Sumas Prairie between Vedder and Sumas Mountains appears to be a structurally dropped block, known as a *graben*, which is German for "grave." The block lies between the Vedder fault, located close to the northwest side of Vedder Mountain, and a probable northeast-trending fault

View over quarries in early Tertiary rocks on Sumas Mountain, looking south across Sumas Prairie to the long northeast-trending ridge of Vedder Mountain. Higher peaks of the North Cascades in Washington lie in the background. —P. Mustard photo

near the southeastern side of Sumas Mountain. We deduce this from the presence of sandstone and shale beds of Eocene age that dip gently southwestward on the southwestern ends of both mountains. You can see the westernmost exposures of these beds in the low roadcuts in the northeast quadrant of the Abbotsford cloverleaf. They also form the low cliff on Sumas Mountain farther east that parallels the rise of the ridge and is visible from Highway 1. Near the Chilliwack cloverleaf, about 25 kilometres northeast of Abbotsford, where the road from Chilliwack to Sardis and Vedder Crossing intersects Highway 1, an exploratory oil and gas borehole drilled in the 1960s found Tertiary strata at a depth of about 300 metres, which is several kilometres below the projected base of the southwest-dipping Eocene rocks on Sumas and Vedder Mountains. The faults bounding Vedder and Sumas Mountains appear to be responsible for down-dropping the Tertiary rocks to their present position.

Because the Vedder and Sumas faults apparently offset the Eocene beds, we know they are younger than about 35 million years old. They are probably the same age as many other young northeast-trending faults in the southern Coast and Cascade Mountains. Near the northeast end of Harrison

Lake is an elongated, northeast-southwest-trending, 24-million-year-old granitic body that may have intruded along a normal fault. North of this, a series of volcanic centres aligned in a northeast direction are about 14 million years old and also may have been emplaced along a northeast-trending normal fault system. Near Highway 5 at the eastern margin of the Coast Belt, volcanic rocks dated at 22 million years old appear to have been faulted on a northeast-trending fault, in the upper Coquihalla River valley, against a 48-million-year-old Eocene pluton. Although we cannot say for sure, the faulting that formed the Sumas Valley graben probably occurred between 24 and 14 million years ago, during the early stages of the Cascade magmatic arc.

Floods in Sumas Valley

Between kilometres 92 at Abbotsford and 109, where Highway 1 crosses the Vedder Canal, the highway follows Sumas Prairie, an unusually level surface. The Sumas River flows only about 2 metres below the valley floor, and both rise almost imperceptibly southwestward between Sumas and Vedder Mountains to a point that lies only about 3 metres above and a few hundred metres northeast of the Nooksack River in Washington. The flat surface of Sumas Prairie is the ancient floodplain of the Nooksack River, which once flowed northeastward to the Fraser River but later adopted its present westward course. In November 1990, the Nooksack River overflowed its banks and spilled water over the imperceptible drainage divide and into the Sumas River.

Sumas Prairie was at one time occupied by a shallow, 6-by-10-kilometre lake, which was drained in 1923 to provide new agricultural land. Sumas Lake stood only 4 metres above mean sea level, so at times its outlet stream would ebb and flow with tides in the Strait of Georgia, 75 kilometres to the west. The lake was bounded on the northeast by land that rises, again almost imperceptibly, to the north end of Vedder Mountain, where the Chilliwack River leaves the Cascade Mountains. On its southwest side, the lake lapped onto the former floodplain of the Nooksack River. Low, timbered ridges, up to 15 metres high, marked its shores and were made in part of sand carried by the wind to form dunes, and in part of gravelly beach materials piled up by waves on the lake and by the push of lake ice in winter. Buildings sit on one such ridge east of the highway at kilometre 97, taking advantage of the good drainage and freedom from floods. Perhaps the builders remembered that in June of 1894 the Fraser River, swollen by late spring snowmelt, sent floodwaters up Sumas Lake to a point near the international boundary.

Chilliwack River Alluvial Fan

From near the Vedder Canal at kilometre 108 eastward to the Chilliwack River Road overpass at kilometre 119, Highway 1 traverses the lower part of the large alluvial fan of sand and gravel deposited by the Chilliwack River where it flows out of the mountains. On an alluvial fan, it is common for a stream to deposit so much sediment that it chokes its own channel and has to seek another course, which geologists call a *distributary channel.* The miniature meandering creek that Chilliwack River Road follows was the main channel of the Chilliwack River in the 1860s. The main stream today, the Vedder River, was only a minor stream then, fed by seepage and perhaps periodic overbank flooding. It passed westward from the apex of the fan through Mr. Vedder's farm into Sumas Lake. In the 1870s, more and more water from the Cascade Mountains tended to follow the Vedder

Shaded-relief image of the Chilliwack River alluvial fan (light brown) *between Sumas Mountain and Chilliwack.* AZ, *Atchelitz Creek;* CC, *Chilliwack Creek;* LK, *Luckakuck Creek.* —Image constructed using Canadian Digital Elevation Data obtained from GeoBase

distributary channel, and floods along the former lower Chilliwack River became correspondingly reduced. Now, the Vedder distributary accepts all the flow of the Chilliwack River.

The shift from the Chilliwack to the Vedder River was not a simple one. For a time two intermediate distributaries, Luckakuck Creek, which Highway 1 crosses 1 kilometre west of kilometre 119, and Atchelitz Creek, which is 2 kilometres farther west near the Luckman Road overpass, carried most of the water from the Chilliwack River. Settlers on lower Luckakuck Creek attempted to minimize damage to their land by chopping out logs obstructing the lowest part of the Chilliwack River, now called Chilliwack Creek. At the same time, First Nations people also living on the lower Chilliwack River were encouraging the diversion by felling more trees along the same channel.

In the past few decades, the upper part of Vedder River has shown signs of again bursting its banks in spite of efforts to contain it. Though intervention may delay a shift to still another course, the Chilliwack River's virtually unending contribution of sediment to the streams at its fan head makes a change inevitable.

(7) *Bedrock in the Southwestern Coast Mountains*

Between Horseshoe Bay in West Vancouver and Harrison Lake, most of the bedrock is granitic rock of middle Jurassic to mid-Cretaceous age, about 170 to 95 million years old. It intruded and metamorphosed volcanic and sedimentary rocks ranging from Triassic to early Cretaceous age. From near Dewdney, just east of Hatzic Lake, to the Harrison River valley, the rocks exposed along Highway 7 at the foot of the mountains are mainly quartz diorite and granodiorite of middle Jurassic age, about 167 to 164 million years old. These granitic rocks intrude Harrison Lake volcanic rock of early to middle Jurassic age that typically weathers to a rusty colour. This rock, which is greenish grey on fresh surfaces, is exposed in the roadcuts on the south side of Mount Woodside, 6 to 8 kilometres east of Harrison River. The volcanic rocks form the mountains southwest of Harrison Lake and Echo Island in the lake, and are tilted and in places faulted but not tightly folded. They may be related to the youngest parts of the latest Triassic and early Jurassic Bonanza volcanic rocks in Wrangellia on Vancouver Island. Farther north, along the west side of Harrison Lake, the volcanic rocks are overlain by fossiliferous sandstone, shale, and volcanic rocks of middle Jurassic to early Cretaceous age. These same rocks are also present on the east side of the lake, but there they are highly deformed and metamorphosed.

On Highway 7 just east of Agassiz, the roadcuts are in light-coloured Miocene granitic rock, about 19 million years old, and about 12 kilometres

The molds of early Cretaceous marine fossil clams of the genus Buchia *in sandstone from the west side of Harrison Lake, about 30 kilometres north of the Fraser River.*

northeast of Agassiz the roadcuts are in Oligocene granitic rock, about 24 million years old. These granitic rocks are some of the youngest known in the Canadian Cordillera and represent older and deeper parts of the Cascade magmatic arc. They are the northern extension of granitic rocks passed on Highway 1 near the Wahleach powerhouse. On the east side of the Oligocene granitic intrusion is its exposed intrusive contact with highly deformed metamorphic rock that possibly is derived from calcareous shale and sandstone equivalent to the fossiliferous early Cretaceous sandstone and shale west of Harrison Lake.

Bedrock in the Cascade Mountains

Between Chilliwack and Hope, Highway 1 climbs from the Fraser River floodplain or river terraces onto the lowermost slopes of the Cascade Mountains, where it traverses debris fallen from the steep mountainsides and roadcuts in bedrock. The mountains south of Highway 1 are slightly metamorphosed late Paleozoic and early Mesozoic volcanic rock, sandstone, shale, and limestone of the Chilliwack terrane. They were folded and thrust-faulted in a northwest direction in mid-Cretaceous time, about 95 million years ago, during Coast Belt mountain building. Near exit 138,

Mount Cheam rises 2,000 metres above the Fraser River; the highway at its base is roughly 50 metres above sea level. On the north face of the mountain, but difficult to recognize because of the dense vegetation, the same section of Permian limestone is repeated three times by folding and thrusting. The lowest body of limestone is passed near exit 138 and shale and sandstone weathering to a rusty colour overlook the highway farther east.

Near exit 146 to Herrling Island, granitic rocks of the Mount Barr pluton are about 18 million years old—some of the youngest-dated granitic rocks known in the Canadian Cordillera. Other granitic rocks to the south and east belong to the Chilliwack batholith, which ranges from 35 to 23 million years old. All of these granitic rocks cooled and crystallized at depths of no more than a few kilometres. At one place near the international boundary, the molten rock apparently broke through to the Miocene surface and created an explosive volcano where a thick accumulation of rhyolitic ash erupted and cooled 12 million years ago. The granitic rocks all belong to the older part of the Cascade magmatic arc, here uplifted and then exposed by very late Tertiary erosion.

Cheam Slide

Between 6,000 and 5,000 years ago, the Cheam Slide swept northwestward off the western slopes of Mount Cheam above Bridal Falls, crossed the site of the future highway, and deposited its debris, forming the hummocky topography between exits 135 and 138 on Highway 1. The debris covered a 1.5-by-2.5-kilometre area and extended at least as far north as the Fraser River. The alluvial fan built by Bridal Creek partly buries the slide scar and testifies to its antiquity. The rough topography of the landslide apron indicates that movement at the foot of the slope was fast—sliding probably was completed in minutes.

Creep at the Wahleach Power Station

About 7 kilometres east of the Cheam Slide, the blocky, windowless concrete structure by Highway 1 is British Columbia Hydro's Wahleach powerhouse, built in 1951 to 1952. The mountainside above the powerhouse is creeping slowly downslope, though the movement probably would have gone undetected were it not for construction and operation of the generating system.

To bring water down to the powerhouse generator, a horizontal tunnel was driven eastward behind the powerhouse for about 300 metres and then extended upward at an angle of 48 degrees until it broke through to the surface about 650 metres above sea level. Another east-west tunnel, about 3.2 kilometres long and 600 metres above sea level, linked the upper part of the inclined tunnel with the storage reservoir on Jones (or Wahleach) Creek. No slope movement was recognized when the tunnels were driven, but such descriptions as "soft and blocky" and "very soft rock" appeared in tunnelling records.

In January 1989, an unexpected flow of water from the upper tunnel down the mountainside to the highway prompted an investigation. It was discovered that outward movement of the rocks caused a 15-centimetre-wide rupture in the steel lining near the western portal of the upper tunnel. Other surveys showed that there was a slow downslope movement of up to a few centimetres per year. Recognizing the potential threat to the road and rail transportation corridor below this hillside, British Columbia Hydro installed a variety of ground-movement indicators, added a computerized warning system tied to the company microwave network, and developed a response plan should the downslope creep show any marked increase in pace. Field surveys showed that the area involved was about 1 kilometre in extent from top to bottom and up to 400 metres wide, and the fastest motion was a fairly steady 6 centimetres per year. The rate of movement diminishes gradually with depth, dying out at a maximum depth of about 120 metres. In all, about 60 million cubic metres of rock are moving. A continuing serious question is how many more hillsides in British Columbia have a cover of creeping rocks like this but remain unrecognized?

Water flow to the powerhouse was restored in September 1989 by way of the original tunnels with no obvious effect on the rate of creep, but a new tunnel system has been developed deeper in the mountainside to minimize the hazard.

⑦ *Suture in the Southeastern Coast Mountains*

A *suture* is the boundary between crustal blocks, terranes, or continents that at one time were separated by an ocean basin and are now together. Sutures are generally marked by big faults along which may be found fragments of former ocean floor rocks and scraps of ultramafic rock such as serpentinite. In the area between Harrison Lake and the Fraser River, and dispersed between the abundant granitic intrusions, are ultramafic rock and gabbro, metamorphosed basalt with recognizable pillows in places, phyllitic quartzite that probably once was chert, and metamorphosed shale and sandstone, all rock types characteristic of the floors of former ocean basins.

Near Johnsons Slough, about 14 kilometres east of Agassiz and 18 kilometres west of Hope on Highway 7, is a body of dark, foliated gabbro and amphibolite, as well as soapy-feeling talc schist derived from ultramafic rocks, and rare blocks of chert. Similar rocks are also found about 15 kilometres north of Highway 7 and from 5 to 10 kilometres east of Harrison Lake. Other scraps of similar rocks are present south of the Fraser River near the mouth of Jones (or Wahleach) Creek and in the high, rugged Cheam Range. These shreds and scraps of rock may be remnants of an ocean floor that in Jurassic time lay between the island arc rocks of Wrangellia to the west and the western margin of the continent to the east.

Spuzzum Batholith

⑦ The dark granitic rocks of the 95-million-year-old Spuzzum batholith are present on both sides of the Fraser River for a distance of about 14 kilometres downstream from Hope. The rock, which weathers to a rusty colour, is mainly quartz diorite and diorite. It is low in silica and relatively rich in aluminum, iron, and magnesium, so it contains more dark minerals than most of the other granitic rocks in the vicinity. You can see it along Highway 1 between exits 153 and 168 and along Highway 7 east of Ruby Creek, almost as far east as Hope.

Intrusive rocks of this age are associated with the highest grade of metamorphic rocks in the southern Coast Belt. In British Columbia they occur northwest of Hope, between Harrison Lake and the Fraser River. They also occur in the North Cascades in Washington, where they similarly are associated with high-grade metamorphic rocks. Metamorphic minerals in the country rock around the Spuzzum batholith north of the Fraser River suggest the batholith was intruded at depths of 25 to 30 kilometres in the crust during the earliest stages of Coast Belt mountain building.

Rocks of the Spuzzum batholith occur immediately east of Ruby Creek along Highway 7 about midway between Agassiz and Hope. High-grade metamorphic rocks are exposed a short distance up the gravel road that runs up the east side of Ruby Creek north of the highway. The "rubies" for which the creek is named are mostly small red crystals of the mineral garnet that grow in the metamorphic rocks.

HIGHWAY 1

Hope—Lytton

108 kilometres

Hope is located on the site of Fort Hope, a Hudson's Bay Company fur trading post built in 1848–49 at a major bend in the Fraser River. North of Hope, the Fraser River flows in a roughly north-to-south direction from Prince George, a distance of about 500 kilometres. Between Hope and

Lytton, about 100 kilometres to the north, Highway 1 follows the course of the Fraser River, much of which is in a deep canyon incised in the eastern part of the Coast Belt.

At Hope, the Fraser River turns and flows generally westward into the Strait of Georgia, about 140 kilometres away. The abrupt change in the course of the river at Hope may be largely related to faults that have weakened the bedrock, making it susceptible to erosion. North of Hope, the

Shaded-relief image of the Fraser River canyon between Hope and Lytton.
—Image constructed using Canadian Digital Elevation Data obtained from GeoBase

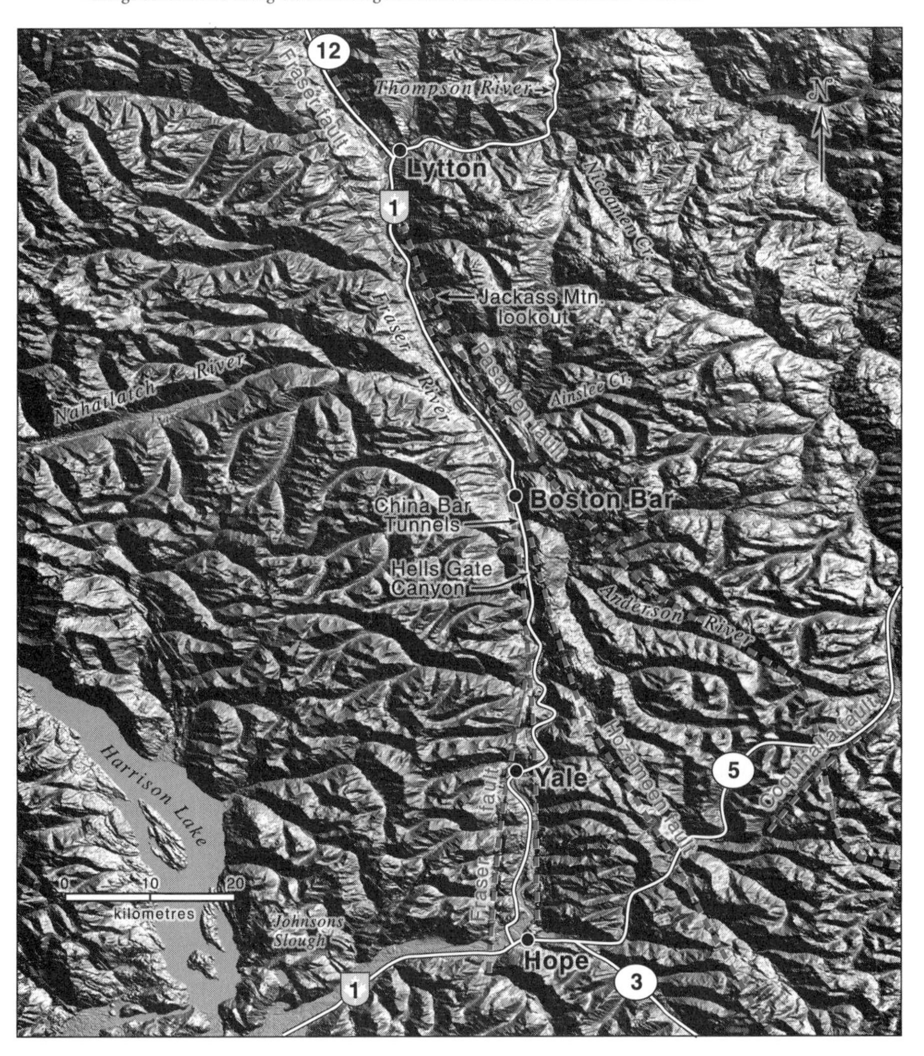

Fraser River follows the roughly north-south-oriented Fraser fault, which was active between about 45 and 35 million years ago. South of Hope, the granitic rock of the Chilliwack batholith, about 35 to 23 million years old, intrudes and obliterates the Fraser fault. The Fraser River southwest of Hope may follow the trend of northeast-southwest faults related to the 25- to 14-million-year-old Sumas Valley graben east of Abbotsford, although the similar geology on both sides of the Fraser River just west of Hope precludes any major offset across any faults in that area.

Fraser Fault System

About 2 kilometres west of Hope, a striking greenish grey cobble conglomerate is exposed on the south side of Highway 1 and in a railway cut across from it on the north side of the highway. Here, the western strand of the Fraser fault system separates the conglomerate from the Spuzzum batholith to the west. Originally, the conglomerate was thought to be Cretaceous, but a rare mudstone bed within it yielded Eocene pollen. Intermittent exposures of conglomerate, aligned in a north-south direction, occur above the east side of the Fraser River valley for 25 kilometres north of Hope, and similarly aligned conglomerate, intruded by the Chilliwack batholith of Oligocene age, can be followed southward across the border into Washington. The conglomerate probably accumulated in a down-dropped block during early movement on the Fraser fault. The trace of the western strand of the Fraser fault system also is marked by a series of aligned notches cut in the ridge crests on the west side of the Fraser River northwest of Hope. One of the notches is visible on the skyline across the Fraser River due north of the conglomerate outcrop along Highway 1.

The same conglomerate, but there reddish brown, forms the western abutment of the bridge over the Fraser River on the north side of Hope. The cobbles in the conglomerate include a wide variety of well-rounded granitic and metamorphic rocks, generally closely packed in a sandy matrix. The conglomerate probably was originally red, as in the exposure by the bridge abutment; the greenish grey colour in the exposure along Highway 1 west of Hope may be due to alteration caused by its proximity to a younger, probably Oligocene, intrusion just to the south.

In places there is just one Fraser fault, but elsewhere, as near Hope, a system of interconnected faults forms a zone up to about 3 kilometres wide. Fault movements shattered and crushed the rocks caught in the fault system, making them readily susceptible to erosion and stream cutting. The Fraser River north of Hope took advantage of this weakness.

The Fraser fault is an enormous, relatively straight structure. It can be traced more or less continuously for a distance of about 325 kilometres

Eocene conglomerate in a railway cut adjacent to Highway 1, 2 kilometres west of Hope, accumulated in a down-dropped block along the Fraser fault. The steep groove across the rock surface is a drill hole made for blasting.

A series of aligned notches in ridges on the west side of the Fraser River valley northwest of Hope mark the trace of the western strand of the Fraser fault system.

northward from Hope, along the roughly north-south valley of the Fraser River, to near the town of Williams Lake in central British Columbia. North of Williams Lake, the fault disappears beneath a cover of Miocene lava flows, although linear anomalies on aeromagnetic and gravity maps indicate that it may continue beyond Prince George, over 500 kilometres north of Hope. Eventually it may link up with the Northern Rocky Mountain Trench fault, which forms the boundary between the Omineca and Foreland Belts. South of Hope, it is partly obliterated by the Chilliwack batholith, but it does continue south of the batholith in Washington, where it is called the Straight Creek fault.

In the 1880s, G. M. Dawson of the Geological Survey of Canada recognized the fault-controlled nature of the Fraser River valley. However, he thought that the sedimentary rocks along the valley were down-dropped on bounding normal faults that brought them against flanking granitic rocks. Not until the 1950s did researchers on the Straight Creek fault in Washington recognize that the predominant motion on the fault was right-lateral strike-slip. *Right-lateral* means that if you face the fault, rocks on the opposite side of the fault have moved sideways to your right. *Strike-slip* means that most movement on the fault was horizontal. Rocks on the western side of the Fraser fault have moved northward with respect to those on its eastern side. In addition, the great length and relative straightness of the fault suggest the strike-slip displacement is of considerable magnitude.

Estimates of the amount of offset across a fault are made by identifying formerly continuous rock units and structures that now are on different sides of the fault and separated by some distance. Estimates of the amount of offset across the Fraser fault range from 80 to 190 kilometres, depending on rock units or structures chosen. The smallest numbers, between 80 and 110 kilometres, restore the continuity of the Bridge River and Methow terranes on both sides of the fault, and if a geological map is cut along the line of the fault and the west side is moved south, this looks at first like the best match. However, 140 kilometres of offset matches a belt of Eocene granitic intrusions just south-southeast of Hope with intrusions of similar age near Lillooet. This match places the prefaulting location of the town of Lillooet close to Hope. In addition, the 140-kilometre restoration brings Permian granitic rocks in the Chilcotin River canyon west of Williams Lake near Permian granitic rocks between Lytton and Lillooet. The 190-kilometre estimate of offset comes from matching distinctive metamorphic rocks east of Harrison Lake with similar rocks in the core of the North Cascades east of Seattle. The current "best estimate" of the offset is between 140 and 190 kilometres.

Significant movement on the Fraser fault probably began during the accumulation of the Eocene conglomerate in a down-faulted block. The fault

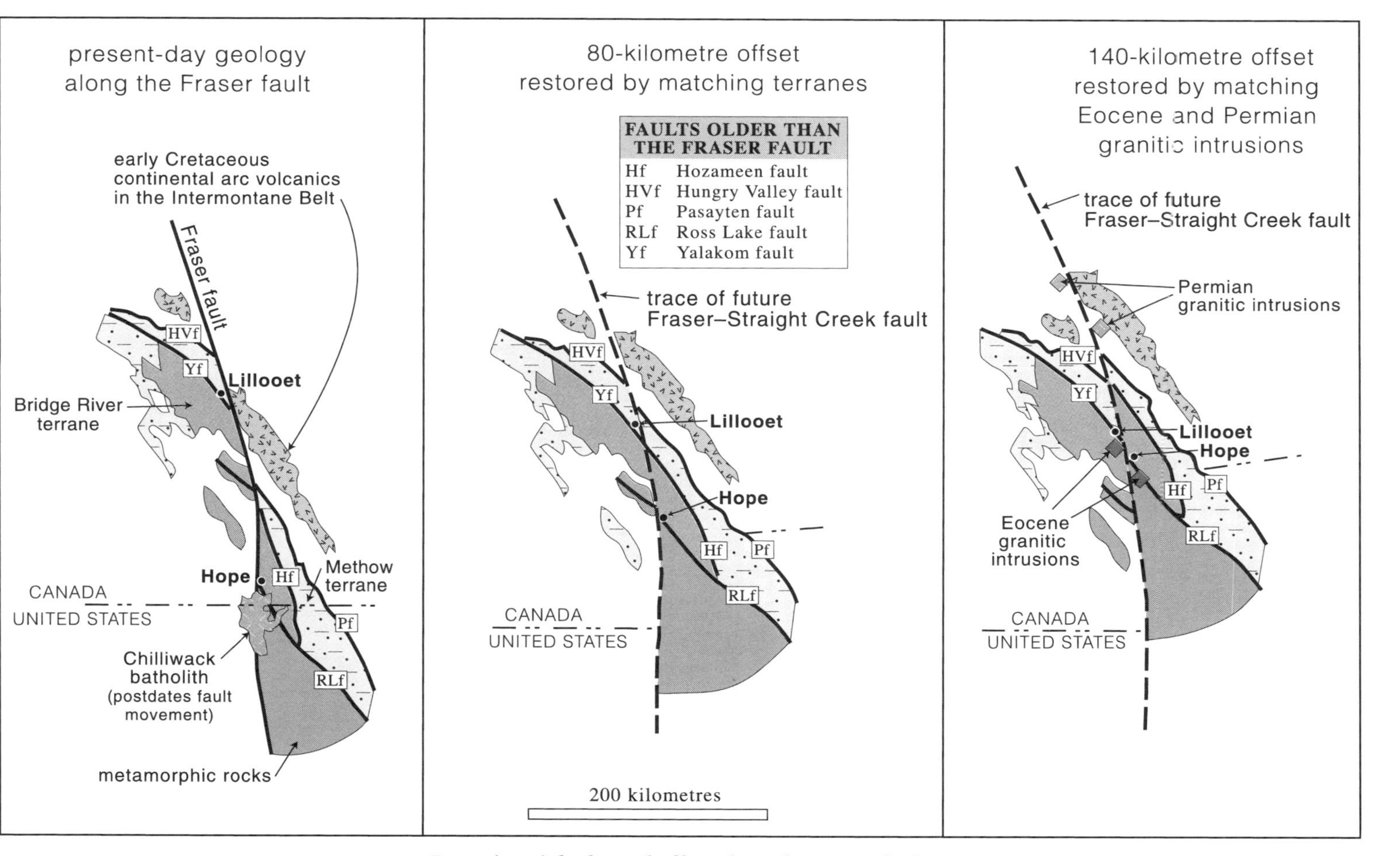

Restoring right-lateral offset along the Fraser fault.

offsets granitic rocks isotopically dated at as young as 45 million years old, indicating that movement was later. Movement on the fault was essentially completed before the Chilliwack batholith intruded across it between 35 and 23 million years ago, because different parts of the batholith are not offset by the fault.

Lake of the Woods Area

Highway 1 north of Hope climbs a hill up the remains of a large prehistoric rockslide to the Lake of the Woods. Roadcuts just north of the Lake of the Woods are greenschist of the Bridge River terrane, here called the Hozameen group, which weathers to a rusty or brownish colour and here forms the eastern wall of the Fraser River valley above an elevation of about 700 metres. The rocks low in the valley here are down-faulted on a fault on the east side of the Fraser fault system. North of this are low roadcuts in light-coloured rock containing dark layers. Originally the rock was gneiss, made of quartz and feldspar with dark hornblende layers, but it has been crushed and fractured, altered by hot water percolating along the Fraser fault, and weathers to a white colour with darker bands.

Hills Bar and the 1858 Gold Rush

On the east side of the Fraser River across from Yale, and 1.6 kilometres downstream, is Hills Bar, an important place in the 1858 gold rush. The great California gold rush of 1849 soon passed to the stage of organized corporate mining activity with little opportunity for individual reward, and many prospectors and miners were still carrying dreams of success. Their hopes were encouraged by rumors of new goldfields, notably one near Fort Colville, a former Hudson's Bay Company post in northeastern Washington, and another on the lower Thompson River. The Fort Colville find was said in 1857 to yield miners $10 to $40 per day. Prospectors on the Thompson River also found gold on the Fraser River, catalyzing another gold rush. In the spring of 1858, some twenty thousand enthusiastic gold seekers arrived in Victoria headed for the goldfields. They travelled by boat to the head of navigation on the Fraser River, which was at or below Yale depending on the state of the river and the vessel, and continued on by foot. Hills Bar was one of the first discoveries, in March 1858, and probably the most productive.

The gold at Hills Bar was of a very small particle size, referred to as "flour gold," which was thought to indicate lengthy transport from its lode source. The gold particles were within 60 centimetres of the upper surface of the bar and could be concentrated with a gold pan, or in greater volume with a rocker. During spring runoff, from mid-May to July, most of the bar was submerged, but as the river level dropped, progressively more of the bar became accessible. Though the high water interfered with mining operations, it seemed to bring in a new supply of gold from sources upstream. In September 1858, about nine thousand men were at work and their efforts for the year yielded slightly more than $500,000. These numbers seem inconsistent with a statement made at the time that the nine thousand men were making good wages, averaging $8 to $13 per day! Perhaps it is not surprising that in August alone almost four thousand gold seekers gave up and returned through Victoria on their way back to San Francisco.

Serious prospectors were not so easily discouraged. In July 1858, upon learning of Governor Douglas's proposal to build a pack trail to the upper goldfields, five hundred miners volunteered their services without pay for the summer. In addition, each agreed to pay a $25 bond for good conduct, which was to be returned at the end of the summer work in the form of provisions from the local store at Victoria prices. The trail was built from the head of Harrison Lake across several portages to Lillooet. Though it was soon abandoned in favor of another trail up the Fraser River and then along the Thompson River, it succeeded in reducing packers' rates to Lillooet from $1.00 per pound to 18 cents per pound. Flour that formerly had sold in Lillooet for $1.25 per pound would have fallen to perhaps 40 or 45 cents.

Hells Gate and the Salmon

Between the Yale and China Bar highway tunnels, Highway 1 and the Canadian Pacific and Canadian National Railways follow a particularly constricted 30-kilometre section of the Fraser Canyon, most of it in granitic or gneissic rock. Along much of the canyon this rock is highly broken and crushed by movement on the Fraser fault, but near Hells Gate, at the upstream end of this narrow section, is a small granitic body that intruded the fault in mid-Tertiary time, after major movement on the fault had ceased. Although jointed, the massive rock of this body resists erosion, and Hells Gate has the most constricted channel anywhere along the lower Fraser River. At low water, the entire flow of the river is confined to a channel 35 metres wide and 26 metres deep, and at flood peaks it is almost twice that depth. With a drop of 2.1 metres in the cascade passing through the channel gap and a velocity of up to 6.1 metres per second (22 kilometres per hour), Hells Gate presents a formidable obstacle to salmon swimming up the river to their spawning beds in the last stages of their four-year life cycle.

In historic times, back eddies about halfway up the constriction gave the struggling fish an opportunity to rest and recover their strength so they

The constricted Fraser River at Hells Gate. A tramway connects the level of Highway 1 with the river viewpoint.

could overcome the second part of the cascade. In 1913, waste rock excavated during construction of the Canadian National Railway along the eastern side of the canyon was dumped into the Fraser River, changing the configuration of its channel and eliminating one or more of the back eddies that had provided resting sites. Hells Gate now became an obstacle that only a handful of the fish could overcome, and most of these had become so exhausted that they were unable to get past the next rapid. Sockeye salmon by the millions choked the river downstream, struggling and dying without a chance of reaching their spawning grounds.

Fisheries officials took steps to alleviate the disaster by blasting out some of the debris from railway construction, but as the river level fell more debris was exposed, and the steps taken to clear the channel had to be repeated. Then on February 23, 1914, the problem was exacerbated by a rockslide from above the new railway grade. This not only blocked a partly excavated tunnel but also spilled into the river, narrowing its channel to 23 metres and adding about 100,000 tons of rock to the obstruction of the previous year. It was only with a monumental effort that the riverbed was sufficiently cleared in time for the 1914 salmon runs.

For a time in 1941, Hells Gate once again proved to be a nearly insurmountable barrier to the fish even though no new debris had choked the river. Careful study of this blockage and a search of old records convinced fisheries biologists that the salmon could cope with the low flows of the river and with some of the very high flows, but at the intermediate stages the currents were too fast or too continuous. The remedy was a pair of fishways designed to be passable at a range of stream flows. They were installed in 1944–46 under the auspices of the International Pacific Salmon Fisheries Commission, established by treaty between the United States and Canada, who share the salmon catch.

Bridge River and Methow Terranes

Between Yale and about 2 kilometres northeast of the northernmost highway tunnel at China Bar, the steep-walled gorge of the Fraser Canyon is in variably fractured and broken granitic and gneissic rocks. In fault contact with these rocks on the eastern side of the canyon, and mostly well above the road, are black and grey slate, chert, and basalt of the Bridge River terrane. North of the northernmost tunnel, the canyon widens and the road turns eastward for a short distance. Where it turns north again, a few blocks of greenish serpentinite crop out along the highway and a notch appears in the ridge to the south. These mark the place where the mainly

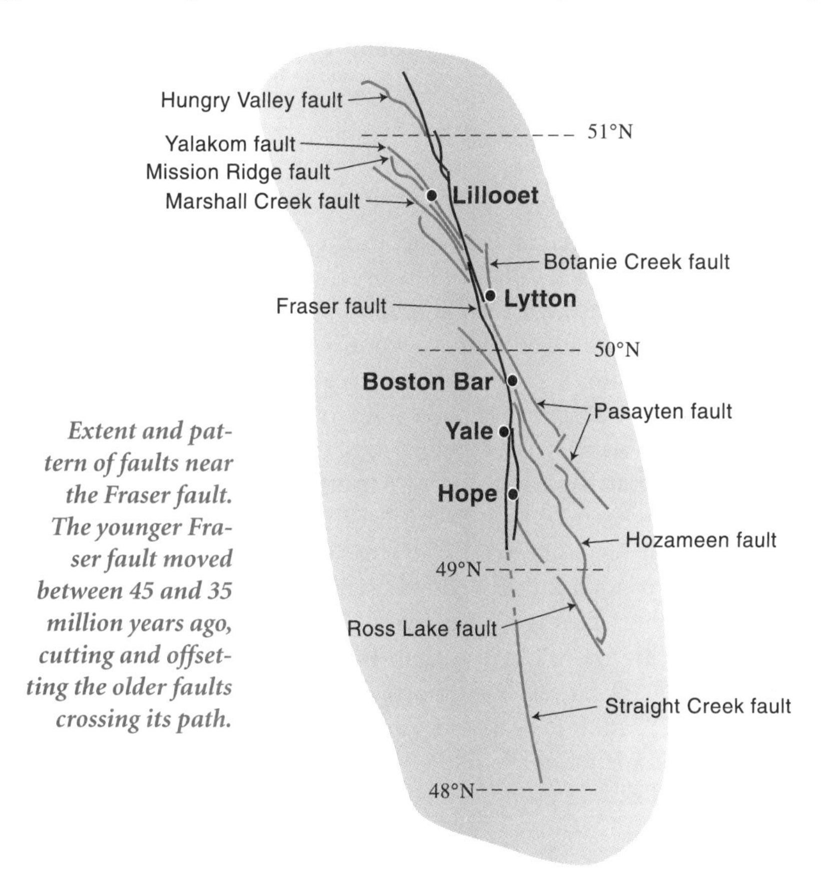

Extent and pattern of faults near the Fraser fault. The younger Fraser fault moved between 45 and 35 million years ago, cutting and offsetting the older faults crossing its path.

Thin-bedded darker layers of hard shale and lighter layers of volcanic ash–bearing siltstone of the Ladner group of the Methow terrane, 15 kilometres north of Boston Bar.

north-northwest-trending Hozameen fault smears into alignment with the younger, north-trending Fraser fault. The Hozameen fault marks the eastern limit of rocks belonging to the Bridge River terrane.

East of the Hozameen fault and north along Highway 1 are rocks of the Methow terrane. Between the Anderson River and north as far as Lytton, Highway 1 passes through early Jurassic shale and tuffaceous siltstone of the Ladner group (that is slightly metamorphosed to slate and phyllite, in places, as near Boston Bar) and early Cretaceous sandstone, shale, and conglomerate of the Jackass Mountain group. In the mountains to the south-southeast of the highway the Ladner group unconformably overlies pillow basalt and associated gabbro and serpentinite of an ancient ocean floor, probably of Permian age.

From Boston Bar to north of the Highway 1 bridge across the Anderson River, the rocks are thin-bedded argillite and siltstone with a strong slaty and phyllitic cleavage. The cleavage surfaces, commonly shiny and lustrous, here are more obvious than the bedding. Higher up the valley side, the rocks are far less metamorphosed and contain middle Jurassic fossils about 175 million years old. Early Jurassic fossils about 190 million years old occur farther north along the highway. The few fossils are mostly ammonites, an extinct group of marine animals related to modern squid and cuttlefish.

An Uncommon Valley Form

Ainslie Creek, about 10 kilometres north of Boston Bar, is spanned by a very high bridge that gives a superb view of the stream canyon's Y-shaped cross section, which is eroded in phyllite of the Ladner group. The deeper part of the canyon, corresponding to the stem of the Y, records an attempt by Ainslie Creek to catch up with the rapid lowering of the Fraser River during postglacial time. Parking space is available on the west side of the southbound approach to the bridge, but northbound vehicles must enter and leave this with caution!

Jackass Mountain

On July 20, 1871, the colony of British Columbia entered into confederation with the Dominion of Canada, which at that time was made up of five eastern provinces, from Nova Scotia to Manitoba, plus the vast tract of land that had been Hudson's Bay Company territory. One of the terms of confederation was that "Canada will assume and defray the charges for . . . the Geological Survey." In keeping with this term, the director of the Geological Survey of Canada, Alfred Selwyn, travelled from Montreal to Victoria in time for the proclamation, and then he, with an assistant and a photographer, spent the rest of 1871 in a reconnaissance of the geology of southern British Columbia. He travelled up the Fraser Canyon, then along the Thompson River and up its north fork to the uppermost part of the Fraser River in the Rocky Mountain Trench. He turned back only a day's march from Yellowhead Pass, which marks the Continental Divide near 53 degrees latitude in the Canadian Rocky Mountains. Taking advantage of the wagon road from Yale north through the Fraser Canyon, which had been completed only seven years earlier, Selwyn described and named several stratigraphic units, such as the Jackass Mountain Conglomerate group and Cache Creek group, using local place names. These stratigraphic names survive today, albeit in the case of the former with the word "conglomerate" deleted so that other rock types can be admitted to the unit.

About 25 kilometres south of Lytton, Highway 1 climbs high above the east side of the Fraser River and crosses a shoulder of Jackass Mountain. As the highway climbs the hill from the north or south, you can see roadcuts and outcrops of mostly greenish sandstone, which weathers to a tan colour, and shale, rock types that are much more common than conglomerate in the Jackass Mountain group. The Jackass Mountain conglomerate is conspicuous in the roadcuts at the crest of the long hill. A lookout here has easy access for southbound traffic, but northbound travellers would have to make an illegal crossing of the highway at a blind corner, so they should stop on the shoulder south of the crest of the hill.

The Jackass Mountain group is the youngest rock unit in the Methow terrane. Fossils, mainly ammonites, are rare but show that the Jackass Mountain group is marine and was deposited in early Cretaceous time, between about 130 and 100 million years ago. The conglomerate is made up of rounded cobbles and boulders embedded in a matrix of green, well-cemented sand. The composition of the sandstone and conglomerate here indicates that their source was a region containing granitic and volcanic rocks, plus some chert. This range of rock types is found in the southern Intermontane Belt, and recent isotopic dates from detrital zircons in the sandstone can be matched with the ages of similar rocks in the Intermontane Belt.

The rounding of the cobbles and boulders indicates that prior to deposition they were transported by vigorous streams heading in steep terrain or by strong wave action on an exposed coast. From there, the sediments were rapidly deposited by slumping and sliding into fairly deep water.

The Jackass Mountain conglomerate, across Highway 1 from the lookout (26 kilometres north of Boston Bar and 18 kilometres south of Lytton) features massive beds of conglomerate interlayered with greenish sandstone and grey shale. The rocks probably were deposited in submarine fans that descended the early Cretaceous continental slope from the east.

A submarine fan complex, such as that forming today on the continental shelf and slope oceanward of the Juan de Fuca Strait, is the most likely depositional setting. This submarine fan formed between a high-standing continent and a deep marine basin and probably marks the early Cretaceous continental margin of North America. The uppermost beds of the Jackass Mountain group contain plant material and probably were laid down near shore or even in a freshwater environment.

From the lookout, the trace of the Fraser fault lies directly below in the bed of the Fraser River. The western side of the valley is dark, silica-rich schist derived from chert and argillite in the Bridge River terrane. A little farther north across the river, the schist is intruded by white, granitic rocks that crystallized 64 million years ago, in earliest Tertiary time.

Fraser Fault at Lytton

Highway 1 continues north along the eastern side of the Fraser River as far as Lytton. The road lies within a fault sliver about 2.5 kilometres wide composed of Jackass Mountain sandstone, shale, and some conglomerate. West of the Fraser fault, whose trace is visible as a series of aligned notches in the ridges that descend the west side of the Fraser River valley, are the high Coast Mountains, here made mainly of light-coloured, late Cretaceous and early Tertiary granitic rocks that intrude the Bridge River terrane. Above Highway 1 to the east, and separated from the Jackass Mountain rocks by the Pasayten fault that runs through Lytton, are more light-coloured granitic and gneissic rocks, but these are of latest Paleozoic and early Mesozoic age and the westernmost rocks of the Intermontane Belt in this area. Here, the typically north-northwest-trending Pasayten fault that separates these older granitic rocks from the Jackass Mountain group is smeared into a more northerly alignment with the Fraser fault.

HIGHWAY 3

Hope—East Gate of Manning Provincial Park

84 kilometres

East of Hope, Highway 3 climbs some 1,300 metres to the summit at Allison Pass in the Cascade Mountains and then descends, following the eastward-flowing Similkameen River, a tributary of the Okanogan River in Washington, in turn a tributary of the Columbia River. Most of the route is

Shaded-relief image of topography along Highway 3 between Hope and the east gate of Manning Provincial Park. The topography becomes somewhat more subdued east of the Pasayten fault, which marks the geological boundary between the Coast and Intermontane Belts but lies within the topographic transition from the Cascade Mountains to the Interior Plateaus. —Image constructed using Canadian Digital Elevation Data obtained from GeoBase

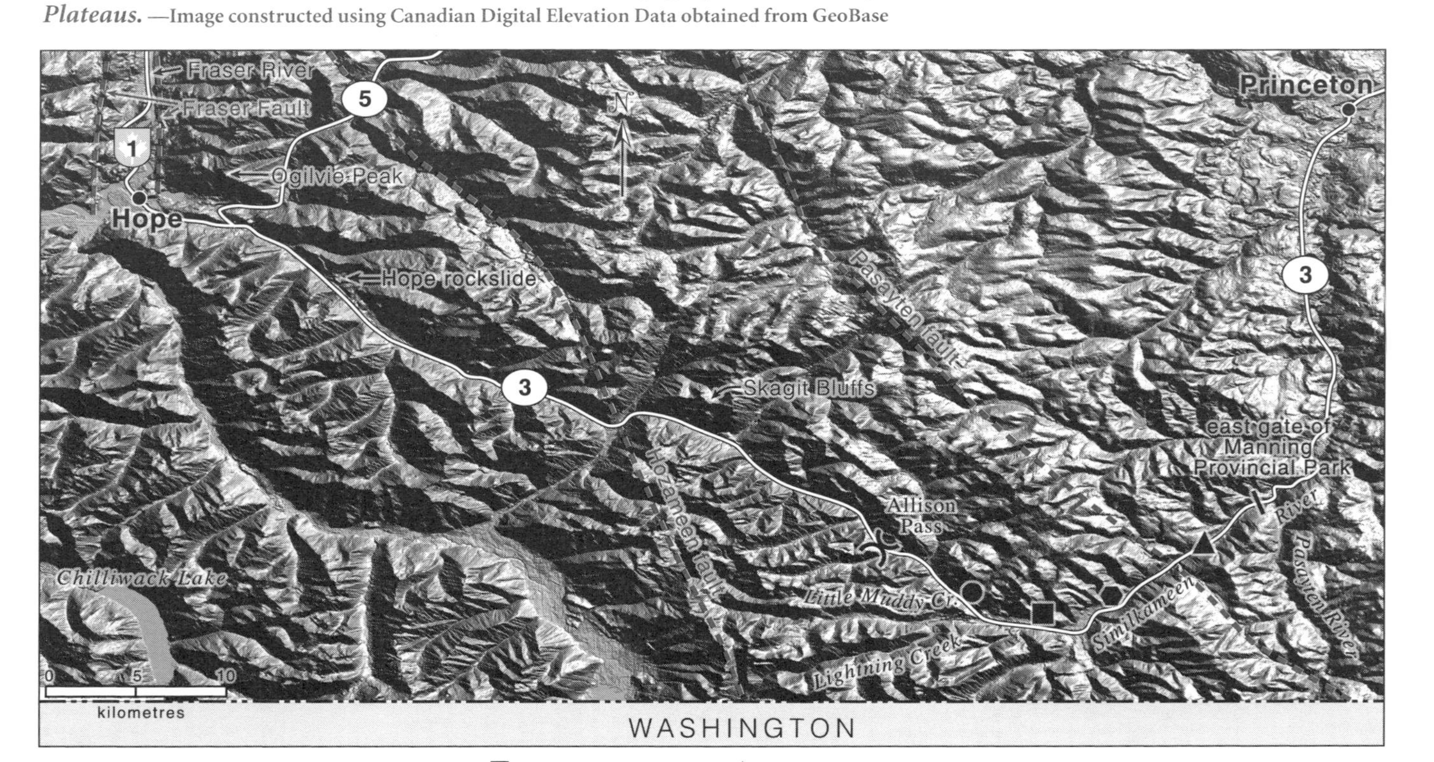

within Manning Provincial Park and crosses rocks of the Bridge River and Methow terranes. The Pasayten fault, about midway between park headquarters and the east entrance, marks the eastern geological boundary of the Coast Belt, but the mountainous topography of the Cascade Mountains continues for more than 60 kilometres farther east, nearly as far as the Okanagan Valley.

Bridge River Terrane

Near the junction of Highways 3 and 5, just east of Hope, look north across the Coquihalla River to the south side of Ogilvie Peak. There, at about the 1,100-metre contour, is the intrusive contact between light-coloured early Tertiary granitic rock, mostly lower down, and the hosting country rock, which is dark-coloured schist derived from argillite and chert and fine-grained amphibolite derived from basalt. The dark rocks belong to the Hozameen group and are part of the Bridge River terrane. From near Hells Gate on Highway 1, the terrane extends south-southeastward for 140 kilometres into Washington in a fault-bounded belt up to 20 kilometres wide. Between Hells Gate and Hope, the western boundary of the terrane is the Fraser fault; the Hozameen fault bounds its eastern side.

Most of the Bridge River rocks near Highway 3 are metamorphosed in low grades but have a strong foliation and their original nature is not obvious. However, on many of the peaks and ridges high above the highway, basalt is interlayered with chert, argillite, and local bodies of limestone, and although highly crumpled and broken, the rocks generally lack the foliation present lower down and are even less metamorphosed. Grey or white chert commonly occurs in irregular layers about 10 to 20 millimetres thick separated by shale a few millimetres thick. In many places, the siliceous shells of radiolarians, minute single-celled organisms, are present. Modern analogues of these rocks are the radiolarian oozes that accumulate on the deep ocean floor. Radiolarians in the Bridge River terrane lived during an interval that spans Carboniferous to Jurassic time, about 330 to 160 million years ago. The rocks of the Bridge River terrane, except for the limestone, are consistent with deposition on a deep ocean floor located far from a major source of terrestrial sediments.

Glacial Deposits

About halfway up the 4-kilometre uphill grade between the bridge over Nicolum Creek and the Hope rockslide viewpoint, the highway bends to the west around a rocky promontory. The bedrock immediately southeast of this promontory is overlain by coarse gravel containing boulders as much as 2 metres in diameter. Such large boulders could not have been transported far and may have been dumped directly off an ice tongue that once filled the valley to the northwest. The coarse gravel is overlapped in turn by much finer gravel and sand exposed in the roadcut for the next few hundred metres. How far northwest the ice front had retreated during the time that meltwater was depositing these higher gravels and sands is not known.

Hope Rockslide

In the early hours of January 9, 1965, a large snowslide blocked Highway 3 about 18 kilometres southeast of Hope. A few hours later a very large rockslide brought down 130 million metric tonnes of rock from the mountainside northeast of the highway. Four people attempting to extricate a car from the snowslide were killed by the rockslide.

From air photos taken of the slide area before the event and a new set taken afterward, the British Columbia Ministry of Transportation and Highways prepared detailed maps showing the changes. Most of the rock was lost from the uppermost one-third of the slide scar, which shed a lens-shaped block about 800 metres in diameter, mostly more than 30 metres thick, and in places almost 150 metres thick. The rock was mostly massive

Hope rockslide from Highway 3. —R. Turner photo

to slightly foliated greenstone of the Bridge River terrane but included minor light-coloured sheets of granitic rock. The latter dipped about 30 degrees to the southwest, roughly parallel to the surface of the present slope. The failure is believed to have taken place along zones of weakness concentrated along the top or bottom contacts of these granitic sheets.

What triggered the slide? High groundwater pressure at the time of the slide is considered to have been unlikely, because meteorological records indicate that below-freezing conditions prevailed for more than three weeks prior to the sliding, and that the only precipitation was snow. Two minor earthquakes of magnitudes 3.2 and 3.1 occurred at this site on the same morning at about 4 AM and 7 AM respectively. The time of the first earthquake was close to the time of the snowslide, and the time of the latter close to the time of the rockslide. It has been suggested that the earthquakes were the triggers for sliding, although a much larger earthquake in the general area on December 14, 1872, with an estimated magnitude of 7.5, failed to initiate major sliding at this site. Some seismologists suggest that the earthquakes in 1965 *record* the slides and were not the triggers. Pre-1965 aerial photographs show linear trenches near the head of the slide. New studies show that the deformation of the trench fillings is gradual, not episodic, and that long-term downhill creep of the mountainside slope probably caused its stability to deteriorate until it collapsed.

An ancient landslide at this site covered almost the same area as its later counterpart. It was derived from the middle slopes on the northeast wall of the valley. Wood from a root under this old slide gave a radiocarbon date of 9,680 years, dating the ancient slide back almost to the time of deglaciation.

The Hope slide makes us more aware of geological hazards, and we've learned that one big landslide may not eliminate the chance of a future landslide at the same site. It has also emphasized the problem of prediction, since the trigger for a slide is so hard to recognize.

Hozameen Fault

About 19 kilometres southeast of the Hope rockslide viewpoint, Highway 3 turns abruptly to the northeast and climbs a steep hill to Rhododendron Park. Near the turn, the highway crosses the Hozameen fault, which is not visible at road level. It separates Bridge River oceanic rocks to the west from the different but partly contemporaneous sequence of rocks to the east that form the Methow terrane, a name derived from a town in Washington. The fault is a big one and can be traced south-southeastward for at least 120 kilometres from 10 kilometres north of Hells Gate in the Fraser Canyon south into Washington. It is marked by discontinuous bodies of green, silica-poor and magnesium-rich ultramafic rocks called *serpentinite*, which together make up the Coquihalla serpentinite belt. Serpentinite forms when hot water alters ultramafic rocks such as peridotite, made of olivine and pyroxene, which are iron- and magnesium-rich minerals. Such rocks probably were upper mantle material that became incorporated in oceanic crust. One body of serpentinite is exposed on the hillside 1 kilometre northwest of the bend in the road and 300 to 800 metres above the valley floor.

Other ocean floor rocks found just east of the Hozameen fault include pillow basalt, whose composition is like basalt that emerges from present-day mid-ocean ridges, and also gabbro. These former ocean floor rocks are probably of Permian age, about 280 million years old, and they are the oldest part of the Methow terrane, much of which is well exposed just to the east near Skagit Bluffs.

Methow Terrane at Skagit Bluffs

The westernmost roadcuts at Skagit Bluffs expose steeply east-dipping, dark, siliceous fine-grained and thin-bedded argillites and interbedded, light grey fine volcanic ash, or *tuff*, and sandstone made of volcanic fragments. The rocks are equivalent to those along Highway 1 near Boston Bar. In places, there are angular and rounded fragments of volcanic rock up to about 10 centimetres across, as well as angular sedimentary fragments. The sedimentary rocks were deposited in deep water, probably on a submarine slope that dived westward and thus promoted downslope slumping and transport of pebbles, cobbles, and blocks in turbidity currents. Slump structures are present on bedding surfaces of some of the coarser strata.

The ridgetop about 6 kilometres south of Highway 3 exposes a 2,250-metre-thick pile of strata; its upper part corresponds with the beds at Skagit Bluffs. Fossils from the ridgetop include *ammonites* and *belemnites*, extinct marine animals related to modern squid and cuttlefish. The fossils are of early and middle Jurassic age, between about 185 and 170 million years old. These rocks overlap in age with younger parts of the Bridge River terrane. During deposition, presumably an open ocean, whose floor is represented by rocks in the Bridge River terrane, lay west of the submarine slope. Along the regional trend, the rocks near Skagit Bluffs are exposed more or less continuously for a distance of about 300 kilometres, in a belt that in places is over 30 kilometres wide.

Immediately east of the strata at the west end of Skagit Bluffs is a section of rocks about 300 metres thick. These rocks were deposited about 155 million years ago, in late Jurassic time. The younger rocks are dark and look very similar to those west of them, and the contact between the two can be found only with careful searching. The rock types are volcanic sandstone and some conglomerate with volcanic pebbles, siltstone, and sandy mudstone. Marine fossils, notably *Buchia*, a bottom-dwelling clam related to modern oysters, are relatively common, and their presence indicates a shallower water environment than the older Methow rocks. A short period of nondeposition or erosion may separate these beds from their predecessors to the west, but the strata of the two units are essentially parallel, implying that there was no big episode of folding and faulting in this time interval.

The easternmost beds in Skagit Bluffs are of early Cretaceous age, about 120 million years old, and assigned to the lower part of the Jackass Mountain group. The Jackass Mountain group along Highway 3 in Manning Provincial Park may be as much as 5,000 metres thick, though measuring its thickness is difficult because the rocks are folded and faulted. Sandstone, siltstone, argillite, and in places pebble conglomerates make up the group. The resistant and thick-bedded pebble to cobble conglomerate, similar to that along Highway 1 at Jackass Mountain, is visible in places along Highway 3 between a stream crossing about 13 kilometres east of Skagit Bluffs (4 kilometres west of Allison Pass) and the headquarters of Manning Provincial Park. West of this, the conglomerate unit is thinner and not as coarse grained.

The pebbles and cobbles were eroded from a wide variety of rock types. Roughly one-third of the cobbles are volcanic rocks, another one-third are granitic rocks, including a boulder isotopically dated at 156 million years old, and the remaining one-third includes chert, argillite, and metamorphic rocks. Fossils from west and northwest of the park headquarters are almost exclusively marine, but some from the north and east are plant materials, indicating a nearby shore. Marine fossils from the youngest Jackass Mountain beds are from very late in early Cretaceous time, only a little more than 100 million years old.

East of the park headquarters, nonmarine mid-Cretaceous sedimentary rocks of the Pasayten group are exposed along Highway 3. Up to 3,000 metres thick, the strata include conglomerate, siltstone, shale, and sandstone. The latter is composed of quartz, feldspar, and mica, indicating it was eroded from an area rich in granitic rocks. Many of the sandstone beds are crossbedded and were probably deposited by westward-flowing torrents. Near the top of the beds are several hundred metres of very distinctive red to maroon sandy siltstones, pebbly sandstones, and greywacke. Although outcrops of the red beds are few, the colour in the soil overlying them is diagnostic. Look for it in the ditch along Highway 3 about 500 metres east of the road to Hampton Campground.

The lack of any angular unconformities in the Methow rocks shows that there was no significant deformation during Jurassic and early Cretaceous time. The basin in which these rocks were deposited evidently lay west of the continental margin until mid-Cretaceous time, about 100 million years ago. Most of the folding and faulting came during or after mid-Cretaceous time, when Wrangellia and its associated island arc collided with the edge of the North American continent. In this area we know the deformation was largely completed before the intrusion of 89- to 88-million-year-old granitic rocks across the border in Washington because these intrusions crosscut the folds and faults.

Lightning Lakes

Gravel deposited by glaciers and meltwater streams form low hills rising above the valley floor midway between Coldspring Campground and the park headquarters. When the highway was first built these hills were much higher, but most of the gravel has since been extracted for road construction. A half kilometre west of Highway 3, across the upper Similkameen River and north of Little Muddy Creek, more gravel forms a *kame* terrace, a flat-topped deposit that slopes gently westward but has a steep south-facing slope down to the valley of Little Muddy Creek. The terrace was deposited by a glacial stream that flowed westward between the valley wall on the north and a relic ice tongue occupying the central and southern part of the valley floor. As the ice melted away, the westward-flowing stream shifted south and built a lower deposit, or terrace, at the foot of the earlier kame terrace.

Terraced gravel on the valley floor of Little Muddy Creek partly or completely surrounds kettles, depressions with or without water in the bottom, formed when stagnant blocks of buried glacial ice melted, permitting the collapse of the gravel cover.

When the ice front occupied the Similkameen Valley to the east, the upper Similkameen River and meltwater from the receding glacier flowed southwest down the Lightning Creek valley. The water deposited gravel near the ice, and then, free of its sediment load, it eroded and deepened the Lightning Creek valley. As the ice retreated back down the Similkameen Valley, a lake developed between the ice front and a spillway farther west. When meltwater overflowed the spillway, it drained westward into the Skagit River valley to the west that today discharges into Puget Sound in Washington. The receding glacier eventually exposed the lower escape route south of Princeton, so the water changed course and flowed eastward down the modern Similkameen Valley, and eventually into the Columbia River in Washington.

When the water ceased to flow into Lightning Creek, the creek's tributaries brought local debris to their mouths faster than the shrunken creek

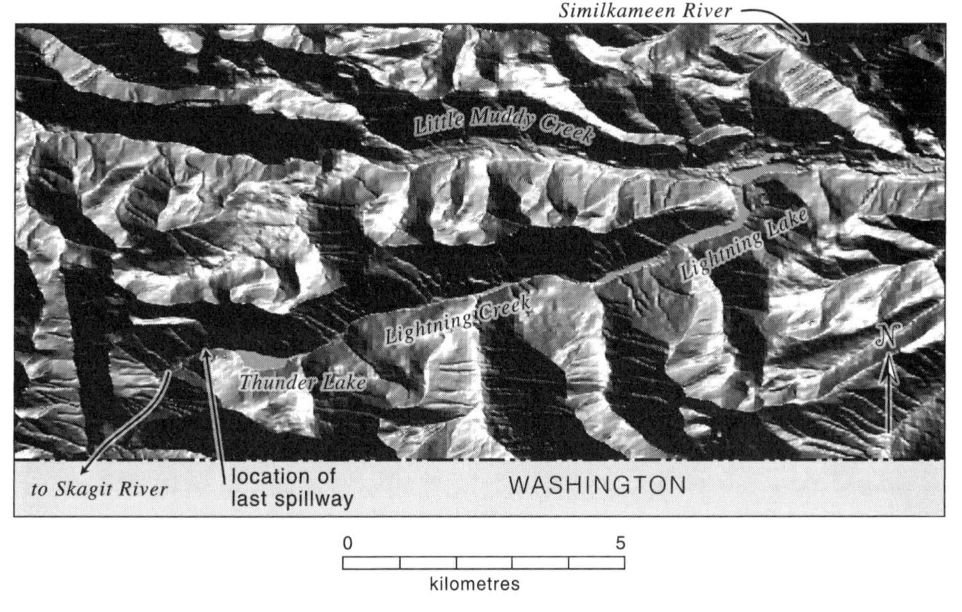

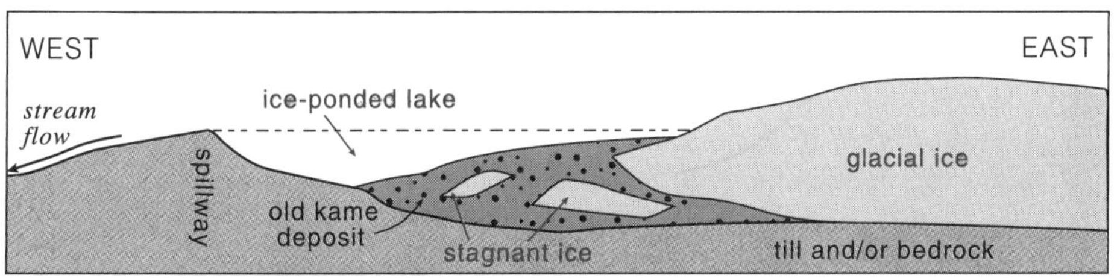

Water in the Lightning Lakes chain flowed southwest to the Skagit River until glacial ice receded from the Similkameen drainage.

could carry it away. Dams developed, ponding a chain of lakes about 12 kilometres downstream from the present head of Lightning Creek. The width and curvature of the Little Muddy Creek channel, now the north end of Lightning Lake, are much like those of the present Similkameen River south of Princeton, reflecting the fact that a much larger river once flowed through it.

Floodplain of the Similkameen River

A strip of alluvium rarely more than 600 metres wide floors the Similkameen River valley between the park headquarters and the east gate. The alluvial gravels, and probably some lakebed deposits hidden beneath them, were deposited by water flowing from the receding glacier. The river has rarely cut more than 10 or 15 metres below the top surface of the alluvium, but there is no record of the river overtopping this surface in historic time. The river is close to the hypothetical condition of equilibrium, in which the stream flow and its gradient are just enough to carry away any debris delivered to it but not enough to cut its channel deeper into the alluvium.

Pasayten Fault

The Pasayten fault, which marks the eastern edge of the Coast Belt along Highway 3, trends in a south-southeast direction for nearly 250 kilometres, from Lytton on Highway 1, where it smears into north-south alignment with the younger Fraser fault, to northern Washington. The Pasayten fault is near what was probably the location of the western edge of this part of the North American continent in early Cretaceous time. Though not exposed along Highway 3, we can infer the fault's presence from a big change (but hard-to-see from a car) in the geology between the park headquarters and the east gate.

About 2 kilometres west of Mule Deer Campground, nonmarine sandstone weathering to a buff colour is made mainly of grains of quartz and feldspar eroded from granitic rock and contains coaly layers. It belongs to the Pasayten group of mid-Cretaceous age and overlies the Jackass Mountain group at the top of the succession included in the Methow terrane. The bedding dips west at about 40 degrees. About 2 kilometres east of Mule Deer Campground and extending for another 2.8 kilometres eastward, granitic

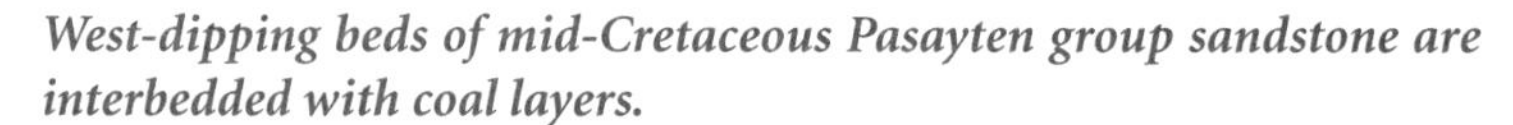

West-dipping beds of mid-Cretaceous Pasayten group sandstone are interbedded with coal layers.

and gneissic rocks are exposed intermittently in a series of roadcuts. The dip of the foliation of the gneiss is like that of the bedding in the sandstone, and the colour is similar, so that driving along the highway it is hard to distinguish the easternmost exposures of sandstone from the westernmost exposures of gneiss. The similar dips of the bedding and gneissic layering might suggest that the sedimentary rocks were deposited on gneiss with a horizontal foliation and both were later tilted westward. However, isotopic dating of zircons from the gneiss shows it to be of latest Jurassic age, about 150 million years old, so it formed at a time when there is little record of deformation or metamorphism in the Methow sedimentary rocks. This indicates that the gneiss and Methow strata may have been widely separated in late Jurassic time, and that the contact between them is a fault on which there probably was considerable movement. The gneissic and granitic rocks form the Eagle plutonic complex, part of a narrow belt of once deep-seated rocks.

HIGHWAY 5

Hope—Coquihalla Pass

45 kilometres

Highway 5 branches from Highway 3 just east of Hope, heads up the Coquihalla River valley, and climbs steeply to cross the crest at Coquihalla Pass at an elevation of 1,244 metres. Parts of Highway 5 are built on the original bed of the old Kettle Valley Railway, which ran from Hope to Princeton, Merritt, and other towns in the interior of British Columbia. In spite of magnificent roadcuts through bedrock in many places, Highway 5 presents considerable difficulty for those wishing to stop and examine the rocks. Concrete barriers on both sides of the highway, as well as along parts of the median, few turnouts, and the speed of the traffic preclude safe examination of the bedrock unless you leave the highway at off-ramps and drive on local roads. These generally are unpaved and narrow and some are used by logging trucks, so make sure that, even there, you park well off the road.

The westernmost kilometre of Highway 5 crosses greenish altered gneiss, which is exposed in abutments of the overpass that carries Highway 5 over Highway 1 on the south side of Hope. Only at and beyond the divergence of Highways 3 and 5 is rock well exposed along the roadside, in this case light-coloured early Tertiary granitic rock. Similar granitic rock is exposed 7 kilometres northeast of Hope on Ogilvie Peak, where it visibly intrudes darker rocks of the Bridge River terrane near the 1,000-metre contour.

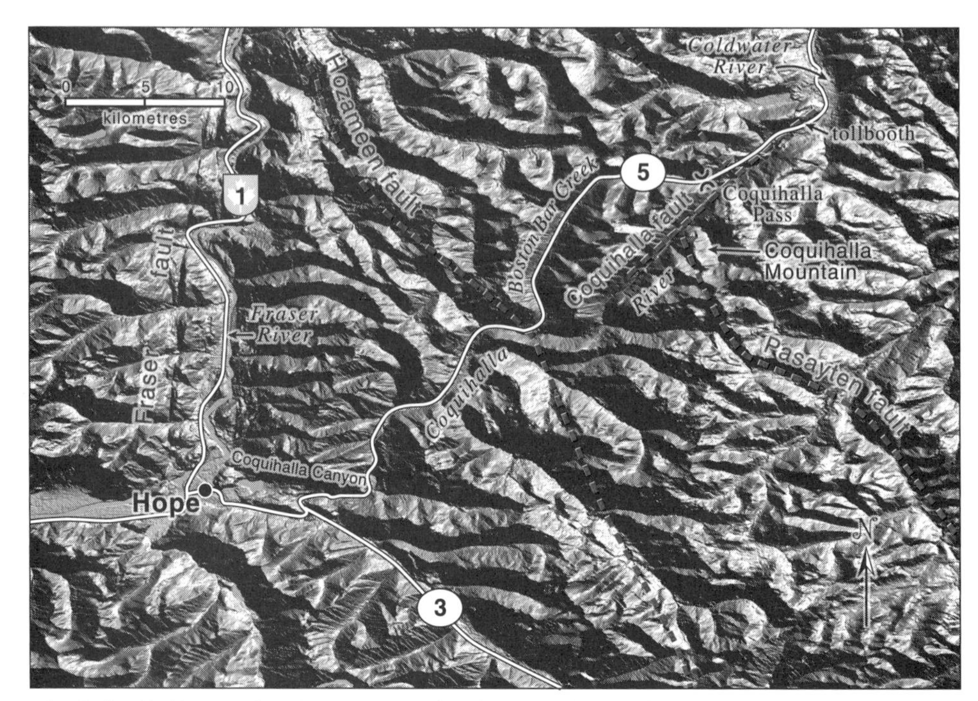

Shaded-relief image showing topography along Highway 5 between Hope and Coquihalla Pass. —Image constructed using Canadian Digital Elevation Data obtained from GeoBase

Coquihalla Canyon

During Pleistocene time, glacial ice in the Fraser River valley blocked the Coquihalla River, diverting it southward to its present position. At the point of diversion, about 7 kilometres from its mouth, the Coquihalla River eroded a bedrock saddle, cutting a spectacular, winding canyon 100 metres deep through the early Tertiary granitic rocks.

The canyon presented a challenge to the engineers building the Kettle Valley Railway in 1914–15. The chief engineer, Andrew McCulloch, negotiated the canyon by means of five tunnels and two bridges in close succession. The operation of the Kettle Valley Railway through Coquihalla Pass began in 1916, but snowslides and rockfalls shut it down for weeks on end during winter months, and the line eventually was abandoned in 1959. Later, the bridges were removed from the lower canyon but have since been replaced by two footbridges in what is now the Coquihalla Canyon Recreation Area. You can reach the area by taking the local road from Hope to Kawkawa Lake and Othello, a former station on the Kettle Valley line. A footpath following a historic cattle trail offers an alternative to returning through the tunnels to the parking area.

Coquihalla Serpentinite Belt

East of the roadcuts in granitic rock and upstream of the canyon, Highway 5 follows the bottom of a steep-walled valley that contains the Coquihalla River. It is cut in metamorphosed chert, argillite, and greenstone of the Bridge River terrane of Carboniferous to Jurassic age. About 20 kilometres east of Hope, and just before the highway turns and starts to climb steeply, a significantly broader part of the valley heralds a change in the type of bedrock. Roadside outcrops are scarce, but you can see the rocks in an abandoned quarry by taking the Caroline Mine Road off-ramp, then turning west across the old mine road and then south for about 100 metres.

The bedrock here includes serpentinite and gabbro that probably once formed parts of the floor of a Permian ocean basin. Serpentinite, a black to bright green rock with a greasy luster, is derived by the addition of hot water to the iron and magnesium silicate minerals olivine and pyroxene. The olivine and pyroxene crystals may be remnants of old upper mantle material or else were concentrated in the crust by settling out from a molten magma of basaltic composition. As the accumulated iron and magnesium silicates cooled, a reaction took place between the minerals and hot seawater circulating through fractures in the rock, converting the olivine or pyroxene to serpentine. In places, such as near the Caroline Mine and visible in the banks of the Coquihalla River at its confluence with Dewdney Creek, pillow basalt is associated with the serpentinite and gabbro. The chemistry of the basalt is similar to that of basalt forming today on mid-ocean ridges, supporting the theory that this was once part of the floor of an ocean.

The serpentinite body crossed by Highway 5 here is part of a belt that is 40 kilometres long, up to 2 kilometres wide, and exposed from valley bottom to mountaintop over a vertical range of more than 1 kilometre. It extends discontinuously from the vicinity of Skagit Bluffs, on Highway 3, north-northwestward to Highway 1 south of Boston Bar. Most of its contacts are steeply dipping faults, the major one being the Hozameen fault, which separates the Bridge River terrane to the west from mostly sandstone, shale, and volcanic tuff of the Methow terrane to the east. The Fraser fault offsets the serpentinite belt, and a very large ultramafic and gabbro body in the Shulaps Range, about 50 kilometres northwest of Lillooet, is probably the offset northwest continuation of the Coquihalla serpentinite belt. Although the age of the serpentinite, gabbro, and basalt of the Coquihalla serpentinite belt is not known here, rocks in the northern body have been dated as 280 million years old, or early Permian.

Jurassic sedimentary rocks of the Methow terrane unconformably overlie the serpentinite belt. A line of about thirty gold prospects in the sedimentary rocks parallels the serpentinite belt on the east. Of these, only the

Caroline Mine has had a significant production, and operating from 1982 to 1984, it yielded slightly less than 50,000 ounces of gold worth about $16 million. Nearly all the gold was either hosted by quartz veins or occurred in areas of sulfide alteration within the base of the Jurassic sedimentary rocks and within 200 metres of the eastern boundary of the serpentinite belt.

The Jurassic sedimentary sequence east of the serpentinite belt is the Ladner group, named from Ladner Creek, crossed on the lower part of the long and fairly steep hill on Highway 5. The rocks are dark, steeply dipping, locally tightly folded, mostly thin-bedded argillite and tuffs.

Needle Peak Pluton

Near Shylock Road (the railroad engineer was a Shakespeare aficionado), Highway 5 turns northward up the valley of Boston Bar Creek, a tributary of the Coquihalla River. Near the turn, you can see dark argillite and tuff of the Ladner sedimentary rocks intruded by light-coloured granitic rocks of the 48-million-year-old Needle Peak pluton. The pluton is about 25 by 15 kilometres in extent, and the highway climbs through its middle, passing out of it into a very narrow band of rocks of the Methow terrane just west of Coquihalla Pass. The granitic rocks support some of the steepest slopes in the region and have smooth glaciated surfaces capped by sharp alpine peaks, known as "horns."

Avalanche Alley

During railroad days, the upper part of the Coquihalla River valley, about 5 kilometres east of Highway 5, was a winter nightmare, with rockslides, debris flows, and snow avalanches diverting train traffic for weeks at a time. Highway 5 follows the valley of Boston Bar Creek, which may have fewer rockslides than the upper Coquihalla River valley, but has far more snow avalanches. Strips of lush but low brushy vegetation extending straight downslope through the dense coniferous forest mark the tracks of avalanches, which occur with a frequency exceeded in only a few other mountain valleys in southern British Columbia. Most of the avalanche tracks extend to the valley floor, and some even continue for a short distance up the opposite slope.

Because there was no route along Boston Bar Creek that could avoid crossing the potentially hazardous avalanche tracks, the British Columbia Ministry of Transportation and Highways adopted a series of remedies. Bulldozed dikes across the lower part of the avalanche chutes stop smaller snowslides, and patterns of bulldozed mounds achieve much the same effect. In potentially more serious places, snowslides are triggered while still of manageable size and cleared if necessary before permitting traffic

Yak Peak, just west of Coquihalla Pass on Highway 3. The smooth surface was glacially carved in the Eocene granitic rock of the Needle Peak pluton. —R. Turner photo

to pass. Mortars mounted on platforms shoot explosive charges into the avalanche source areas, and cables hoist explosives over the source areas of two avalanche chutes. A snowshed diverts the Brown Bear snowslide, on the most severe avalanche track, over the highway. As a final protection for travellers, gates have been installed to close the highway when the conditions are extremely hazardous.

Eastern Boundary of the Coast Belt

Highway 5 crosses a divide, Coquihalla Pass, at the head of Boston Bar Creek and emerges into a more open area above the headwaters of the Coquihalla River. The dark, folded, metamorphosed sedimentary rocks in this area probably belong to the Methow terrane. These rocks bound the northeast side of the Needle Peak pluton; a contact between the two rock types is exposed close to the sign marking Coquihalla Pass.

The upper Coquihalla River valley, 3 kilometres south of the pass on Highway 5, follows the northeast-trending Coquihalla fault. The northwest side of the valley consists of the 48-million-year-old Needle Peak pluton. Coquihalla Mountain on the southeast side is underlain by

22-million-year-old volcanic rocks that are the easternmost rocks of the Cascade magmatic arc in southern British Columbia. In places, the volcanic rocks extend down to the valley floor, where the fault appears to cut them. The fault may be contemporaneous with and related to those faults bounding the similarly trending Sumas Valley graben, near Highway 1 east of Vancouver.

A short distance north of Coqhihalla Pass and the narrow belt of dark rocks of the Methow Terrane are granitic rocks that weather to a tan colour, and farther north, near the tollbooth, is grey, banded, gneissic granodiorite. The tan rocks are 105 million years old, and the gneissic rocks are of latest Jurassic age, about 150 million years old. These rocks underlie the southeasternmost part of the Intermontane Belt and are part of the Eagle plutonic complex. The boundary between the Coast and Intermontane Belts probably lies between the dark Methow rocks and tan-weathering granitic rock, and although elsewhere the boundary is the Pasayten fault, the fault surface is not obvious here.

HIGHWAY 12
Lytton—Lillooet
64 kilometres

Highway 12 follows the east bank of the Fraser River between Lytton and Lillooet and runs along the boundary between the Coast and Intermontane Belts. Lytton sits at the confluence of two major rivers, the Fraser and the Thompson. Highway 12 crosses the Thompson River just above the confluence, where its clear green water flows into the muddy Fraser River water.

Faults at the Belt Boundary
South-southeast of Lytton, the north-northwest-trending Pasayten fault separates rocks of the Intermontane Belt from those of the Coast Belt. Near Lytton, the Pasayten fault swings into alignment with the younger, north-trending Fraser fault, that major system of interconnected faults that the Fraser River follows nearly continuously for more than 300 kilometres between Hope and Williams Lake. The bridge over the Thompson River at Lytton crosses the lower end of a canyon cut in latest Paleozoic and Triassic granitic and gneissic rocks of the Mount Lytton complex, here the westernmost component of the Intermontane Belt. A little farther north along Highway 12, fractured granitic rocks, weathering to pink, of the Mount

Lytton complex occur in roadcuts, and below the road at river level, near the eastern terminus of the small Fraser River ferry, are dark outcrops of Jackass Mountain sandstone, the youngest rocks in the Methow terrane of the Coast Belt. Here, the position of the fault separating the Coast and Intermontane Belts can be pinned down to within 100 metres or so. On the west side of the river, a series of aligned notches in the ridges about 900 metres above river level traces the western strand of the Fraser fault system. You can see the latest Cretaceous and early Tertiary granitic rocks, weathering to a white colour, above and west of the Jackass Mountain rocks, which weather brown.

The Mount Lytton complex, the root of an early Mesozoic arc in the Quesnel terrane, is between 250 and 225 million years old. The Jackass Mountain sedimentary rocks of the Methow terrane are between 130 and 110 million years old. These two bodies of rock were faulted together around 100 million years ago, during the initial phases of mountain building in the

Shaded-relief image of topography along Highway 12 between Lytton and Lillooet. —Image constructed using Canadian Digital Elevation Data obtained from GeoBase

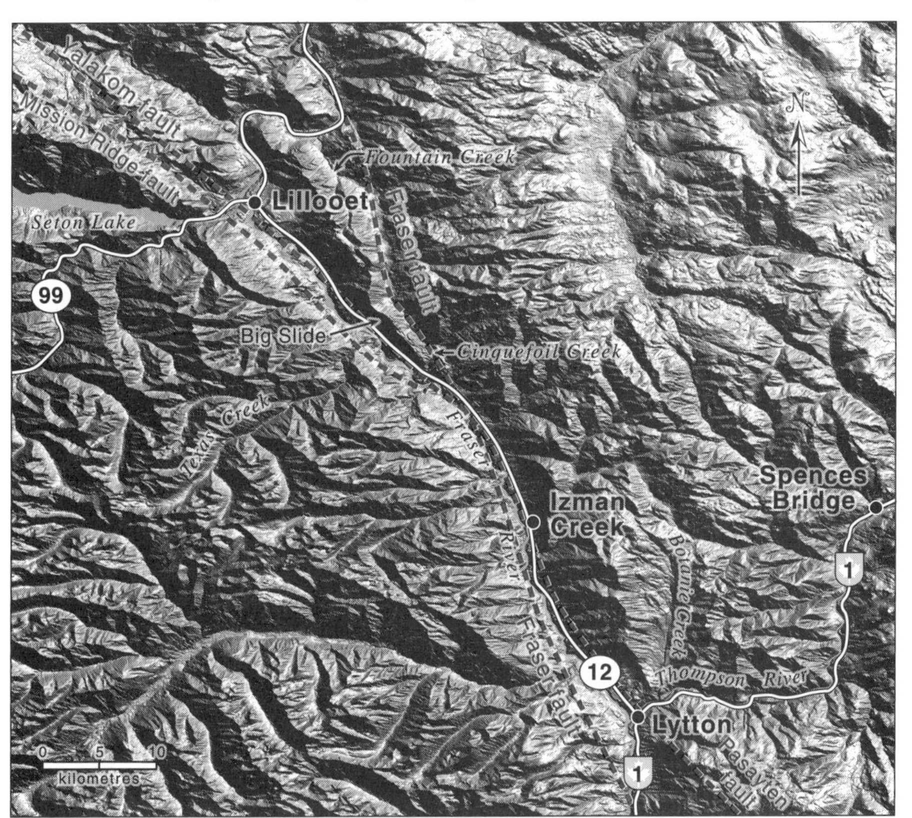

Coast Belt. We know this because distinctive nonmarine sedimentary rocks overlie both terranes and contain fossil pollen of mid-Cretaceous age, about 100 million years old. This is just slightly younger than the youngest known Jackass Mountain rocks.

You can see this sandstone along Highway 12 about 10 kilometres north of Lytton, where roadcuts expose highly fractured red, buff, and grey sandstone and shale, with coaly layers, as well as conglomerate. The sandstone is distinctive, containing grains of black chert, white quartz, feldspar, volcanic rock, and shiny white mica. In places, the grains are coated with *hematite*, a red iron-oxide mineral. No fossils other than plant material represented by very thin coal layers have been found in these roadcuts, but similar rocks down-faulted into Mount Lytton granitic rocks in Botanie Creek, about 5 kilometres east of Highway 12, contain mid-Cretaceous fossil pollen. The highly variable grain size of these sediments and the plant fossils, coal layers, and red coatings suggest that streams probably deposited the sandstone, conglomerate, and shale in fault-bound basins within a continent, not in a sea.

These rocks appear to be the oldest sedimentary unit that overlaps terranes of both the Coast and Intermontane Belts. Similar sandstone of the same age also occurs in the Coast Mountains northwest of here, where it overlies the Bridge River terrane and is called the Silverquick formation. It is also exposed along Highway 99 between Lillooet and Cache Creek, where it overlies the Cache Creek terrane. The sandstone evidently postdates the final closure of a marine basin that once separated the early Cretaceous continental margin from Wrangellia.

Deposits at the Edge of the Ice

Between the sandstone outcrops and Cinquefoil Creek, about 37 kilometres north of Lytton and 25 kilometres south of Lillooet, Highway 12 crosses, for the most part, unconsolidated sediments deposited at the end of the last glaciation when the last tongues and patches of ice were disappearing. At that time, the hillsides would have been plastered loosely with wet unconsolidated till that had been dropped from the ice. Vegetation was probably still sparse and provided little protection from bad weather and periodic mudslides. Landslides and debris flows, which are slurries of mud carrying coarser fragments, would have been very common. The Fraser River was probably unable to carry away the sediment as fast as it came down, so the debris accumulated in the valley bottoms. This environment probably persisted for centuries after the glaciers melted away and before more vegetation grew and stabilized the slopes. Even today, after rare prolonged periods of rain, debris flows at times block Highway 12.

The last stages of a debris flow that blocked Highway 12 in June 1980, when angular rubble was carried downhill in a slurry after a period of heavy rain in this normally arid region.

In places, the road crosses incised ravines hosting small tributaries that contributed to the flood of debris at the end of the last ice age. The road angles down into the ravines and up the far side, giving excellent views of the materials brought down. Highway 12 has been built well above the Fraser River, near the boundary between the unconsolidated deposits and the granitic bedrock, because it is easier to traverse the ravines near their heads.

Blue Quartz Eyes

Bedrock exposed in low cuts just north of Izman Creek, about 20 kilometres north of Lytton, consists of greenish Mount Lytton granitic rock. In places, the granitic rock has become so highly sheared and recrystallized that it has turned into *mylonite*, a finely laminated rock that in places contains bluish quartz "eyes." Shearing and rolling of the resistant quartz grains in the granitic rock has produced a circular outline to which may be attached two triangular shadows resembling the corners of an eye. The shearing also produced minute fractures within the quartz grains, which scatter light, creating the bluish tinge.

Fraser Fault Temporarily Leaves the Fraser River

At Cinquefoil Creek, about 37 kilometres north of Lytton and 25 kilometres south of Lillooet, the Fraser fault leaves the Fraser River valley and runs in a more northerly direction up Cinquefoil Creek, rejoining the river about 10 kilometres northeast of Lillooet. The Fraser fault runs near the bottom of the north-south-oriented valley whose southern part contains south-flowing Cinquefoil Creek and whose northern part contains north-flowing Fountain Creek. West of the valley the rocks are late Jurassic and early Cretaceous marine shale, siltstone, sandstone, and local conglomerate of the Jackass Mountain group and underlying strata, also of the Methow terrane. East of the valley are 105-million-year-old mid-Cretaceous volcanic rocks of the Spences Bridge group that overlie the Mount Lytton plutonic complex and, farther north, the Cache Creek terrane.

Yalakom Fault

The Fraser River north of Cinquefoil Creek follows the Yalakom fault, which splays off the Fraser Fault in a north-northwest direction. The Yalakom fault separates the Methow terrane east of the river from little-metamorphosed Bridge River terrane rocks exposed low on the west side of the river. This fault continues northwestward, for about 300 kilometres, through Lillooet and into the Yalakom River valley. The Yalakom fault, slightly older than the Fraser fault, is a right-lateral strike-slip fault that moved in latest Cretaceous and earliest Tertiary time. The estimates of offset across it range from 115 to 150 kilometres. South of Hope, the south-southeast-trending Ross Lake fault has somewhat similar characteristics and may actually be a segment of the Yalakom fault, offset from it by movement on the younger right-lateral Fraser fault. Forming the steeper slopes above the west side of the Fraser River and across the Mission Ridge fault (discussed below), metamorphosed Bridge River rocks contain 45-million-year-old granitic intrusions that are very similar to the rocks on the eastern Fraser River valley wall east of Hope. We can bring the two bodies of granitic rock into alignment by removing about 140 kilometres of right-lateral movement on the Fraser fault. Movement on the Fraser fault is younger than 45 million years, the age of the offset intrusion, and older than 35 million years, the age of the oldest part of the Chilliwack batholith, which intrudes the fault south of Hope. Thus, if we combine the amounts of movement on both the Fraser and Yalakom faults, and we could have stood on the Mount Lytton complex back in late Cretaceous time and looked westward, the rocks of the Coast Mountains probably lay some 250 to 300 kilometres south of where they are today.

Mission Ridge Fault

The steep, upper western wall of the Fraser River valley as far north as Lillooet is hard, silica-rich schist and fine-grained amphibolite that is metamorphosed Bridge River terrane. It contains many small bodies of 45-million-year-old early Tertiary granitic rock that weathers to white. These rocks are separated from the unmetamorphosed Bridge River rocks at the base of the slope by yet another fault, the Mission Ridge fault, whose east-northeast-dipping surface probably lies close to the present ground surface of the steep valley wall across the Fraser River between a point west of the end of Cinquefoil Creek and Lillooet. Unlike the strike-slip Yalakom and Fraser faults, this fault is a normal fault on which the unmetamorphosed rocks to the east have slipped down against the metamorphic rocks to the west. Vertical movement on the fault may be as much as 15 kilometres, judging from the difference in metamorphic grade of the rocks on opposite sides of the fault.

The roadcuts along Highway 12 north of Cinquefoil Creek and on the east side of the Yalakom fault are in shale, siltstone, and minor greywacke that are part of Methow terrane. A boulder probably derived from these rocks contains latest Jurassic or earliest Cretaceous fossils. To the north, opposite Lillooet, these strata are overlain by sandstone and conglomerate of the Jackass Mountain group. On the west side of the Fraser River and sandwiched between the Mission Ridge and Yalakom faults, the small knolls at the base of the steep western wall are fractured and broken but unmetamorphosed bodies of chert, gabbro, and serpentinite of the Bridge River terrane.

The Big Slide

Highway 12 crosses an area known as the Big Slide about 6 kilometres northwest of Cinquefoil Creek and 16 kilometres south of Lillooet. There, the Fraser River is deflected against the eastern side of the valley by the fan of Texas Creek, which flows through a narrow gap in the western wall of the Fraser River valley, and by old rockslides derived from the east wall of the valley. Highway 12 crosses the lower part of the slide source area, which is made of argillite, siltstone, and minor intrusive rocks. The bedding in the sedimentary rocks dips into the hillside nearly at right angles to the surface slope and thus played no part in the sliding. The bedrock failed along fractures nearly parallel to the hillslope.

At the Big Slide the road is narrow, clinging to a slope that drops precipitously to the river. The road surface may be littered with debris from small rockfalls, and the overall slope of the hillside is close to 35 degrees, which is the angle at which most rocks will start to slide over one another. If this makes you uneasy, it is a sign that you are becoming attuned to at least one type of geological hazard in mountainous regions!

Slide debris visible on the opposite side of the Fraser River records two old landslides. River terraces eroded in older debris are in turn covered by younger debris. The volume of the earlier rockslide was about 45 million cubic metres, and that of its successor about 7 million cubic metres. Both were big enough to temporarily block the river, which ceased downcutting above the blockage, changing the character of the river valley. The Fraser River valley floor between Cinquefoil Creek and Fountain Creek, northeast of Lillooet, is much wider than the stretch between Cinquefoil Creek and Lytton. At Lillooet, the valley floor below the 400-metre contour is 5 kilometres wide, whereas only 7 kilometres to the south it is barely 2 kilometres wide and both banks of the river are cut in the bedrock.

Based on rates of river downcutting, geologists estimate the first slide occurred between 5,000 and 2,000 years ago. No carbon has been analyzed from this slide to provide a more precise date. The later slide may be slightly less than 1,200 years old based on carbon isotopes from wood in the slide. The salmon-dependent economy of the prehistoric Lillooet Culture collapsed around 1,100 years ago, possibly as the result of the second slide. Archeologists learned this from studying the remains of a large pithouse village on a terrace of the Fraser River north of Lillooet.

HIGHWAY 99
Vancouver—Lillooet and Beyond
250 kilometres

Fraser River Delta

South of Vancouver, Highway 99 crosses the flat, low-lying country of the Fraser River delta, the largest delta in western Canada. It extends southwestward for about 25 kilometres from New Westminster, in the southeastern part of Greater Vancouver, and its perimeter along the Strait of Georgia is about 40 kilometres long. Sediments in the delta have a maximum thickness of just over 200 metres, all of which accumulated in the past 10,000 years. On deltas, rivers often overfill their channels with sediment, and new channels, called *distributaries*, branch off from the original channel. The big distributaries of the Fraser River that cross the delta are the Main Channel and the North Arm.

The Fraser River brings just over 17 million tonnes per year of fine sand and mud to the delta, which is mostly transported during high-water

Shaded-relief image of topography along Highway 99 between Vancouver and Pemberton. —Image constructed using Canadian Digital Elevation Data obtained from GeoBase

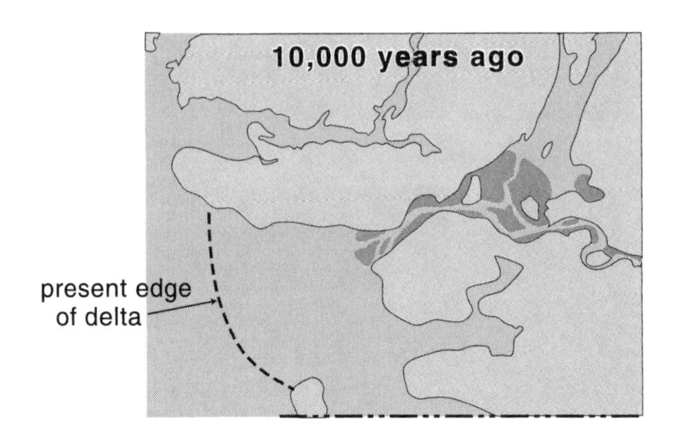

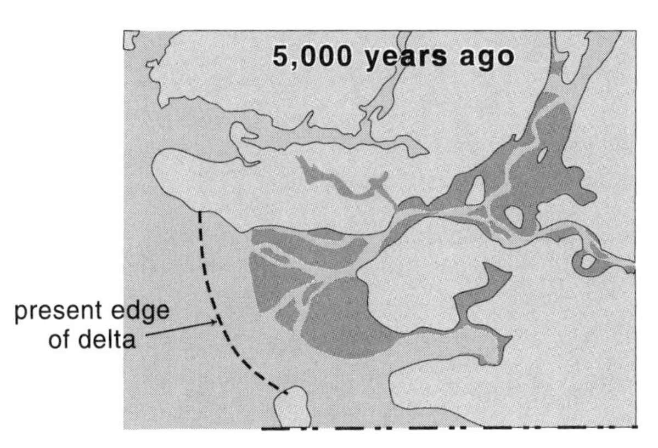

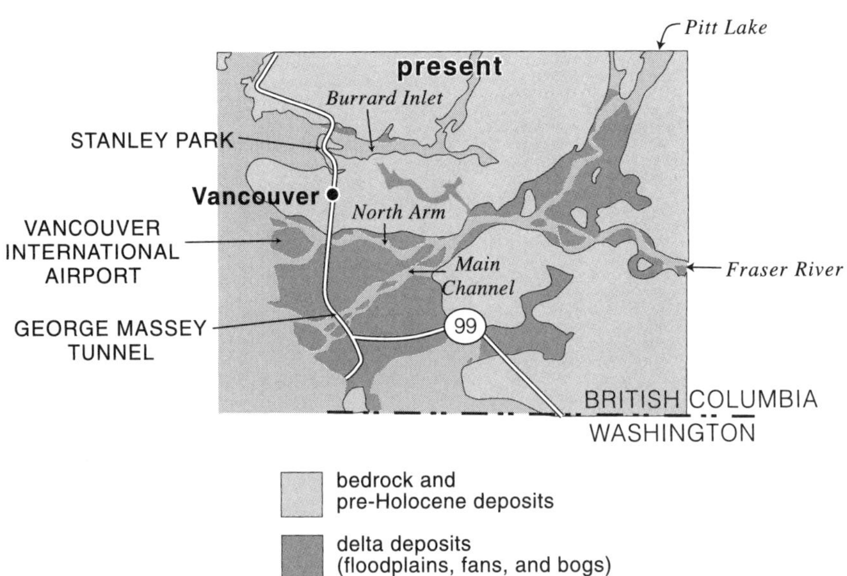

The Fraser River began building its delta 10,000 years ago, after glaciers from the last ice age retreated from the region.

flow in late spring and early summer. During high tides, denser seawater intrudes the delta via the deeper channels, and fine-grained sediment is carried in suspension seaward over the top of the saltwater intrusion. Sediments are deposited in bogs (mainly in the delta's eastern parts) and on the floodplain, estuarine river channels, tidal flats, and the slope below the seawater of the Strait of Georgia.

Deltas exhibit two or three parts, from top to bottom. Uppermost is the *delta platform*, built mainly just above the contemporaneous sea (or lake) level in floodplains, marshes, and river channels. Its top surface and internal bedding layers slope gently seaward at angles of a few degrees, an angle established by the flow rate and load of the input stream. Second is the *delta front*, or slope, whose bedding may dip as much as 23 degrees. It is made of gravel and coarse sand that rolls and bounces along the riverbed, and upon reaching the sea, cascades down the steep slope. The third part is an outer *apron* of fine sediment carried in suspension in the river and possibly deposited well out into the sea. Silt- or clay-sized material tends to slip off sea bottom slopes that are more than a few degrees, so these apron deposits may be completely isolated from their delta platform and slope counterparts.

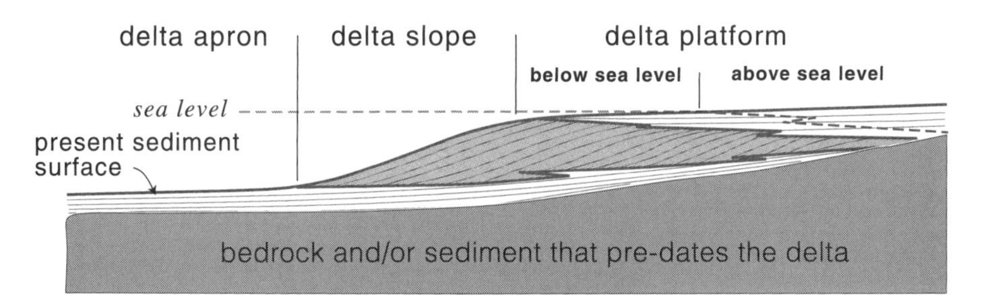

Components of a typical delta

The Fraser River delta is an important agricultural and waterfowl area, and salmon must cross it during their upstream migration. However, in recent years, its natural state has been greatly modified by rapid urban and industrial growth, including at the Vancouver International Airport and major docks along the Main Channel.

In parts of the delta, sediments saturated with water overlie consolidated glacial and preglacial strata. Major earthquakes can convent the water-saturated layers to a fluidlike mass, evidenced by sand dikes on the delta. These small sandy intrusions penetrate the peat and clayey material overlying the water-saturated sediment. Earthquakes may also cause slumping of the submarine delta front. These potential hazards have raised concerns about the safety of buildings on the delta. Major buildings are supported on pilings that penetrate the water-rich layer and others, like the airport terminal, are constructed after preloading the area with sand heaped on the surface to squeeze excess water out of the underlying sediments before construction.

George Massey Tunnel

The George Massey Tunnel, formerly called the Dease Island Tunnel, carries the four-lane Highway 99 beneath the Main Channel of the Fraser River, which is used by cargo vessels to access docks farther upriver near New Westminster. The tunnel consists of six straight sections that were built in a nearby excavation on the north bank of the channel and kept free of water by diking and pumping. When the tunnel sections were complete, the dike was breached and the basin was flooded; it now serves for holding ships of the British Columbia Ferries fleet during off-seasons. Then the tunnel sections were floated and towed one by one into position over a trench excavated in the riverbed. There they were sunk, jacked into position, and finally buried under steel plates and sand.

This maneuver was possible only when the river bottom sediment was not in motion and the trench could remain open for several weeks. A preliminary experiment was undertaken to establish if and when this condition existed. A line of pits was excavated on the riverbed at the tunnel site and surveyed at weekly intervals. The experiment demonstrated that there were negligible changes in the riverbed during early spring when stream flow was still low. However, once snowmelt in the mountains had begun in earnest and the annual runoff was underway, the pits in the riverbed and presumably the future tunnel trench as well had no chance of survival.

Another consideration that dictated the season for placing tunnel sections was the intrusion of a wedge of salty water from the Strait of Georgia at times of high tides and low river flow, particularly in mid- to late winter. The rapidly changing buoyancy caused by the interaction of denser salty water and fresh river water could compound the difficulties of placing tunnel segments should the wedge reach and pass up beyond the tunnel line. These considerations meant that tunnel sections had to be placed in March and April.

Geology beneath Vancouver

Bedrock directly beneath Vancouver, and exposed in excavations for major buildings downtown, mostly consists of sandstone and shale, with some plant fossils. Natural exposures are on the shore between Kitsilano Beach and Spanish Banks, in cliffs around the Stanley Park seawall, and on the south side of Vancouver Harbour. The sedimentary rocks are early Tertiary in age, about 50 million years old, except for those underlying the northern half of Stanley Park and a few places at the foot of the north shore mountains, which are late Cretaceous, about 80 million years old, and related to the Nanaimo group exposed mainly on the western side of the Strait of Georgia. The Cretaceous strata overlie a south-dipping, weathered surface on Cretaceous and older granitic rocks of the Coast Mountains, and the contact is exposed in the Capilano River valley, which is north of Lions Gate Bridge, the Highway 99 crossing of the entrance to Vancouver Harbour. The late Cretaceous to early Tertiary strata thicken to the south, attaining a thickness of over 4,000 metres in a borehole drilled near the international boundary at Point Roberts, just south of the Tsawwassen ferry terminal.

In places, the sedimentary rocks are cut by small, dark, near-surface basaltic intrusions, and in places extrusions of middle Tertiary age (34 to 31 million years old). In Vancouver, these rocks are exposed in Queen Elizabeth Park in Vancouver, in the south abutment of the Lions Gate Bridge in Stanley Park, and along Sixth Avenue east of Main Street. These volcanic rocks are considered to be the westernmost exposures of older parts of the Cascade magmatic arc.

East Side of Howe Sound

North of Vancouver, Highway 99 follows the steep eastern side of Howe Sound as far as Squamish, at its head. Howe Sound is a fjord whose steep walls and a U-shaped cross section were carved by a glacier. Please note that this section of Highway 99 has a high accident rate because it is a busy road with only limited opportunities for passing or stopping. Compensate by displaying a high degree of caution. The highway presents many interesting engineering challenges that must be addressed before the 2010 Winter Olympics in Whistler.

For the first 10 kilometres or so north of Horseshoe Bay, the highway traverses 118-million-year-old granitic rocks that intrude fine-grained amphibolite, gneiss, diorite, and pegmatite of uncertain age and origin. They may be the metamorphosed equivalents of rocks of the latest Triassic to middle Jurassic Bonanza arc on Vancouver Island. For the next 15 kilometres or so beyond this, Highway 99 crosses slightly metamorphosed volcanic and sedimentary rocks, weathering to a rusty colour, in which a single fossil, an ammonite, has been found. The fossil is of early Cretaceous age, about 110 million years old. Younger granitic rocks exposed higher up the eastern slope show that these strata form a narrow screen along the east side of Howe Sound. Similar strata occur on some of the islands to the west in Howe Sound.

The east side of southern Howe Sound is a spectacular place to live, but this comes at great private and public cost. For the first 30 kilometres or so north of Horseshoe Bay, the many streams descending the steep side of Howe Sound have long and full records of flash floods, many of which carry enough rocks and soil to qualify as debris flows—avalanches or slurries of debris carried by water. Steep hillsides, a shallow layer of loose rocky soil, and torrential rain, much of it concentrated in storm cells no more than 1 kilometre across, contribute to the hazard and make forecasting difficult. Logging practices have probably added to the dangers. The death toll associated with the debris flows since the highway was opened in 1958 has grown to about fifteen people, including nine who plunged into M Creek after its bridge had been carried away in a flood. Extensive and costly public engineering near Lions Bay to control the debris flows includes channelling of creek beds and building check dams to collect debris during floods. Warning systems now reduce the hazards to drivers and those living on the steep and bouldery alluvial fans.

Porteau Cove Provincial Park

Bedrock at Porteau Cove Provincial Park is beautifully exposed on the east side of the highway at the foot of the grade leading down to the level of the British Columbia Rail line. The pale grey, 96-million-year-old granitic rock in places contains dark, finer-grained inclusions 1 to 2 centimetres thick and ten to a hundred times as long from end to end. The inclusions dip about 60 degrees to the east. You can best see them on a surface inclined at a high angle to them. The much less common surfaces nearly parallel to the inclusions show that they are pancake shaped. The dark inclusions were fragments of the country rock incorporated into the granite and softened, squeezed, and stretched before the granitic body finally crystallized.

Fracture systems, which are sets of nearly parallel breaks in the granitic rock, are conspicuous on the cliff overlooking the highway and railway just north of the park. The most significant set, spaced at intervals from 1 to 8 metres, dips toward the road at about 50 degrees, an angle less steep than the cliff face. The fractures present a major engineering challenge. In November 1964 and January 1969, slippage along the westward-dipping fractures produced rockfalls. Remedial work included the installation of rock bolts 1 to 9 metres long, some blasting and scaling, diversion of runoff from heavy rain or snowmelt away from the top of the cliff, and boring horizontal drainage holes at the foot of the face. To provide data for a warning

Light-coloured granitic rock near Porteau Cove Provincial Park contains numerous slightly flattened and elongate dark inclusions of country rock.

system, movement gauges were set across open fractures at the top of the slide, and listening devices were set at the bottom.

Glaciers have polished the bedrock surface directly east of the Porteau Cove Provincial Park entrance. The surface was buried under a former cover of weakly cemented gravel, protecting it from weathering until construction of the highway in the late 1950s uncovered it. Features on the polished surface indicate ice flowed south at the time they were formed. Numerous striae, nearly parallel scratches, run north to south across the face, and these formed when rock fragments embedded in the glacial ice were dragged across the bedrock. The edge of small steps in the bedrock are rounded where they face north and sharply broken off where they face south. Depressions about the size of dinner plates deepen southward gradually, then terminate abruptly at smoothly curved surfaces facing northward.

A Submerged Moraine

West of Porteau Cove, a submerged moraine of magnificent proportions is hidden from sight beneath the waters of Howe Sound. It extends from the east shore of Howe Sound to the rocky islets on the far side. Water depths reach 285 metres several kilometres north of here but shoal to as little as 13 metres at the crest of the moraine and then drop off southward, rapidly at first, then more gradually, to a depth of almost 250 metres. Sediment sampling near the crest brings up stony, sandy mud, the kind of material that a glacier would deposit.

The moraine may have formed in one of several ways. One possibility is that during retreat of the last ice sheet, a short-lived return to a cool, wet climate created an environment favorable for glacial development. This may have halted the retreat of the Howe Sound ice tongue. Just such a halt, or even a minor readvance, for several centuries about 11,000 years ago, is recorded in the Fraser Lowland and in other parts of the world. A similar date has been obtained from glacially deposited gravel not far north of Porteau Cove Provincial Park. Sediment carried in the ice could have been deposited in one place during this time, creating the moraine.

The moraine may have formed for other reasons. Alaskan glaciers that terminate in tidewater tend to halt and produce moraines at their leading edge because cold but not frigid salt water promotes the melting of ice and the calving of icebergs. The place where an advancing ice tongue can widen or deepen is a likely site for the advance to halt because the significant increase in the area of ice in contact with the water encourages melting and calving. In addition, tidewater glaciers tend to bulldoze fjord-bottom sediments, forming big moraines in front of them that minimize the contact between seawater and ice. In turn, this enables the glacier to advance, pushing its shield ahead of it until, for example, it reaches a broader part of the fjord, where its moraine armour becomes attenuated and the ice front halts. As long as the ice tongue terminates near the moraine crest it can survive, but even a slight retreat of the ice front into deeper water behind the crest may lead to a serious imbalance of supply and demand, and a catastrophic retreat may follow. The Columbia Glacier, near the Alaskan oil-shipping port of Valdez, is one such body of ice reaching tidewater that during the last two decades has passed from an advanced but stable phase in contact with its moraine crest, to the early part of a catastrophic retreat. The Columbia Glacier now produces many more icebergs than before, increasing the threat to nearby tanker traffic. Perhaps significantly, Howe Sound widens about midway along its length, near the position of the moraine.

Furry Creek, 3 kilometres north of Porteau Cove Provincial Park and now a golf course and housing development, was formerly an abandoned gravel pit. Marine shells from the gravel pit have provided radiocarbon dates of 11,300 years, and wood fragments have given 10,690 years, showing that the gravels were deposited shortly after the last ice age. Four kilometres north again, the road crosses the broad floor of a former gravel pit that once was a raised marine delta from which 44 million tons of sand and gravel, worth about $90 million, were extracted.

Britannia Mine

The small settlement of Britannia Beach hosts the British Columbia Mining Museum. Directly north of the museum is the building that formerly housed the concentrator for the Britannia Mine. The 1894 discovery of a surface exposure of copper mineralization led to the development of the mine, which came into production in 1905. It operated until 1974, producing more than 500 million kilograms of copper, 15 million grams of gold, 180 million grams of silver, 125 million kilograms of zinc, and lesser amounts of lead and cadmium.

The orebodies that were mined occurred within a narrow northwest-southeast-trending zone of intensely sheared, foliated, early Cretaceous *pyroclastic rocks*, angular fragments exploded from a volcano. The rocks are now greenschist, having undergone low-grade metamorphism, and can be seen, along with equipment used in the mine, in the car park just south of the museum. The ore, which occurred over a vertical range of 1.3

The abandoned concentrator of the former Britannia Mine at Britannia Beach. —R. Turner photo

kilometres and a horizontal distance of 3 kilometres, was in massive sulfide deposits consisting of the iron-sulfide mineral pyrite with lesser amounts of chalcopyrite (copper-iron sulfide) and sphalerite (zinc sulfide), and also concentrated along irregular quartz veinlets.

The delta of Britannia Creek, below the mill, was the site of the administrative offices for the mine, housing for many of the workers, and loading facilities and the dock for the community, which in its early days was accessible only by water. In an effort to reclaim more flat land in the 1920s, a retaining wall was built of mine waste across the intertidal zone and then was backfilled with mill tailings. Before long, the reclaimed land had slumped beneath the waters of Howe Sound, and mill tailings have since been found on the fjord floor. This was probably the first recognized example on the British Columbia coast of a submarine delta-front collapse induced by human activities.

Squamish Granodiorite

Roadcuts along Highway 99 on lower parts of the hill just north of Britannia Creek are in the same early Cretaceous pyroclastic rock as at the Britannia Mine, but here the rock, although metamorphosed, is not sheared, and you can clearly see the angular fragments within it.

Stawamus Chief, viewed from the northwest, is a large mass of granodiorite.

Toward the top of the hill there is an intrusive contact with sparsely jointed and relatively coarse-grained granitic rock. Similar rock flanks the roadside between here and Squamish. The rock crystallized about 100 million years ago. Unlike most other granitic rocks of the Howe Sound area, it contains a small amount of pink, potassium-bearing feldspar, which makes it granodiorite. Its sparse jointing has made it resistant to erosion, and the relatively few fallen fragments are unusually large. Bold bluffs of pale rock are commonplace, with the most spectacular being the bluffs of the Stawamus Chief (more commonly known as Squamish Chief) just southeast of Squamish. Rising 600 metres in precipitous cliffs, it challenges the boldest rock climbers. These granitic rocks, like those at Porteau Cove, are the deeper and youngest parts of the middle Jurassic and early Cretaceous magmatic arc built on the eastern parts of Wrangellia.

Shannon Falls Provincial Park

About 3 kilometres south of Squamish, Shannon Creek drains a hanging valley that heads 6.5 kilometres southeast of the highway and spills over a sheer cliff of granodiorite in a spectacular waterfall about 300 metres high. A raised delta on the west side of the road here has yielded marine shells recording a late-glacial sea level about 30 metres above its present position and a radiocarbon date of 10,600 years old. At its maximum, the ice sheet depressed the land surface by more than 200 metres, so this site tells us that most of the rebound of the Coast Mountains occurred before this site was exposed from beneath the ice cover.

Mount Garibaldi

Mount Garibaldi, about 20 kilometres northeast of Squamish, is a late Pleistocene volcano rising 2,678 metres above sea level that is visible in good weather from several places on Highway 99 just south of Squamish. Mount Garibaldi and other young volcanic centres in this part of the Coast Mountains, including Mount Cayley west of Whistler and Meager Mountain northwest of Pemberton, are the northernmost volcanoes of the Cascade magmatic arc.

The Mount Garibaldi volcano grew during and following the climax of the last regional glaciation. The ice had withdrawn from its upper limit, here about 1,800 metres, but was present in the valleys below the 1,325-metre level. The Garibaldi eruptions built a broad cone with a summit elevation of about 2,750 metres, centred at the pyramidal south peak of Mount Garibaldi. The eastern part of the cone was built on high, ice-free ground and retains the original 12- to 15-degree slope of the cone. However, the western half of the cone was built on the ice sheet. It collapsed as

Mount Garibaldi, half of a Cascade volcano, viewed from Highway 99 south of Squamish. Parts of the pyramidal south peak formed on the ice during the last glaciation. The western half of the cone collapsed when the ice it was built on melted.

the ice melted, creating debris flows that were funnelled down the Cheekye River into the valley of the Squamish River. The slide scar left behind on the present Mount Garibaldi exposes a magnificent cross section of a volcano. This scar is partly covered at its northern end by younger lava flows.

Cheekye River Fan

Highway 99 continues northward from the traffic light at the Squamish turnoff and crosses the floodplains of the Mamquam and Squamish Rivers for about 10 kilometres. The route then begins a gradual climb up the alluvial fan of the Cheekye River, a tributary of the Cheakamus River. The fan rises 150 metres in the 3.6 kilometres to its apex, where Highway 99 crosses the 10-metre-deep canyon of the Cheekye River cut into granitic bedrock. The alluvial fan was built in postglacial time from debris mostly derived from the collapse of the western part of Mount Garibaldi.

Proposals to build a housing development and a hospital on the fan raised concerns about geological hazards. The unstable northern slope and crest of Cheekye Ridge potentially could produce a landslide that would temporarily block the Cheekye River. If the blocked water overtopped the dam formed by the landslide, a devastating flood could extend onto the fan. In 1991, provincial and local authorities sponsored a geotechnical investigation into the hazards of the fan and the housing development was at least temporarily put on hold.

Shaded-relief image of the Squamish area. Note that the valley containing the Stawamus River and the southwest side of the Tantalus Range west of Squamish appear to be aligned along the Ashlu Creek fault. (1) ***Daisy Lake,*** (2) ***Black Tusk,*** (3) ***Rubble Creek,*** (4) ***the Barrier,*** (5) ***Garibaldi Lake,*** (6) ***a young, postglacial lava flow, heading in a small cone, that flowed down the Mamquam River valley,*** (7) ***Cheekye River fan,*** (8) ***Stawamus Chief,*** (9) ***Shannon Falls Provincial Park.***

The Mamquam River, east of Squamish, contains a lava flow that heads in a small cone east of Mount Garibaldi. The flow is postglacial and younger than 6,670 years old.

Granite of the Cheakamus Valley

For about 16 kilometres north of the bridge over the Cheekye River, Highway 99 passes through a glacially carved landscape of rounded hills and smaller rocky knobs of granite separated from one another by narrow chasms. Along this part of the route, the Cheakamus River lies to the west, hidden from the highway in the bottom of a deep valley.

Granitic rocks exposed in roadcuts north of the Cheekye River bridge range from about 155 to 145 million years old, late Jurassic time. These rocks look a bit different from the generally clean-looking 100- to 95-million-year-old mid-Cretaceous granitic rocks along Howe Sound with their white feldspars with dark brown and black hornblende and biotite crystals and few dikes. In contrast, the Jurassic granitic rocks look messier, with greenish altered hornblende and biotite, and they typically are cut by numerous dark green dikes. Most of the dikes probably were conduits or "feeders" to the early Cretaceous volcanic rocks, such as those near Britannia Beach, but rare dikes, which weather to brown, may have fed the very young volcanic rocks, such as those forming Mount Garibaldi.

A large, dark greenish grey dike cut through somewhat altered late Jurassic granitic rock in a roadcut north of the Cheekye River.

VIEWPOINT GEOLOGY

Pullouts on both sides of Highway 99 where it emerges high above the Cheakamus River, roughly 6 kilometres north of the bridge over the Cheekye River and 14 kilometres south of the Rubble Creek dam, provide viewpoints of the spectacular scenery and the bedrock geology.

To the southwest, the spectacular Tantalus Range marks the boundary between Jurassic and Cretaceous granitic rocks. The Tantalus Range is mainly made of late Jurassic granitic rock. Out of sight on the southwest side of the range, the rocks are faulted along the Ashlu Creek fault against a narrow sliver of early Cretaceous volcanic rocks. Cretaceous granitic rocks, which form much of the western side of Howe Sound, intrude the volcanic rocks, and very locally, the Jurassic granitic rock.

To the southeast, the country east of Squamish shows the geological relationships of the Tantalus Range in a cross section. On the skyline above the sheer granitic cliffs of the Stawamus Chief, early Cretaceous volcanic strata form Goat Ridge, with gently west-dipping layers more or less parallel with the skyline. Volcanic rocks also form the upper part of Sky Pilot Mountain,

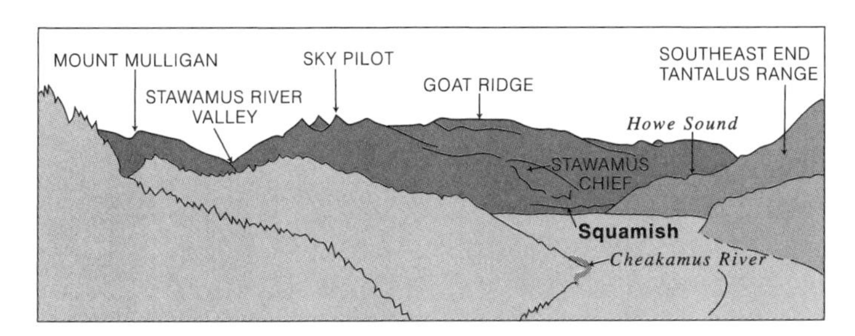

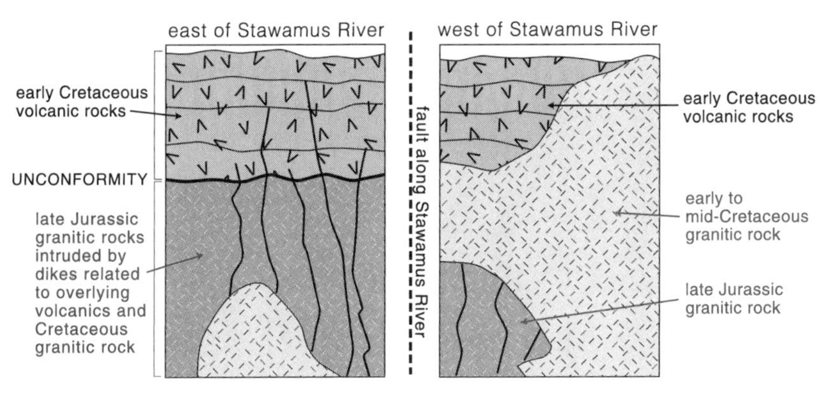

Schematic view to the south from viewpoints along Highway 99 about 6 kilometres north of the Cheekye River bridge. Intrusive and stratigraphic relationships east and west of the Stawamus River are shown.

east of Goat Ridge. Lower parts of Goat Ridge and Sky Pilot Mountain are made of early Cretaceous granitic rock that is physically continuous with that forming Stawamus Chief Mountain. It intrudes the slightly older volcanic rocks above, as seen along Highway 99 north of Britannia Beach. The steep valley to the east of Sky Pilot Mountains contains the Stawamus River and probably follows the trace of the Ashlu Creek fault whose northwesterly continuation runs along the southwest side of the Tantalus Range. East of the fault, late Jurassic granitic rocks on Mount Mulligan and Alpen Mountain (the latter just out of sight) are continuous with those crossed between the Cheekye River and the viewpoints. The Jurassic granitic rocks are intruded by abundant green dikes and are overlain by early Cretaceous volcanic rocks.

Sheared Granitic Rocks in Cheakamus Canyon

Between the top of the descent of Highway 99 to the level of the Cheakamus River (between about 26 kilometres north of the Cheekye River bridge and 5 kilometres south of Rubble Creek), the highway crosses a shear zone in the late Jurassic granitic rocks, exposed in continuous roadcuts along Highway 99 for a distance of about 3 kilometres. The normal granitic texture of randomly oriented crystals changes abruptly to one with a foliation formed by shearing and growth of new crystals. The foliation dips to the

Mylonitic late Jurassic granite by the Cheakamus River.

northeast and the width of the zone of sheared rock, measured at right angles to the foliation, is about 1.5 kilometres. In some exposures, accessible in two pullouts alongside the river on the west side of the road, examination of the rock with hand lens shows that platy or tabular mineral grains of mica and feldspar have been rotated into rough parallelism, probably while the rock was still partly molten and the crystallizing minerals were still free to assume their characteristic forms. In other exposures, some grains have been bent as well as rotated, and in still others, the weaker minerals have been ground up and recrystallized in varying degrees to form mylonites, which are very fine-grained, foliated rocks.

"Pristine" granitic rocks have undeformed crystals of feldspar (white), *quartz* (gray) *and biotite* (black).

In sheared granitic rock, the coarser part has a poor foliation, but the finer-grained, well-foliated part starts to approach true mylonite. The small circled crystal of feldspar has "tails," showing that the rock to the left of it moved up with respect of that to the right.

Rubble Creek Rockslide from the Barrier

About 1 kilometre south of the hydroelectric dam below Daisy Lake and about 3 kilometres above the head of Cheakamus Canyon, Highway 99 crosses Rubble Creek. South of Rubble Creek you can see dead trees standing upright in the bouldery bed and banks of the Cheakamus River. The trees were killed and partly buried in the fall or winter of 1855–56 when a major rockslide swept down Rubble Creek and dammed and diverted the Cheakamus River.

The source of the rockslide is a cliff called the Barrier. You can see the cliff either from the northwest end of the Cheakamus hydroelectric dam or from the end of the 2.8-kilometre paved road that leaves the east side of Highway 99 and leads to the trailhead at Black Tusk Meadows in Garibaldi Provincial Park. The Barrier exposes black, grey, and reddish lava of andesitic composition that was extruded at the very end of the last glaciation, about 10,000 years ago. The lava emerged from a volcanic centre on Clinker Peak and in its descent gave rise to four lobes of lava. One lobe reached the north wall of the valley now containing Rubble Creek and formed a dam 250 metres high. This natural dam impounds Garibaldi Lake, which is about 10 kilometres east of Highway 99 at an elevation of 1,500 metres. Another lobe, one that was involved in the landslide, also reached the north wall of Rubble Creek and spread westward, making contact with the ice that occupied the lower Rubble Creek valley. Another of the lobes also is believed to have made contact with the ice and today forms the spectacular 450-metre-high north-facing wall visible south of the landslide scar on the Barrier. This face displays a high proportion of lava columns oriented roughly perpendicular to the present cliff face, a feature that indicates the surface of the face is parallel with the surface on which the lava cooled against the ice. Before the winter of 1855–56, the western end of the collapsed lobe probably looked like this ice-contact face.

The rockslide generated by the collapse of the lobe spread swiftly down the valley and reached the present site of the Cheakamus dam before it came to rest. The valley floor was covered with a layer of coarse bouldery debris. The lower slopes of the Rubble Creek valley show a scattering of angular andesite fragments, or in places a levee of andesite debris, and an absence of trees more than 140 years old below a well-defined limit. Above this limit, there are no andesite fragments on the ground surface and old trees survive. The lowermost of these trees exhibit bark scars on the up-valley sides of their trunks, and two trees have been found with andesite fragments embedded in the wood. Tree-ring counts show that the damage occurred after the summer growth of 1855 and before the spring growth of 1856. Careful mapping of the limits of debris and damaged trees shows

View of the Barrier at the head of Rubble Creek. The Black Tusk is the peak in the background.
—C. Hickson photo

that the rockslide swept higher on the outside of the curves in the valley than on the inside. From the measurements of the different elevations of the damage on inner and outer curves, it is possible to calculate that the velocity of the rockslide must have reached 30 to 40 metres per second, or about 108 to 144 kilometres per hour.

In 1970, a subdivision had been started west of Highway 99 and immediately south of Rubble Creek. The appropriate government departments of highways, education, and health approved the initial development. After completion of this first phase, the developers applied for permission to subdivide the area directly east of Highway 99, and being confident of approval, they put in roads and water lines. This time, however, the hazard of a future rockslide was raised and approval was denied. A court hearing supported the denial on the grounds that the hazard should be considered over the life span of the community that could arise here, rather than just over the life span of an individual resident. A panel of engineers later found evidence of multiple rockslides in this area and recommended limiting any housing development on and adjacent to the rockslide, as well as around the perimeter of the reservoir and along the river banks for 3 kilometres downstream from the Cheakamus dam. In response to these recommendations, the provincial government set aside a fund to buy back property within the designated area. In addition, the high-water level in the Daisy Lake hydroelectric reservoir has been lowered by 2 metres as a precaution against wave damage in the event of a new slide.

Lava Esker near Brandywine Falls

About 7 kilometres north of the Cheakamus dam, you can view Brandywine Falls by walking east across the railroad bridge over Brandywine Creek and then south along the footpath to the falls lookout, where the creek drops about 70 metres in a single jump. Extending from the south side of the parking lot for Brandywine Falls Provincial Park, lava forms a conspicuous sinuous ridge that rises above the late-glacial or postglacial alluvium of Brandywine Creek. The ridge extends southward for several hundred metres but is best exposed in the railroad cut near its northern end. The term *esker* is usually applied to an elongate, sinuous, steep-sided

Exposed in a railway cut near Brandywine Falls is a cross section of a sinuous ridge of basalt with columnar jointing. The lava flowed into a tunnel or trench in an ice sheet and solidified. Its formation corresponds to that of an esker made of gravel deposited by a stream flowing in a tunnel beneath the ice.

ridge of gravel built by a glacial stream that flowed within a tunnel or trench beneath a glacier. Here, the esker is made of basaltic lava! Many basalt columns in the lava are nearly perpendicular to the present top, side, and bottom surfaces of the flow. Because columns form perpendicular to the quickly cooling outer surfaces of a lava flow, these present-day surfaces probably correspond to the original outer surfaces of the flow. At the base of the flow, a layer of granulated basaltic glass formed by rapid quenching of the lava. The linear esker form, the columnar fracture pattern, the quenched substratum, and the absence of surviving material that may have confined the lava laterally all point to a lobe of lava that flowed through a trench or tunnel in the ice.

Between Brandywine Falls and Pemberton

Visible on top of the eastern ridge from parts of Highway 99 north of Brandywine Falls is a spectacular, 2,500-metre-high black peak shaped like a shark's fin, which in north-south profile is narrow enough to resemble its name—the Black Tusk. The volcanic rocks forming this peak are isotopically dated at between about 200,000 and 90,000 years old.

End-on view of the Black Tusk from the ridge to the north of it.

Yet another manifestation of Cascade arc magmatism is a basaltic lava flow seen in roadcuts and natural exposures along the highway above the hydroelectric dam on Daisy Lake. It emerges from the valley of Callaghan Creek about 4 kilometres north of Brandywine Falls. The 30-kilometre-long flow has columnar joints and lies on top of till and glacial outwash sediments that yield radiocarbon dates of 34,000 years old.

North of the flow, Highway 99 passes through Whistler, venue for the 2010 Winter Olympics. Much of this part of the highway crosses mainly broken and fractured greenish and buff granitic rocks. In Whistler, roadcuts of greenish chlorite schist, which weathers to a tan colour, and silvery schist probably formed by alteration along yet another shear zone. The original rocks are probably highly sheared equivalents of the early Cretaceous volcanic rocks found higher up on the eastern valley wall below the Black Tusk and crossed well to the south along Howe Sound. A diorite body about 30 kilometres north of Whistler has been dated at 113 million years old.

Pemberton Valley

About 50 kilometres north of Brandywine Falls Provincial Park, Highway 99 descends into the Pemberton Valley. Just east of Pemberton, the road branches. Highway 99 is to the east and continues southeastward along the Lillooet River to Lillooet Lake. The valley containing Lillooet Lake is the northwestward continuation of the valley containing Harrison Lake. These valleys approximately mark the boundary between the different rocks and structures in the southwestern and southeastern Coast Mountains.

Between the coast and the Pemberton Valley, the granitic rocks range in age from middle Jurassic to mid-Cretaceous (170 to 95 million years old), with no clear sideways shift in the axis of magmatic activity with time. Near the coast, the middle Jurassic granitic rocks intrude Wrangellian strata. In several places near Highway 99, early Cretaceous volcanic and sedimentary rocks lie on top of eroded and exposed late Jurassic granitic rocks. Much of the deformation in the southwestern Coast Mountains is concentrated in discrete shear zones, such as those at Britannia Beach, Cheakamus Canyon, and in Whistler, but between those shear zones the rocks are relatively little deformed.

By contrast, east of the Pemberton valley, the granitic rocks become younger from west to east. Their ages range from mid-Cretaceous to Eocene: 102-million-year-old rocks occur near Lillooet Lake, 87- to 85-million-year-old rocks near the summit crossed by Highway 99 between Pemberton and Lillooet, 65- to 63-million-year-old rocks along the descent of the eastward-flowing Cayoosh Creek, and 45-million-year-old rocks near Lillooet.

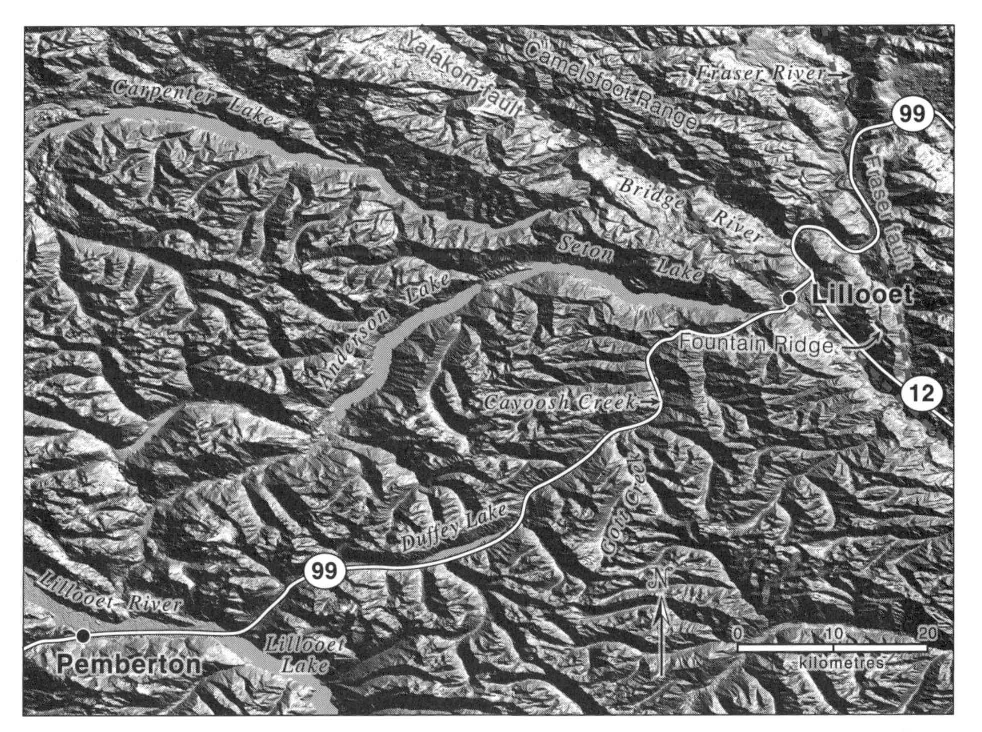

Shaded-relief image of topography along Highway 99 between Pemberton and Lillooet. *—Image constructed using Canadian Digital Elevation Data obtained from GeoBase*

The nongranitic rocks along Highway 99 east of Pemberton are all metamorphosed and have well-developed foliations. The original rocks ranged in composition: chert, shale, sandstone, basaltic volcanic rock, and ultramafic rocks are now respectively metachert, phyllite, metamorphosed sandstone, amphibolite, and rare talc schist. You can find the equivalent rocks, little metamorphosed but highly broken up, about 50 kilometres north of Highway 99 near Carpenter Lake, where they mostly belong to the Bridge River and Methow terranes. Many rock types there are similar to those found on modern ocean floors, and the chert contains microfossils made of silica, called *radiolarians*, that lived between Carboniferous and middle and possibly late Jurassic age, about 350 to 160 million years old. In a few places, late Jurassic and early Cretaceous sandstone, shale, and conglomerate overlie these rocks and are probably lateral equivalents to rocks of the same ages in the Methow terrane.

The older rocks in the southeastern Coast Belt appear to be the remnants of a marine basin floored by oceanic lithosphere that, starting in mid-Cretaceous time, was caught in the jaws of a tectonic vise, squeezed, metamorphosed, and intruded by the mid-Cretaceous and early Tertiary granitic rocks. The eastern jaw of the vise was the early Cretaceous North

American continental margin, represented today by the westernmost rocks of the southern Intermontane Belt. The western jaw comprised Wrangellia and the middle Jurassic to mid-Cretaceous granitic, volcanic, and sedimentary rocks of the southwestern Coast Belt.

Lillooet Lake

Lillooet Lake, which is about 84 kilometres by road southeast of Lillooet, is 34 kilometres long, 1.3 to 2 kilometres wide, and 137 metres in maximum depth. It was first mapped in 1860 by Colonel Moody of the Royal Engineers. The delta of the upper Lillooet River, formed where it discharges into the head of the lake, was mapped in greater detail in 1913, 1948, and 1969. It was found that the delta front advanced into the lake at a rate of 7.8 metres per year prior to 1948 and at more than 20 metres per year from then until 1969. Erosion caused by activities related to logging may have been the cause of the rapid increase. The lake bottom sediments display annual layers, called *varves*, which show that most of the incoming suspended sediment is deposited near the base of the delta at rates of 5 to 8 metres per year, diminishing to 16 centimetres per year at the south end of the lake.

Seismic studies have revealed a prominent layer deep within the bottom sediments, discovered because it reflects seismic energy. The layer may be ash deposited at the time of a volcanic eruption near Meager Mountain on the upper Lillooet River, about 75 kilometres above the head of Lillooet Lake. The eruption produced a widespread fallout of volcanic ash, recognized as far east as the Alberta boundary. This was misnamed "Bridge River ash" long before its source on the Lillooet River was established. Radiocarbon dates of the ash from elsewhere in the region indicate the eruption took place about 2,550 years ago. Assuming the layer in the bottom sediments formed at the time of this eruption, the average prehistoric rate of erosion in the drainage basin of the Lillooet River above the lake was about half the rate prevailing in historic time.

Between Lillooet Lake and Lillooet

On the northeast side of Lillooet Lake, Highway 99 climbs up the steep eastern wall of the valley. For the next 16 kilometres it crosses a 102-million-year-old body of quartz diorite and granodiorite, and northeast of this for a distance of about 10 kilometres is an 87-million-year-old body of quartz diorite.

About 21 kilometres northeast of Lillooet Lake, the highway passes into the watershed of northeast-flowing creeks and crosses the uppermost bridge on Cayoosh Creek. The highway follows this creek almost as far as

View from Duffey Lake, looking east. The high mountains are 102- to 85-million-year-old granitic rocks that intrude schist and phyllite derived from late Paleozoic and Mesozoic strata of the Bridge River terrane.

Lillooet. Along this part of the route are many roadcuts in nongranitic rocks, including dark grey phyllite weathering to a rusty colour, derived mainly from pyritic shale; pale grey to white quartz-rich rock probably derived from chert; pale green schist derived from volcanic rock; minor marble derived from limestone; and rare talc schist derived from ultramafic rock. A good place to examine these rocks is immediately east of where Highway 99 crosses Gott Creek, about 50 kilometres east of Lillooet Lake and 32 kilometres west of Lillooet. Although these rocks are undated along Highway 99, on the basis of the range of rock types geologists believe they belong to the Carboniferous to late Jurassic Bridge River terrane.

Near Boulder Creek, 3 kilometres east of Gott Creek and 29 kilometres west of Lillooet, the road passes through 63-million-year-old granodiorite. Across Cayoosh Creek to the north, the rocks are thinly laminated siltstone and shale of probable late Jurassic or early Cretaceous age. A fault separates the granite from the sedimentary rock, and Cayoosh Creek probably follows the fault, taking advantage of the more-readily eroded rock along the fault.

About 70 kilometres east of Lillooet Lake and 14 kilometres west of Lillooet, the canyon near the Cayoosh Creek bridge is cut through foliated sandstone, siltstone, and phyllite. The rocks are not dated here, but fossil clams in the same kinds of rocks are present about 10 kilometres to the southeast, near the top of Mount Brew, at an elevation of nearly 3,000

White, deformed quartz veins crosscut finely layered siliceous schist derived from Bridge River chert.

metres. The fossils are of earliest Cretaceous age, about 140 million years old, and lived in sand and mud that probably were deposited on top of the Bridge River terrane. The mountains in this area are the highest in the southern Coast Mountains.

A Trail of Distinctive Erratics

Rocks deposited from a glacier that do not come from the local area are called *erratics*. Distinctive erratics of a particular rock type can form trails that show the paths glaciers took during an ice age. A source of a distinctive and attractive quartz-lazulite rock, eagerly sought after by rockhounds, is believed to exist somewhere near the bridge over Cayoosh Creek 14 kilometres west of Lillooet. This location is the highest and northernmost known occurrence of boulders of this rock. The rock contains conspicuous flecks of the azure blue aluminum-, iron-, or magnesium-bearing phosphate mineral lazulite. Pebbles and cobbles with the same unique mineralogy, sometimes containing lilac-coloured dumortierite or the colourless, fibrous mineral sillimanite as well, have been collected at several localities down the Fraser Canyon as far as Hope. Similar pebbles have been found near Chilliwack and southwest from there across northwestern

Washington to Cape Flattery on the northwestern tip of the Olympic Peninsula. Transport as far as Hope could have been the work of the Fraser River, but transport from Chilliwack to the open ocean probably needed glacial assistance.

Seton Lake

The viewpoint from the car park on the west side of Highway 99 at the top of the hill leading down to the outflow of Seton Lake, 6 kilometres east of Lillooet, is of archeological, historical, and geological interest. Near the viewpoint are the remains of several pit dwellings that once housed the original inhabitants of the region. Seton Lake, below the viewpoint, extends westward into the Coast Mountains. In the early 1860s, before a road was built along the Fraser Canyon, the lake was part of the route to the Cariboo goldfields at Barkerville, near Quesnel in the interior of British Columbia. Steamboats carried passengers from New Westminster (now part of Greater Vancouver) up the Fraser River to the north end of Harrison Lake. From there the route was overland to the south end of Lillooet Lake, and from its north end, northeastward over a low pass to Anderson Lake, which is separated from Seton Lake by only a short portage. From the east end of Seton Lake, the route to the Cariboo goldfields was entirely overland. A cairn in Lillooet marks mile 0 on the road to the goldfields.

To the north of the viewpoint, Seton Creek flows eastward out of Seton Lake along the floor of a spectacular steep-walled valley, to join the Fraser River at Lillooet. The origin of this big, deep valley, which today is occupied by a relatively small stream that cuts across the grain of the Coast Mountains, is unknown. It is possible that the valley follows an old river system that was present before the last uplift of the Coast Mountains, and as the mountains rose, the old valley gradually became deeply incised, a feature called an *antecedent drainage.* The lower part of the eastern end of the valley shows the U-shaped profile typical of a glaciated valley, so the valley has been shaped by ice at least in part.

The steep walls of the valley are well-foliated rocks, some similar to those at Gott Creek and others to those by the bridge over Cayoosh Creek, derived from the Bridge River terrane and overlying sediments. Abundant white granitic veins intrude the metamorphic rocks. Some intrusions, called *sills,* lie in the near-horizontal plane of foliation and are themselves foliated and mylonitic. Others, called *dikes,* crosscut the plane of the foliation but appear to be folded, and still other dikes are undeformed and rise steeply up the cliff face. Surprisingly, all the sills and dikes are about 45 million years old, which means that they were intruded during and immediately after the latest deformation.

View eastward from the viewpoint above Seton Lake. Note the glacially carved, U-shaped profile of the lower valley. The dark cliffs are metamorphic rocks intruded by light-coloured, 45-million-year-old granitic sills and dikes. Beyond them is the valley of the Fraser River, here carved in Mesozoic sedimentary rock; and in the background, east of the Fraser River, Fountain Ridge is capped by cliffs of early Cretaceous Jackass Mountain sandstone, shale, and conglomerate.

Farther northwest and southeast are larger granitic intrusions of the same age. Together, these represent the easternmost, youngest episode of magmatic activity in the Coast Belt before the Cascade magmatic arc "fired up" in this region about 35 million years ago. Note that granitic rocks of about the same age also occur just east of Hope. These two areas of similar rock were probably adjacent to each other before right-lateral strike-slip movement on the Fraser fault offset them.

Yalakom Fault

At the eastern end of the Seton Creek valley, the large Yalakom fault runs northwestward away from the Fraser River. At the easternmost Highway 99 bridge over Seton Creek, a large body of serpentinite marks the trace of the fault, which continues northwestward through Lillooet, over the ridge north of town, and into the Bridge River and Yalakom River valleys. The Yalakom fault is a right-lateral strike-slip fault like the Fraser fault whose west side moved northwestward. The displacement on the fault is estimated

The white granitic sills and dikes in the cliff face in the Seton Creek valley were all intruded into the metamorphosed Bridge River terrane and younger rocks 45 million years ago. Note the heads of large talus cones, piles of rock that fall from the cliffs above.

to be between 150 and 155 kilometres, and the fault appears to have been active in latest Cretaceous to early Tertiary time. The Yalakom fault is cut and offset by the younger Fraser fault, and its counterpart on the west side of the fault is probably the Ross Lake fault southeast of Hope.

East of the Yalakom fault are early Cretaceous sandstone, shale, and conglomerate, which form cliffs of greenish buff colour on Fountain Ridge on the west side of the Fraser River. The rocks belong to the Jackass Mountain group, the youngest part of the Methow terrane. The north-trending Fraser fault follows a north-south-oriented valley that runs on the other side of Fountain Ridge.

Air Circulation in Talus below Fountain Ridge

Fountain Ridge, across the Fraser River from Lillooet, is flanked by some of the largest *talus cones* and *talus aprons* in southern British Columbia. These accumulated in postglacial time from innumerable falls of rock from the cliffs above. Small rock fragments come to rest on the 35-degree slope of the talus and a few larger blocks of rock roll to the bottom and out on the valley floor before stopping. Most rockfalls, which are common in this semiarid region, occur in autumn at the time of early frosts and in spring with early thaws.

On hot summer days—and it can get very hot in Lillooet—ice used to build up on the floor of an old root cellar excavated in the toe of the talus from Fountain Ridge, only to thaw again the next autumn or winter. The phenomenon was a response to seasonal exchange of air temporarily held within the talus. In fall and winter, cold air entering the talus displaces the less dense warm air upward out of the talus, whereas in the spring and summer the cold and dense air within the talus drains downward and outward, escaping to the atmosphere at the toe of the talus. The warm air that enters the talus can itself become chilled through contact with the cold rock fragments, and this chilled air in turn will drain downward and outward. Only after nearly all the talus is warmed to the temperature of the outside atmosphere will the cold air drainage cease.

View looking south down the Fraser River valley from its eastern bank; Lillooet sits on the opposite bank. Here the eastern side and valley floor are carved in Jurassic and Cretaceous sedimentary rocks. The 2,200-metre-high ridge of the easternmost Coast Mountains in the background is formed of metamorphic rocks intruded by numerous 45-million-year-old granitic bodies. The valley is much wider here than farther south and features numerous terraces, which formed when slides blocked the river south of here more than one thousand years ago (**see "The Big Slide" in Highway 12: Lytton—Lillooet**). *Highway 99 comes through the big valley southeast of Lillooet.*

At Lillooet, the volume of the talus cones and the size of the rock fragments and the open spaces between them are such that months are required before the chilling effect of the previous winter is lost. The diffuse upward escape of warm air during autumn and winter does little to increase temperatures at the base of the talus. The excavation of a root cellar or short tunnel into the toe of the talus produces a small increase in the flow of the escaping cold air and thus concentrates the chilling. Alas, the owner of the land with the root cellar became so upset at all the visitors coming to see his "ice cave" that he demolished it!

Highway 99 North of Lillooet

Highway 99 crosses the Fraser River at Lillooet on the Twenty-Three Camels Bridge, so named from the brief use of Bactrian camels as pack animals in this arid region in the 1860s. The camels apparently worked well here, but unfortunately the more commonly used pack animals, mules and horses, reacted violently to the camels and caused big problems on steep, narrow pack trails. The camels were let go, and at least one survived for about twenty years after release. The Camelsfoot Range, north of Lillooet, is named for them.

About 7 kilometres north of the Highway 12 junction, just east of Lillooet, the Bridge River flows into the Fraser River from the west. Just north of this, Highway 99 and the Fraser River turn sharply eastward, cutting across the north end of Fountain Ridge. Across the river, you can see massive, eastward-dipping beds of buff-coloured sandstone and darker beds of shale in the valley wall, which is carved in the southeastern end of the Camelsfoot Range. All the roadcuts along this part of Highway 99 are in sandstone, shale, and conglomerate of the Jackass Mountain group. East of this, the valley opens up at the north end of Fountain Valley, east of Fountain Ridge.

About 15 kilometres north of the Highway 12 junction, Highway 99 passes under the British Columbia Railway, which traverses a valley filled with coarse, bouldery gravel. About 0.5 kilometres east of the bridge, Highway 99 crosses the Fraser fault, which here marks the boundary between the Coast and Intermontane Belts. Pink to grey Eocene volcanic rocks are exposed in the road and railway cuts. To get a good overview of the Fraser fault and its relationships, we suggest you continue for another 3 kilometres along Highway 99, where a pullout on the west side of the highway provides a magnificent view of the scenery. For information about the geology along Highway 99 north and east of this pullout, see **Highway 99: Highway 97—Fraser River**.

VIEWPOINT GEOLOGY

Two sketches show views to the south and to the north of the Fraser River valley from the viewpoint on the west side of Highway 99 about 18 kilometres north of Lillooet and 68 kilometres east of Cache Creek. To the south toward the valley of Fountain Creek, the ridges to the east are capped by volcanic rocks of the early Cretaceous Spences Bridge group, about 105 million years old, that unconformably overlie the Cache Creek terrane. Here, both are the westernmost elements of the Intermontane Belt. The knoll to the south in the middle distance is a fault-bound block of Eocene volcanics, about 50 million years old, which is caught up in the Fraser fault, the main strand of which lies along Fountain Valley. Fountain Ridge, on the skyline, is made of Jackass Mountain sedimentary rocks, about the same age as those of the Spences Bridge volcanics, and here the upper part of the Methow terrane, the easternmost terrane in the Coast Belt.

To the west, the lower cliffs across the Fraser River are Eocene volcanics; the main strand of the Fraser fault runs behind the bluff, and above and to the northwest in the Camelsfoot Range are Jackass Mountain rocks. Prior to movement on the Fraser fault, about 45 to 35 million years ago in late Eocene time, Jackass Mountain rocks across the river probably were about 140 kilometres to the south relative to those at the viewpoint.

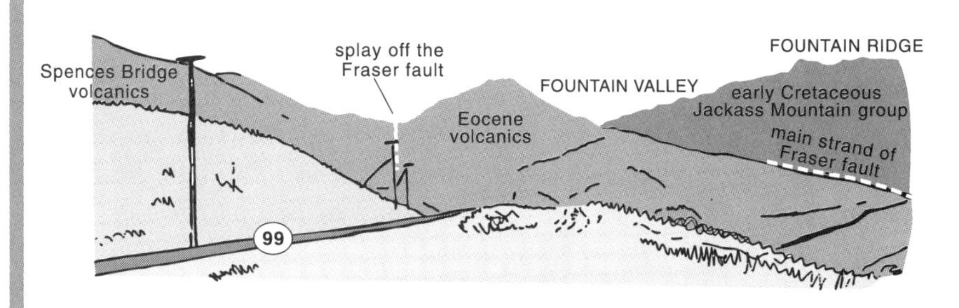

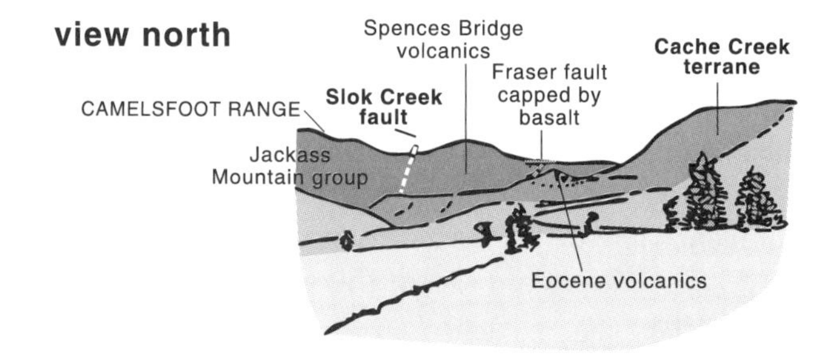

To the north, the west side of the valley is the Camelsfoot Range of Jackass Mountain rocks, here separated from Spences Bridge volcanic rocks by the Slok Creek fault, a splay off the Fraser fault. The main strand of the Fraser fault runs below the flat skyline, which is capped by a 2-million-year-old basalt flow. Cache Creek rocks form the east side of the Fraser River valley, and the knoll in the middle distance is another fault-bound block of Eocene volcanic rocks.

The flat skyline of this area northwest of Kamloops exemplifies the topography formed by the capping Miocene plateau basalt that gave rise to the name "Interior Plateaus." Faulted and tilted Eocene volcanic rocks form the slopes.

INTERMONTANE BELT

Interior Plateaus

The Intermontane Belt occupies the central part of the Canadian Cordillera between south-central Yukon and northernmost Washington, a distance of nearly 1,800 kilometres. It is about 200 kilometres wide in the middle, near Prince George in central British Columbia, but tapers and pinches out at both ends. In contrast with the flanking, mostly mountainous Coast and Omineca Belts, the topography is more subdued. High plateaus, rolling uplands, and deeply incised valleys are typical of the southern Intermontane Belt, although there are mountainous areas within the belt at its southern end near the international boundary, and in northern British Columbia. In addition, metamorphism of volcanic and sedimentary rocks in the Intermontane Belt is generally of very low grade or is absent, again in contrast with that of most rocks in the flanking belts.

About 18 to 10 million years ago, in Miocene time, basaltic lava flowed out onto the surface of this part of southern British Columbia, blanketing the rolling landscape with flat-lying layers of hard rock. Later, erosion removed the Miocene basalt cover in many places, but elsewhere, mainly in the northern part of the area covered by this guidebook, left extensive high-standing, relatively flat areas capped by the lava. The lava forms the plateaus that gave rise to the names "Interior Plateaus" and "plateau basalt."

Seven different names (four plateaus and three highlands) are given to different parts of the interior region in the southern half of British Columbia. They include not only the areas covered with Miocene basalt but also the uplands that approximate the level of the surface upon which the basalt was deposited. For simplicity, we've used the collective name "Interior Plateaus" in this book.

Although the Interior Plateaus are mostly coextensive with the southern Intermontane Belt, in places the topographic and geological boundaries between the two do not precisely coincide. This is because the belt boundaries

Princeton Basin from the southeast. Princeton is out of sight in the basin below the hills in the middle ground. Note the deeply incised, high, rolling upland surface. The snowcapped Coast Mountains are on the left skyline.

are based largely on bedrock geology and were established during the major period of mountain building in later Mesozoic and early Tertiary time, whereas the topography reflects, to some degree, late Tertiary to recent differential vertical movement. For example, the bedrock boundary between the Coast and Intermontane Belts is delineated by the Pasayten fault south of Lytton and the Fraser fault north of Lytton. The faults separate very different kinds of geology from one another, but the topography of the high, rugged Coast and Cascade Mountains declines and gradually merges eastward with that of the more subdued, lower Interior Plateaus. Near the international boundary, the topographic Cascade Mountains extend almost to the eastern side of the Intermontane Belt, but because the bedrock in this area is continuous with that underlying the Interior Plateaus to the north, the region is included in the Intermontane Belt. Similarly, on the eastern side of the southern Intermontane Belt, faults that mark an eastward jump to higher metamorphic grades delineate the boundary between the Intermontane and Omineca Belts, but the topographic change from the Interior Plateaus to the Monashee Mountains is very gradual.

The Interior Plateaus lie in the rain shadow of the Coast and Cascade Mountains and experience more sunshine than other parts of British Columbia. At low elevations, where the annual rainfall is less than 30 centimetres per year, the natural vegetation includes plants typical of a semiarid climate, such as ponderosa pine, sagebrush, and cactus. The area supports an agricultural economy with fruit growing, including grapes, mainly in the Okanagan Valley, and cattle ranching and milk production throughout much of the region. These activities need water for irrigation, which has promoted the widespread construction of water storage reservoirs that take advantage of the numerous hollows already existing on the glaciated upland surface. Rapid urbanization of sites on the valley floors, especially in the Okanagan Valley and around Kamloops, has created new problems in water supply, pollution, waste disposal, and sand and gravel supply for road construction.

At its maximum extent, about 15,000 years ago, the latest Cordilleran ice sheet covered the Interior Plateaus, smoothing off higher parts of the landscape and scouring many valleys to a great depth, a characteristic shared with the fjords along the western side of the Coast Mountains. As the ice receded, glacial lakes ponded behind ice dams that blocked the outlets of many big valleys. Silt and clay settled out of the lake water, and side streams discharging into the lakes deposited deltas of gravel and sand. Today, these deposits form distinctive terraces in the bottom of many major valleys, where the unconsolidated sediments may be hundreds of metres thick. The terraces, well preserved by the relatively arid climate, provide a picture of the late-glacial evolution of the region that is far more complete than that in the flanking mountainous and wetter regions.

Plateau Basalts

The plateau basalts are mostly about 10 million years old, but range in age from 20 to 1 million years old. They are akin in composition and age to the Columbia River basalt flows in central and eastern Washington, although the latter are far more extensive and much thicker. The stack of lava flows here is mostly less than 150 metres thick but in places reaches 350 metres. The greater thicknesses accumulated over topographically low areas, for example, in former river channels or where earlier flows had undergone gentle down-warping.

The interior of southern British Columbia has been exposed above sea level since late Mesozoic time, so erosion, rather than deposition, dominated landscape development prior to extrusion of the basalt flows. In Miocene time, the eroded land surface ranged from about 1,100 to 1,500 metres in elevation, though in places it was higher. Some old hilltops protruded above the lava flood and now stand between 200 and 500 metres above the surrounding region. Today, the Miocene erosion surface gradually increases in elevation as it approaches the Coast Mountains, where it can be traced up to heights of nearly 3,000 metres.

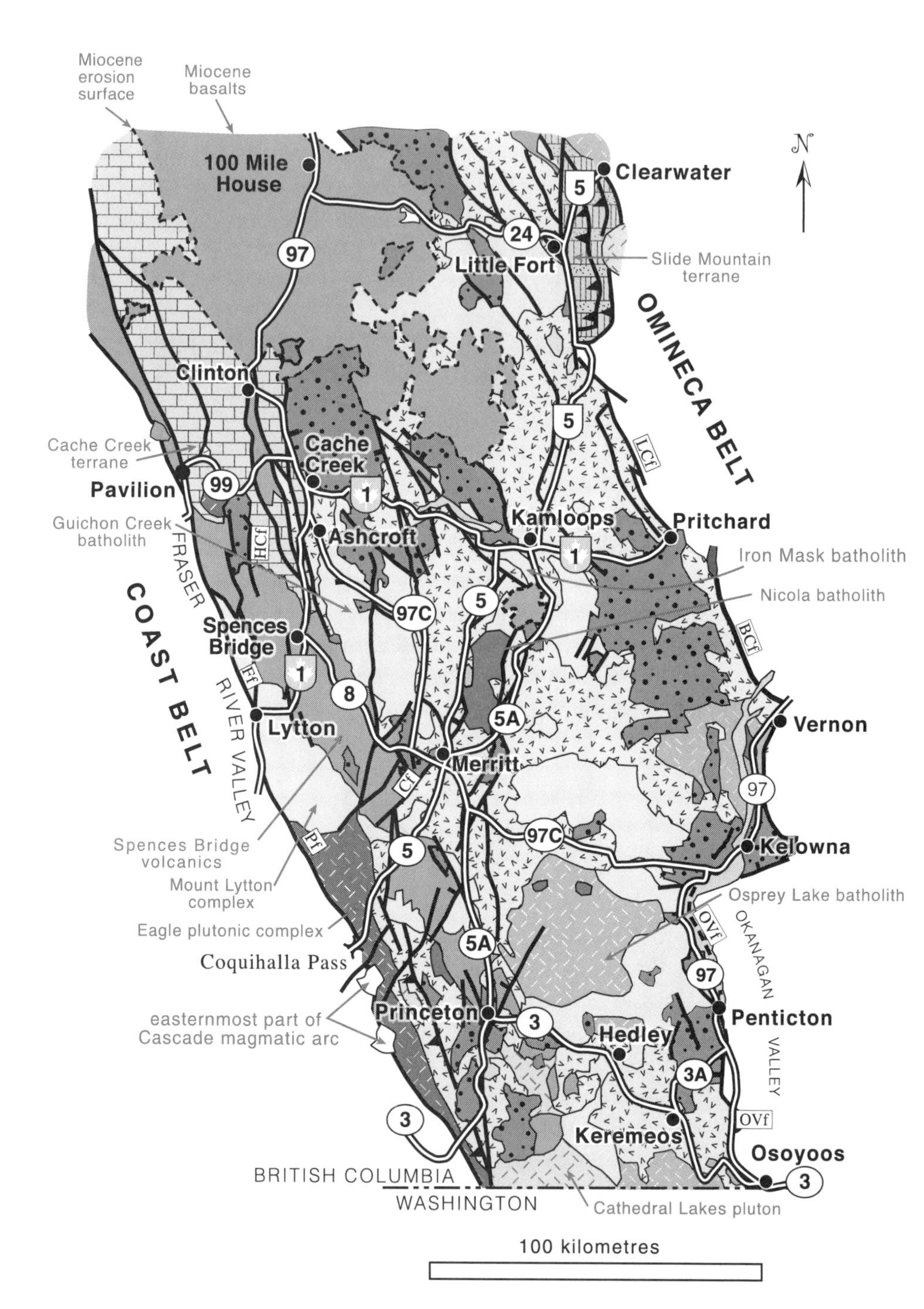

Geological map of the southern Intermontane Belt.

The plateau surface is most clearly seen in the northern part of the region covered by this book. Basalt is exposed at the surface in many places, particularly at scarps and cliffs marking the edges of the plateaus, but more commonly the plateau top is covered by glacially disturbed debris in which basalt fragments dominate. In much of the area covered by this book, the present upland surface probably differs little in its nature from that of the Miocene surface on which the lava was deposited. South of Highway 1, the basalt is only present in a few places, and the upland surface that lay below the flows is incised by relatively steep-walled valleys. Upland surfaces, which were eroded before late Miocene time and reach elevations of well over 2,000 metres near the international boundary, are isolated from one another by this network of valleys.

Older Bedrock of the Intermontane Belt

The oldest rocks of the Intermontane Belt, of late Devonian or Carboniferous through early Jurassic age (about 360 to 180 million years old), belong to the Quesnel, Cache Creek, and Slide Mountain terranes. Their rock types and chemistry suggest they formed on ocean floors and in volcanic island

Rock units on the geological map

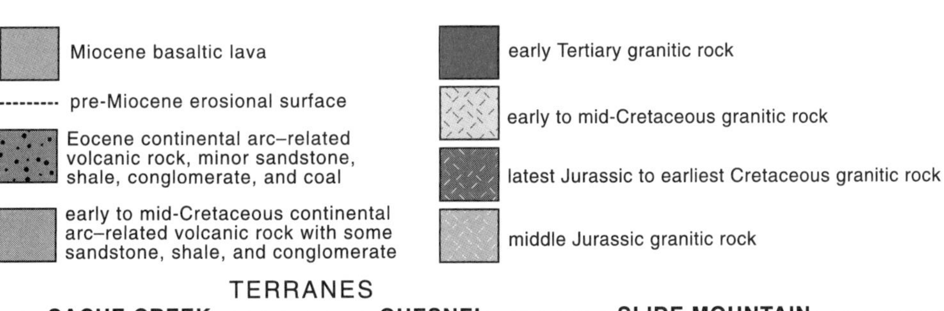

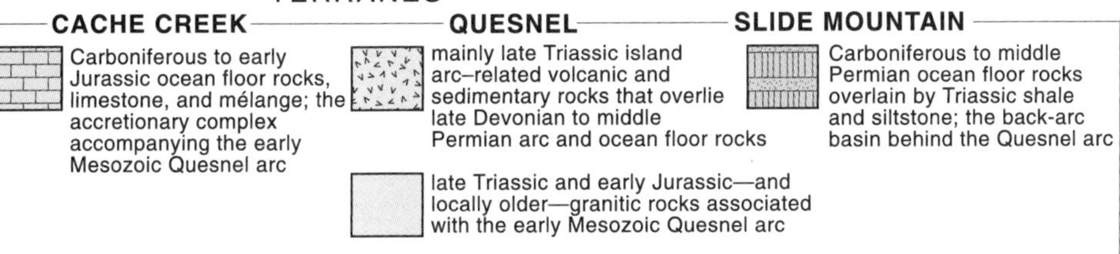

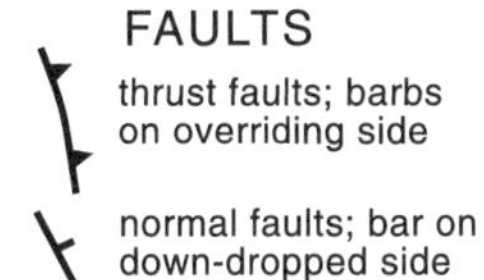

NAMED FAULTS

Cf Coldwater fault
BCf Bolean Creek fault
Ff Fraser fault
HCf Hat Creek fault
LCf Lewis Creek fault
Pf Pasayten fault
OVf Okanagan Valley fault

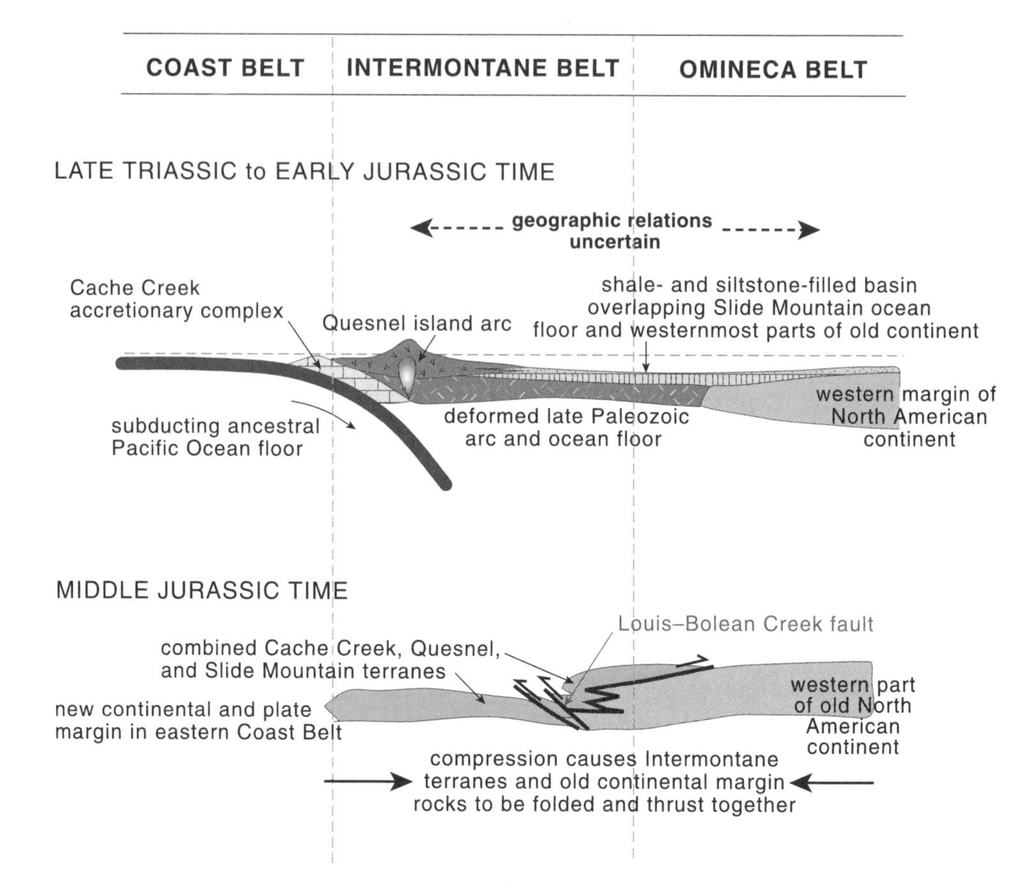

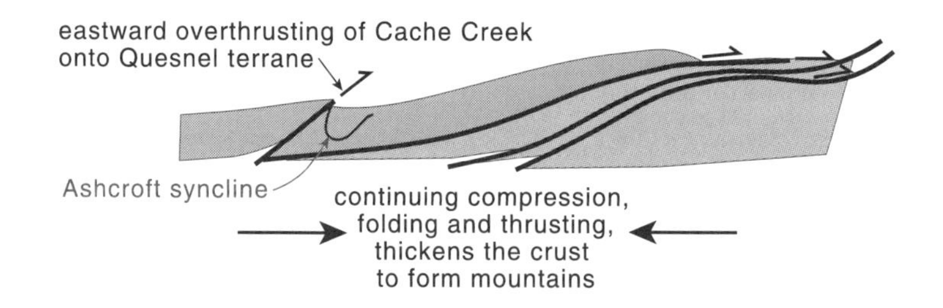

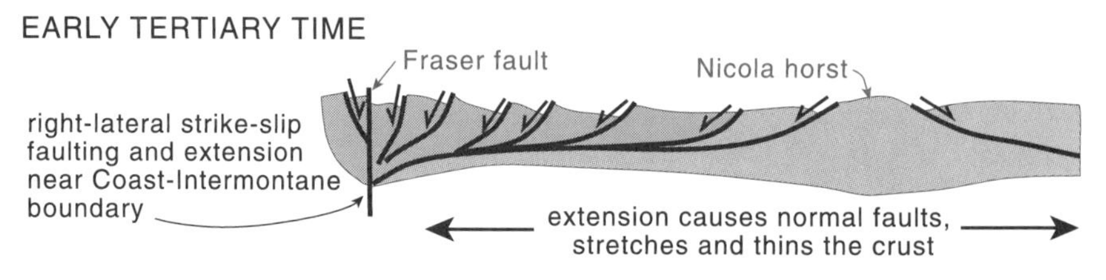

Four cross sections showing stages in the geological evolution of the southern Intermontane Belt.

arcs, well away from the influence of the old continent. The three terranes were accreted to the western edge of the old North American continent in Jurassic time, about 180 million years ago. Subsequently, between about 180 and 100 million years ago, they formed a new continental margin into which younger granitic rocks were intruded and upon which volcanic and sedimentary rocks were laid down, mainly in continental arc settings.

Quesnel terrane. The Quesnel terrane, named for a town in central British Columbia about 300 kilometres north-northwest of Kamloops, is the most widespread older rock package in the southern Intermontane Belt. It extends eastwards into the Omineca Belt and southwards into northernmost Washington. Much of the Quesnel terrane is composed of late Triassic and minor early Jurassic lava, volcanic breccia, and tuff interbedded with marine shale and minor limestone, with numerous granitic intrusions of similar age to the volcanic rocks. The granitic rocks host major copper-gold deposits, and copper mineralization is widespread in the volcanic rocks. Although Triassic volcanic rocks dominate the western part of the Quesnel terrane, east of Kamloops they grade laterally into marine shale, siltstone, and localized volcanic-rich sandstone and tuff. The volcanic and granitic rocks probably formed in an offshore volcanic island arc like the modern Japanese or Philippine island chains, with little or no continental crust below it at the time it formed. It was separated from the ancient continental margin of North America by a basin, containing mud and silt derived both from the arc and the old continental margin. Parts of the basin are preserved east of Kamloops and in the Omineca Belt.

The early Mesozoic volcanic arc rocks overlie at least two different late Paleozoic rock units, a relationship that shows the older rocks came together before the Triassic rocks were deposited on top of them. One unit is of late Devonian through Permian age and is composed of volcanic sandstone and tuff of arc origin with interbedded shale, limestone, and in places chert-pebble conglomerate. The other is basalt, tuff, radiolarian chert, shale, sandstone, and minor ultramafic rocks of Carboniferous and Permian age that probably formed as the floor of an ocean basin and may be part of the Slide Mountain terrane.

Cache Creek terrane. The Cache Creek terrane lies west of the Quesnel terrane. Near the village of Cache Creek, the terrane ranges in age from late Carboniferous to early Jurassic, although elsewhere in British Columbia it is as old as early Carboniferous. It thus overlaps the age of many rocks of the Quesnel terrane but is very different in character. It consists of disrupted and broken radiolarian chert, argillite, volcanic rock including pillow basalt, large distinctive masses of limestone, and locally serpentinite and gabbro. With exception of the limestone, the other rock types are typical of those found on floors of deep ocean basins, and this, together with the jumbled, deformed nature of the Cache Creek terrane, suggests that it is material scraped off an oceanic plate at a subduction zone—part of an ancient accretionary complex.

The presence of blueschist within the Cache Creek terrane near Fort St. James in central British Columbia supports this suggestion. *Blueschist,* a rock containing the bluish sodium amphibole mineral glaucophane, forms under the unusual metamorphic conditions of very high pressure and low temperature encountered in accretionary complexes. The blueschist in central British Columbia is of late Triassic age, about 220 to 210 million years old, indicating that subduction was active at this time. Blue amphibole is known in two places in the south, but no true blueschist.

Although the original relationship between the Quesnel and Cache Creek terranes is obscured by the later faults that separate them today, the geographic distribution and chemical composition of rock types in the Quesnel island arc tells us that its accompanying subduction zone and accretionary complex lay west of the arc. The Cache Creek terrane has the right characteristics, is the right age, and is in the right place to be the accretionary complex that accompanied the early Mesozoic arc of the Quesnel terrane.

Slide Mountain terrane. The Slide Mountain terrane, named for a locality near Barkerville in the Cariboo Mountains about 300 kilometres north of Kamloops, is the easternmost of the accreted terranes and straddles the boundary between the Intermontane and Omineca Belts. Rocks of the Slide Mountain terrane form a series of discrete bodies that extend from near Kootenay Lake in southeastern British Columbia northward into the Yukon. The terrane includes ocean floor rocks of late Paleozoic age that possibly formed in a back-arc basin associated with older parts of the Quesnel island arc. It is overlain by a blanket of Triassic shale and phyllite that probably extended from the old continental margin across to the east side of the early Mesozoic Quesnel island arc.

Fossils in the three terranes suggest that the rocks did not form near their present position relative to the ancient North American continent. Limestone in the Cache Creek terrane contains shells of calcareous single-celled animals called *fusulinids* that are of middle Permian age. Similar fossils of the same age are widespread in Asia, the Middle East, and the Mediterranean region but are unknown in the Americas except in limestone associated with oceanic rocks like those of the Cache Creek terrane.

Limestone in the middle Permian Quesnel rocks contains fusulinids unlike those in the Cache Creek terrane but similar to those found in the Slide

Mountain and Chilliwack terranes as well as the Stikine terrane in northwestern British Columbia and in arc terranes of the same age in California and Nevada. They are not too different from fossils of the same age found on the continental platform in the central and southwestern United States and in parts of Mexico and the Andes.

The Cache Creek fossils with their Asian affinities probably lived in *western* parts of the ancestor of the Pacific Ocean, called Panthalassa, which in Permian time occupied more than half of Earth's surface. As new ocean crust formed along a mid-ocean ridge, the rocks in which the fossils were preserved were carried across the ocean on the oceanic conveyer belt and eventually became lodged in the accretionary complex at the margin of the North American Plate. The animals whose remains occur in the early Mesozoic Quesnel arc and its Paleozoic underpinnings, as well as in the Slide Mountain terrane, probably lived in *eastern* Panthalassa, but farther south than they are today relative to the old continental margin. Today, a distance of about 100 kilometres separates the different Cache Creek and Quesnel faunas, but in middle Permian time, about 265 million years ago, they probably lived on different sides of the enormous Panthalassa Ocean and were separated by many thousands of kilometres.

Continental Arcs of the Intermontane Belt

The thrusting and folding that accompanied accretion of the Cache Creek, Quesnel, and Slide Mountain terranes to the old continental margin of North America, about 180 million years ago, also thickened the crust and eventually raised its surface above sea level. By middle Jurassic time, seawater mostly was restricted to regions to the east, in what is now the Foreland Belt, and to the west, in what is now the Coast Belt. In the southern Intermontane Belt, sedimentary and volcanic rocks younger than middle Jurassic age, less than about 160 million years old, are nonmarine. The middle Jurassic through Eocene granitic rocks represent deeper parts of continental arcs, whose volcanic edifices are preserved in many places. The granitic rocks are found east of the Intermontane Belt in the Omineca Belt, where they intrude rocks that formed on the old continental margin. Furthermore, the chemistry of middle Jurassic and younger granitic rocks in the southern Intermontane Belt shows they were derived from crust that contained some old continental crust, unlike the early Mesozoic granitic rocks. This confirms that these rocks formed after the terranes were accreted to the old continental margin.

A Cretaceous, 105-million-year-old continental arc composed of andesitic and rhyolitic lava, tuff, and breccia is associated with sandstone, shale, and conglomerate deposited in lakes and rivers and is accompanied by granitic intrusions of similar age. These volcanic and sedimentary rocks belong to the Spences Bridge group and were deposited on top of both the Quesnel and Cache Creek terranes.

Eocene continental arc volcanic rocks, about 50 million years old and belonging mostly to the Kamloops group, are interbedded with sandstone, shale, conglomerate, and minor coal beds, as well as small intrusions, and are widespread across the southern Intermontane Belt and flanking belts. The volcanic rocks include flows that range in composition from basalt to rhyolite and associated pyroclastic rocks.

HIGHWAY 1

Lytton—Cache Creek

85 kilometres

See the map on page 119 for the Lytton area.

At Lytton, where some of the highest air temperatures in Canada have been recorded, Highway 1 crosses the Pasayten fault, which marks the boundary between the Coast and Intermontane Belts. Though not visible, the fault can be located near the small ferry landing on the east bank of the Fraser River just north of Lytton. There, it lies between dark outcrops of Jackass Mountain sandstone (the youngest part of the Methow terrane) in the river bank and pink, weathered granite of the Mount Lytton complex in roadcuts on Highway 12 above the ferry's eastern terminus.

Mount Lytton Complex

Highway 1 turns eastward north of Lytton and follows the south side of the Thompson River, which flows into the Fraser River at Lytton. The spectacular cliffs on the north side of the Thompson River, and roadcuts along this part of the highway, are granitic, dioritic, and metamorphic rocks of the Mount Lytton complex, which is of latest Paleozoic and early Mesozoic age. Across the Thompson River, light-coloured gneiss, made mainly of quartz and feldspar, contains thin dark bands of amphibolite. These rocks probably were metamorphosed at depths of about 15 kilometres. This part of the complex is 225 million years old and formed in late Triassic time, but rocks farther north along the Fraser River have been dated at 250 million years, on the boundary between Permian and Triassic time. In places, irregular brown dikes that crosscut the gneiss probably are feeders to the overlying Spences Bridge volcanic rocks of mid-Cretaceous age. Botanie Creek, whose valley is north of the Thompson River about 3 kilometres east of Lytton, contains down-faulted, red continental sandstone and shale from which mid-Cretaceous pollen has been obtained.

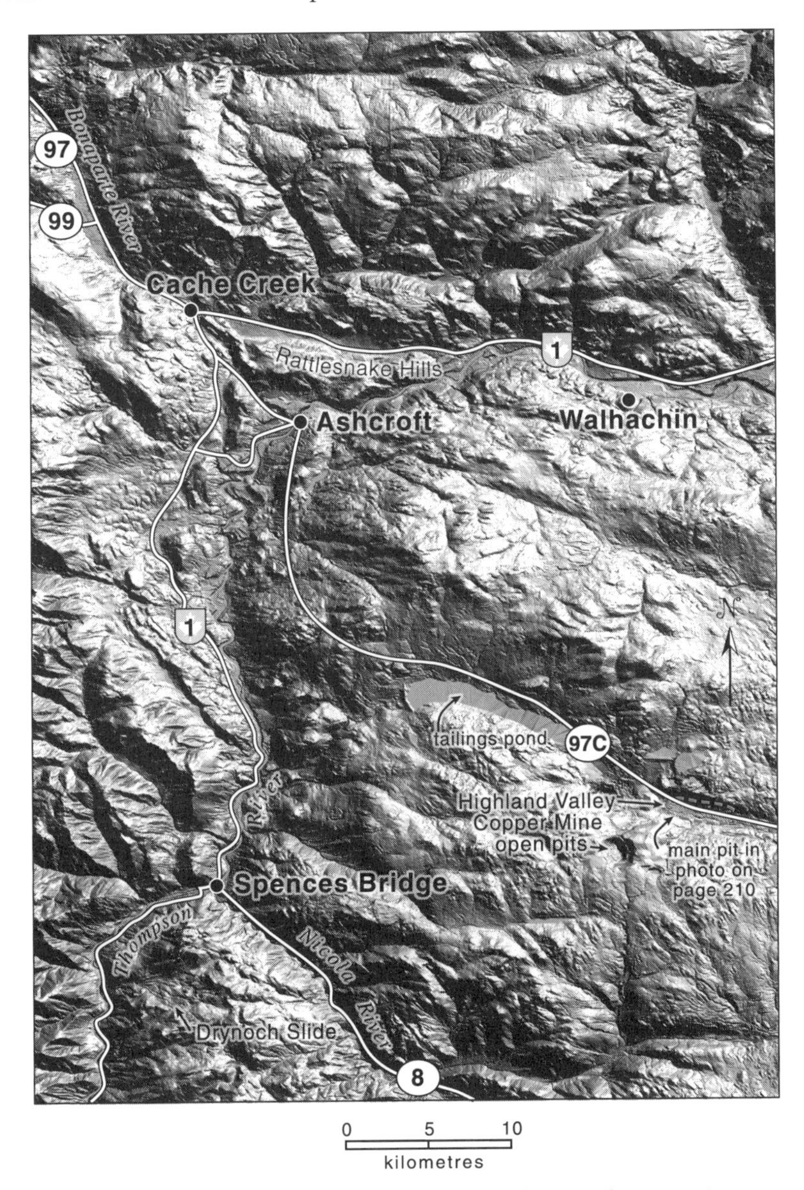

Shaded-relief image of topography along Highway 1 between Spences Bridge and Cache Creek. Note how the topography east of the Thompson River is more subdued than that west of the river. —Image constructed using Canadian Digital Elevation Data obtained from GeoBase

Cliffs on the north side of the Thompson River east of Lytton are composed of altered, banded gneiss with light layers of quartz and feldspar and thin dark bands of amphibolite of the Mount Lytton complex. Small, irregular brownish intrusions that cross the foliation in the gneiss may be feeders to the mid-Cretaceous Spences Bridge volcanics that overlie the complex. Note the shed over the Canadian National Railway track to divert rockslides from the crumbling cliffs. (The older Canadian Pacific Railway track is below Highway 1, on the south side of the river.)

Folded dark amphibolite layer in lighter-coloured granite at the base of cliff on north side of Thompson River.

About 15 kilometres southeast of Lytton and above Highway 1, a large, 212-million-year-old granodiorite pluton intrudes the gneiss. This age is about the same as that of another large granitic body, the Guichon Creek batholith, which is about 50 kilometres northeast of Lytton and intrudes volcanic rocks of the Nicola group that are only slighter older (**see "Guichon Creek Batholith and the Highland Valley Copper Mines" later in this roadlog**). The batholith and volcanics belong to the early Mesozoic island arc that forms much of the Quesnel terrane. The contact between the Mount Lytton complex to the west and the Nicola group and granitic rocks to the east is covered by the 20-kilometre-wide tract of Spences Bridge volcanic rocks. However, the similar age and nature of the granitic rocks in both areas, and their proximity to one another, suggests that the two probably are related and that the Mount Lytton complex is the deeply eroded and exposed root of the Quesnel arc.

About 15 kilometres east of Lytton, where the Thompson River bends northward, Nicoamen Creek enters from the southeast. Gold collected from the creekbed by the aboriginal people in 1855 led to the Cariboo gold rush.

Spences Bridge Volcanic Rocks

The bedrock in the Thompson River valley between Nicoamen Creek and about 7 kilometres north of the village of Spences Bridge is mainly red, brown, and locally buff volcanic flows of the Spences Bridge group. The rocks range in composition from basaltic andesite to rhyolite, and their nature and chemistry suggest that they formed in a continental arc. In places, and best seen on Highway 8 along the Nicola river east of Spences Bridge, the volcanics are interlayered with sedimentary rocks that were deposited in lakes and by rivers and in places contain coaly material and the fossils of freshwater clams. These interlayered rocks are about 105 million years old. They lie in a structural depression that extends south-southeastward from Spences Bridge for about 160 kilometres, almost to the international boundary. Straddling the boundary are granitic rocks of the same age that were the magma chambers related to the volcanics. Similar rocks occur for 50 kilometres to the north-northwest as far as the Fraser fault and, offset by the right-lateral strike-slip fault, continue on its west side for at least 180 kilometres from Lytton. The rocks appear to be remnants of a continental arc built on the edge of the early Cretaceous continent, east of the marine basin containing rocks of the Jackass Mountain group of the Methow terrane.

Relatively flat-lying flows of reddish volcanic rocks of the mid-Cretaceous Spences Bridge group are exposed near Nicoamen Creek about 15 kilometres east of Lytton.

Gravel Terraces

A conspicuous deposit of gravel and sand occurs along the valley bottom of the Thompson River east from Lytton almost to Spences Bridge and, less clearly, southward down the Fraser River valley to Boston Bar. The rivers have carved the gravel and sand into a series of terraces, some matching, others not. The greatest thickness of sand and gravel, several kilometres east of Lytton, is about 150 metres. The deposits rest on till or bedrock, and their uppermost surface forms alluvial fans extending up into the mouths of stream gullies. The gravels include a mixture of cobbles and pebbles of well-rounded and presumably far-travelled granitic rock and less-well-rounded basalt, andesite, and rhyolite from nearby sources.

The deposits formed at the end of the last glaciation, when the last tongues and isolated patches of glacial ice were disappearing, and the steep hillsides would have been plastered loosely with unconsolidated, nearly water-saturated till that had dropped or slid from the ice. Vegetation was probably sparse and provided little protection from the weather. Mudslides, debris flows, and landslides would have been very common, so

Gravel and sand deposited near the margin of glaciers during the last ice age are widespread along the Thompson River valley between Lytton and Spences Bridge. This exposure is on the eastern side of the Thompson River, near Nicoamen Creek and just north of the big river bend about 15 kilometres east of Lytton.

much so that the Thompson and Fraser Rivers might have been unable to carry away the sediment as fast as it accumulated. This environment probably persisted for centuries after the glaciers melted away and before more stable conditions were restored, enabling the rivers to once again get ahead of the sediment supply and carry away the loose material.

Three Big Slope Failures

The first of three landslides of special interest in this part of British Columbia occurred toward the end of the last glaciation. It is buried in the gravel terrace deposits a few hundred metres north of the Canadian Pacific Railway bridge that crosses Nicoamen Creek and Highway 1. Horizontal slabs of brick red to dark-chocolate-coloured volcanic rock, mostly 1 to 2 metres thick, lie within the gravel and sand deposits a few metres above highway level. Similarly coloured volcanic rocks crop out on the opposite bank of the Thompson River and possibly were the source of the slide. The base of the slide throughout its length follows a single stratum in the gravel, and the top surface of the slide was planed off by the meltwater-laden Thompson River, probably within a few hours of the time of sliding.

The second slide, called the Drynoch slide, is 11 kilometres north of the first and 7 kilometres south of Spences Bridge. The slide starts 650 metres above its toe and 5 kilometres east of the highway, ending in the river. Its width is mostly a few hundred metres but it spreads to 1 kilometre at its toe. Overall, it shows a very strong resemblance to a mountain glacier. This may not be an accident: its downhill movement, like that of a glacier, is exceedingly slow. Between 1951 and 1972 the average rate of movement of the slide was about 3 metres per year, somewhat more during the spring months. Remedial measures in the 1960s slowed this rate considerably. Drill hole surveys indicate the slide is moving at depths of 18 metres. With its slow motion and the clay-rich unconsolidated material involved, this landslide is best described as an *earth flow*.

The Drynoch slide was known to be moving in gold rush days and was described in 1871 by Chief Justice Matthew Begbie. Charcoal associated with Indian artifacts buried by this earth flow has been dated at 3,150 years old. The slide presented almost no problem to travel as long as humans moved on foot or horseback along trails, but when the wagon road to the Cariboo goldfields was built, it had a bad habit of moving toward and into the river. The track of the Canadian Pacific Railway on the east side of the river, built across the toe of the slide in the early 1880s, also moved and had to be repeatedly jacked back into place. When the present highway was built in the early 1960s, a serious effort was made to understand and control the slide. A drainage system was built a short distance above the road, and this has done much to eliminate the problem, at least for the time being.

The persistent movement of soil and rocks to the east shore of the river constricted its channel and thereby increased the water velocity. This in turn encouraged undercutting, not only of the east bank but also of the western side. With no room there for the Canadian National Railway to lay its track on the surface, it used three short tunnels and two rock sheds to cross the area. The undermining of a riverbank is often seen opposite a significant source of debris, whether carried by a landslide or a major tributary stream.

The third slide of interest is directly south of the Highway 1 bridge across the Thompson River at Spences Bridge. Here the western bank of the river, composed of till, gravel, and silt, has been the source of no less than three landslides in historic times, with the greatest one in August 1905. The slide plunged into the river, directly killing five people and creating a wave 3 to 5 metres high, which swept upstream into the Indian village near Spences Bridge, there killing another ten people and injuring thirteen. The river was completely blocked for several hours and rose 6 metres before it overtopped the landslide dam and cut a new channel.

Zeolites in Spences Bridge Lava

Outcrops of Spences Bridge lava, best examined on the quiet road just northeast of the village of Spences Bridge (north-bound travellers turn off before crossing the Thompson River bridge), contain *zeolites*, hydrous aluminum silicate minerals with varying and interchangeable amounts of sodium, potassium, calcium, and magnesium. White or pink, relatively soft, low in density, and filling cavities and forming veins in the lava, zeolites are products of shallow, relatively low-temperature metamorphism and were here derived from unstable volcanic glass in the volcanic ash and lava of the Spences Bridge group.

Ashcroft Landslides

Highway 1 crosses the Thompson River at Spences Bridge and passes roadcuts in mainly buff-coloured and greyish green rocks of the Spences Bridge Group. About 15 kilometres north of Spences Bridge, the highway leaves the bottom of the Thompson River valley and climbs up the western side, eventually passing behind a series of low hills and terraces that hide the river from sight. Near the top of the climb a turnout on the east side of the road provides a view across the river and northward into the basin containing the town of Ashcroft. Soon after completion of the Canadian Pacific Railway in 1886, Ashcroft was established as a freight transfer point to serve the Cariboo goldfields nearly 300 kilometres to the north and as an agricultural centre in its own right.

Within a decade, a swarm of landslides from the hillsides above the track just south of town interrupted the rail service. A consultant recommended the railway buy out some of the farmers to curtail the use of newly established irrigation systems. The railway did this, and the resulting decreased soil saturation did indeed reduced the incidence of landslides.

Cache Creek Mélange

Highway 1 descends into the village of Cache Creek in the valley of the Bonaparte River, which flows southward to join the Thompson River just above Ashcroft. The bedrock underlying the grassy, sagebrush-covered hillsides near the village of Cache Creek consists of a soft matrix containing resistant blocks of harder rock. In natural exposures, only the blocks stand out, at first glance resembling either erratics deposited by glaciers or landslide debris. To see the true nature of the bedrock, you must find an artificial outcrop, such as a roadcut.

At the stoplight in Cache Creek, where Highway 97 from the north joins Highway 1, turn west across a small bridge over the Bonaparte River and onto a side road parallel with Highway 1 that is cut into the toe of the hillside

west of Cache Creek. The roadcuts there expose a matrix of contorted, broken, and disrupted black to dark grey graphitic argillite and lighter grey bedded chert in which larger blocks of chert and limestone are embedded. A block of greyish white limestone surrounded by argillite and beds of chert is exposed just northeast of the bridge crossing. The limestone block is of late Carboniferous age. The matrix age at this site is not known, but elsewhere along the roadcuts, the matrix contains Triassic microfossils. About two-thirds of the distance from the north end of the roadcuts, a large block of folded and contorted bedded chert has fallen free of its matrix.

Elsewhere along the Bonaparte River valley are blocks of all sizes and a wider range of rock types. On the east side of the valley north of Cache Creek, pillow basalt forms a "block" several hundred metres long. Along Highway 97 north of its junction with Highway 99, smaller blocks of limestone, gabbro, and amphibolite and shreds of serpentinite occur within the mélange.

Rock formations containing a variety of blocks in a scrambled matrix of argillite, chert, or even serpentinite are called *mélanges*, from the French

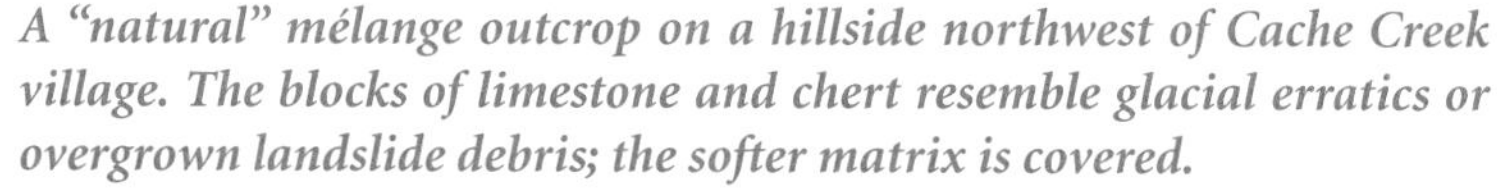

A "natural" mélange outcrop on a hillside northwest of Cache Creek village. The blocks of limestone and chert resemble glacial erratics or overgrown landslide debris; the softer matrix is covered.

A roadcut in the Cache Creek mélange reveals a block of late Carboniferous limestone embedded in a matrix of black, cherty argillite.

word for "mixture." The composition of the matrix and most blocks in the Cache Creek mélange suggests they originated on the floor of an ocean basin. Most mélanges probably form in subduction zones at convergent plate margins when material is scraped off the ocean floor at a trench, folded, faulted, and chaotically broken and disrupted so that a wide range of ocean floor rock types becomes kneaded together to form an accretionary complex. In addition, some mixing may occur even before the rocks enter the subduction zone. Some mélanges probably started out as gigantic submarine rockslides off the slopes of seamounts and atolls in the ocean basin. Such rockslides occur off Hawaii, whose islands are the tops of the highest mountains on earth, rising steeply for more than 9,000 metres from the ocean floor. Eventually, the ocean floor carrying the rockslides feeds into a trench at a convergent plate margin, to be further stirred around and become part of an accretionary complex. Such an origin may account for the presence of blocks of limestone, probably formed in shallow water, in the mélange, which otherwise consists mainly of rocks that originated in or under deep water. The distinctive nature of the rocks exposed in the roadcuts at Cache Creek is typical of that in many accretionary complexes. You can see younger mélanges in Pacific Rim National Park, but sandstone is more abundant and chert less abundant there than in the Cache Creek mélange.

At the south end of the roadcuts on the west side of Cache Creek, beds of chert, each a few centimetres thick, are separated by thin layers of shale. Examination with a hand lens of freshly broken, clean, wetted surfaces of the chert reveals minute dark dots within lighter grey chert. The dots are the shells of radiolarians, which are common today in siliceous oozes on the bottom of deep ocean basins. Wear safety goggles or glasses as you must hammer the rock to get a fresh surface to see the radiolarians, and chert can send off very sharp splinters! Radiolarians have proven invaluable in providing ages of otherwise sparsely fossiliferous rocks in Cordilleran accretionary complexes.

Thin-bedded radiolarian chert is interlayered with argillite by Highway 1 at the south end of the roadcuts on the west side of the village of Cache Creek.

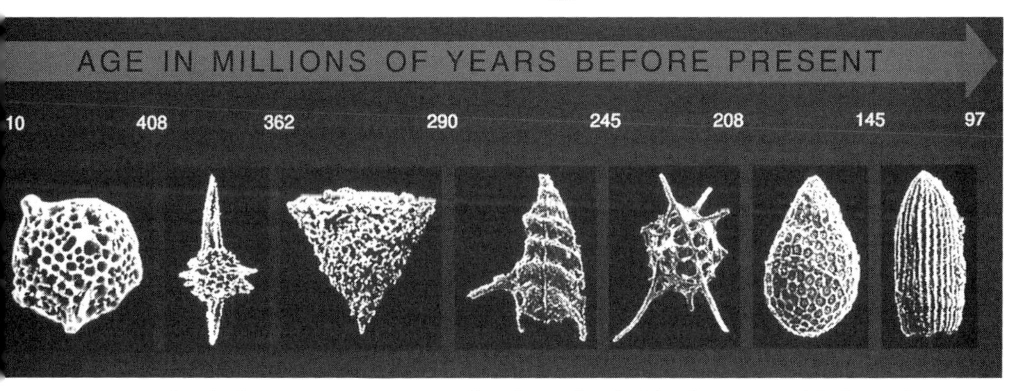

A variety of radiolarians—microscopic fossils made of silica here magnified roughly 100 times—from Cordilleran chert and shale of different ages. They were extracted by dissolving the rock in dilute hydrofluoric acid and repeated washing. —F. Cordey photo

The accretionary complex that is the Cache Creek terrane formed in late Triassic and early Jurassic time, between about 225 and 180 million years ago. However, some rocks in the accretionary complex are older. The block of limestone near the bridge across the Bonaparte River at Cache Creek contains late Carboniferous microfossils, about 310 million years old, and other carbonate blocks farther north along Highway 97 contain early Permian microfossils. The radiolarian chert in the matrix here is Triassic in age, though in a few other places the matrix rock is as old as Permian, and west of Clinton on Highway 97, is as young as early Jurassic.

SIDE TRIP TO ASHCROFT AND BEYOND

A side trip to Ashcroft along Highway 97C provides a good view of the structural relationship between the Cache Creek and Quesnel terranes. You can reach Ashcroft by either of two routes. Leave Highway 1 at Ashcroft Manor, about 10 kilometres south of Cache Creek, or (recommended) leave Highway 1 about 4 kilometres south of Cache Creek and follow the Bonaparte River valley. Cuts along both roads down to Ashcroft expose contorted dark shale and sandstone of the early and middle Jurassic Ashcroft formation. Cross the bridge and continue past the town and southward uphill on Highway 97C, on the east side of the Thompson River valley, to a point 2.8 kilometres south of the bridge, where the highway turns sharply east. Make sure you park your vehicle well off the road because large trucks carry ore from the Highland Valley mines down to the railway at Ashcroft.

The view from Highway 97C displays one of the few big pre-Tertiary structures clearly visible in the southern Intermontane Belt. North of the

viewpoint, at the southeast end of the Rattlesnake Hills to the north of Ashcroft, and in the hills northeast and east of the viewpoint, early Jurassic sedimentary rocks of the Ashcroft formation lie unconformably above an erosional surface on late Triassic volcanic rocks of the Nicola group and on the Guichon Creek batholith, both integral parts of the Quesnel terrane. The Ashcroft beds dip generally westward. The Thompson River valley below the viewpoint is eroded in shale and sandstone of the Ashcroft formation (passed on the descent to Ashcroft), which forms the black cliffs and slumped areas above the western bank of the river. To the northwest, near the top of the hill that lies west of the Thompson River between Highway 1 and Ashcroft, and on the western tip of the Rattlesnake Hills, the unconformity is overturned, with Nicola rocks capping the hill and Ashcroft sediments lower down the slopes. Westernmost, near Highway 1 and forming the hills on the western skyline, are rocks of the Cache Creek terrane, which are faulted against the Nicola rocks. You can see the fault, marked by slivers of greenish grey serpentinite, in the east side of the Bonaparte River valley southeast of Cache Creek, just north of the northern turnoff from Highway 1 to Ashcroft.

A *syncline* is a fold in which the layers forming the opposite sides, or limbs, of the fold dip toward one another so as to make a structural trough. At Ashcroft, the structure is an overturned syncline: the eastern limb of the fold dips westward, the younger rocks lie in the core of the structure, and the western, upside-down limb also dips westward. Above the western limb, the Cache Creek terrane appears to have been thrust eastward over the western edge of the Quesnel terrane, possibly overturning the underlying strata as it did so. The thrust displacement may be large. If the Mount Lytton complex is the exposed root of the early Mesozoic Quesnel arc, then the known western limit of the Quesnel terrane must be near the Fraser River, which is about 40 kilometres west of Ashcroft.

The folding and thrusting took place after deposition of the youngest Ashcroft strata, which are about 158 million years old, and before the 105-million-year-old Spences Bridge volcanics were laid down on top of both the Cache Creek and Quesnel terranes. About 100 kilometres south of here, near the tollbooth on Highway 5 and along the trend of the Ashcroft syncline, are granodiorite with gneissic layering and schist with a parallel fabric formed from the Nicola group. The structures in these rocks may represent the southern continuation of the syncline structure, although there they formed at a much deeper level in the crust. The gneissic granodiorite and accompanying structures are between 157 and 148 million years old, and this late Jurassic age is the current best bet for the time when the Ashcroft syncline formed and when the Cache Creek terrane was thrust eastward over the Quesnel terrane.

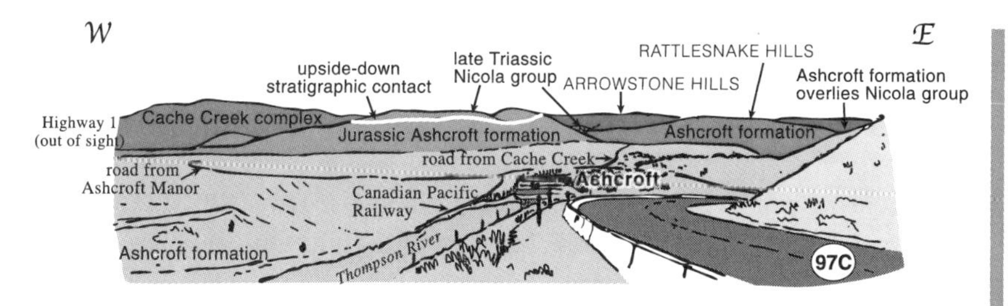

View of the Ashcroft syncline, looking north from Highway 97C, 2.8 kilometres south of Ashcroft.

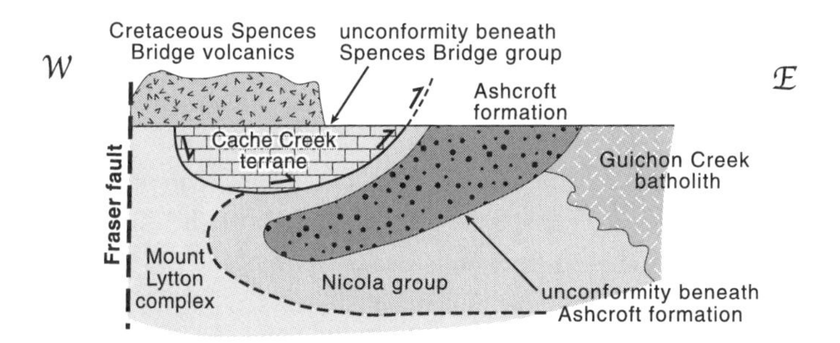

Cross section showing an interpretation of the structure in the region.

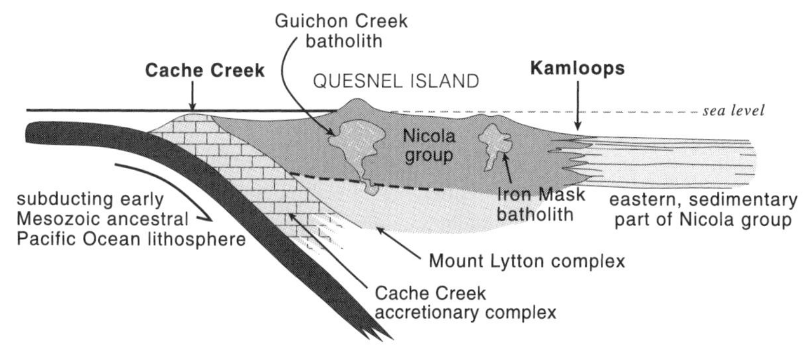

Cross section showing the probable relationships between the rock units in early Mesozoic time.

Guichon Creek Batholith and the Highland Valley Copper Mines

Southeast of and uphill from the last viewpoint, Highway 97C passes some small roadside quarries that expose shale, sandstone, and conglomerate of the Ashcroft formation, and with careful searching, you can find small ammonites, fossil mollusks that lived in Jurassic time. The road then levels out in a high, rolling upland containing the Highland Valley. Here, the roadcuts are in granitic rock of the Guichon Creek batholith, which is about 50 kilometres long in a north-south direction and 30 kilometres across. The 210-million-year-old batholith intruded just slightly older volcanic rocks of the Nicola group at depths of 4 kilometres or less. Together, the volcanic and granitic rocks are integral parts of the early Mesozoic Quesnel island arc.

Because of its economic importance, the Guichon Creek batholith is probably the most carefully studied large granitic intrusion in British Columbia. Mining operations have been continuous in the Highland Valley since they started in the late 1950s. Rocks in the central part of the batholith contain finely disseminated copper sulfide minerals (chalcopyrite and bornite) and molybdenum sulphide (molybdenite), and minor amounts of gold.

The ore production is from enormous open-pit mines, the deepest and newest of which, Highland Valley Copper Mine, is on the south side of Highway 97C. A small road off the north side of the highway about 32 kilometres southeast of Ashcroft leads into a viewing area above the highway. The top of the Highland Valley Copper pit is visible south of the viewing area. In the hillside to the southwest beyond the mine mill is the Lornex open pit. Rock rubble from the now-defunct Bethlehem pit is visible to the east. In 2003, the Highland Valley Copper pit was 2.5 by 2 kilometres in extent at the surface and 600 metres deep; it will be mined to a depth of 800 metres. The Lornex pit has the same surface dimensions, is 400 metres deep, and will be mined to a depth of 500 metres.

The ore is crushed and concentrated at the mines, trucked down the highway to Ashcroft, and from there shipped by rail to the coast. The fine waste from the concentrator is pumped as slurry to the valley west of the mine, where it settles out in the enormous white tailings pond that can be seen in places south of Highway 97C from Ashcroft.

View of the open-pit copper mines in the Highland Valley. The top of Highland Valley Copper pit is on the right; Lornex pit is on the left, behind the mine mill buildings.

HIGHWAY 1

Cache Creek—Pritchard via Kamloops

117 kilometres

See the map on page 197 for the Cache Creek area.

About 2 kilometres east of Cache Creek, Highway 1 crosses the covered boundary between the Cache Creek and Quesnel terranes. North of the highway, bluffs of red, brown, and buff-coloured lava and breccia that are part of a widespread Eocene continental arc lie unconformably on early to middle Jurassic Ashcroft strata and on rocks of the Cache Creek terrane. South of the highway, the Rattlesnake Hills are mainly underlain by soft shale and sandstone of the Ashcroft formation, and at the southeast end of the hills, but not visible from Highway 1, the unconformity between the Ashcroft formation and the underlying Nicola group is exposed. A small cap of Eocene lava perches near the western end of the Rattlesnake Hills.

From Cache Creek boundary, near the western end of the Rattlesnake Hills, to the eastern side of the Intermontane Belt near Pritchard, a distance of about 110 kilometres, most of the volcanic, sedimentary, and intrusive rocks exposed along Highway 1 belong to the early Mesozoic Quesnel island arc. The arc rocks are badly faulted and locally folded, making their stratigraphic thickness hard to determine, but they may be about 3 to 4 kilometres thick. In spite of folding and faulting, we rarely get into overlying or underlying strata in this 110-kilometre stretch of Highway 1, so it appears that at least this part of the southern Intermontane Belt is structurally relatively "flat." This contrasts with the great structural relief in the flanking Coast and Omineca Belts, where rocks of very different ages and metamorphic grades are juxtaposed.

The Importance of Water

The region is arid, with an annual rainfall of about 30 centimetres or less. Finding the water needed to grow crops has long been a problem. An irrigation scheme undertaken by colonists early in the twentieth century was not very successful. The scheme operated in the area about 20 kilometres east of Cache Creek, near Walhachin, with the object of growing apples on the terraces along the Thompson River. Water from an intake on the Deadman River, crossed by Highway 1 just west of Kamloops Lake, was brought to the orchards by gravity flow along a 32-kilometre wooden flume. Early in World War I, 97 of the 107 men working on the project left to serve with the British armed forces and those remaining were unable to keep the flume operating, particularly after it was damaged by a

torrential cloudburst. Its remnants are still visible in places above Highway 1, but few of the fruit trees it was designed to water remain. Availability of metal pipe and relatively low-priced electric power for pumping has significantly changed the economic landscape by making water from the Thompson River more readily available, though perhaps not enough to revive fruit growing in the Thompson Valley.

Shaded-relief image of topography along Highway 1 near the Deadman River valley and Kamloops Lake. Note how deep the meltwater-carved, north-south-oriented portion of the Deadman River is. —Image constructed using Canadian digital Elevation Data obtained from GeoBase

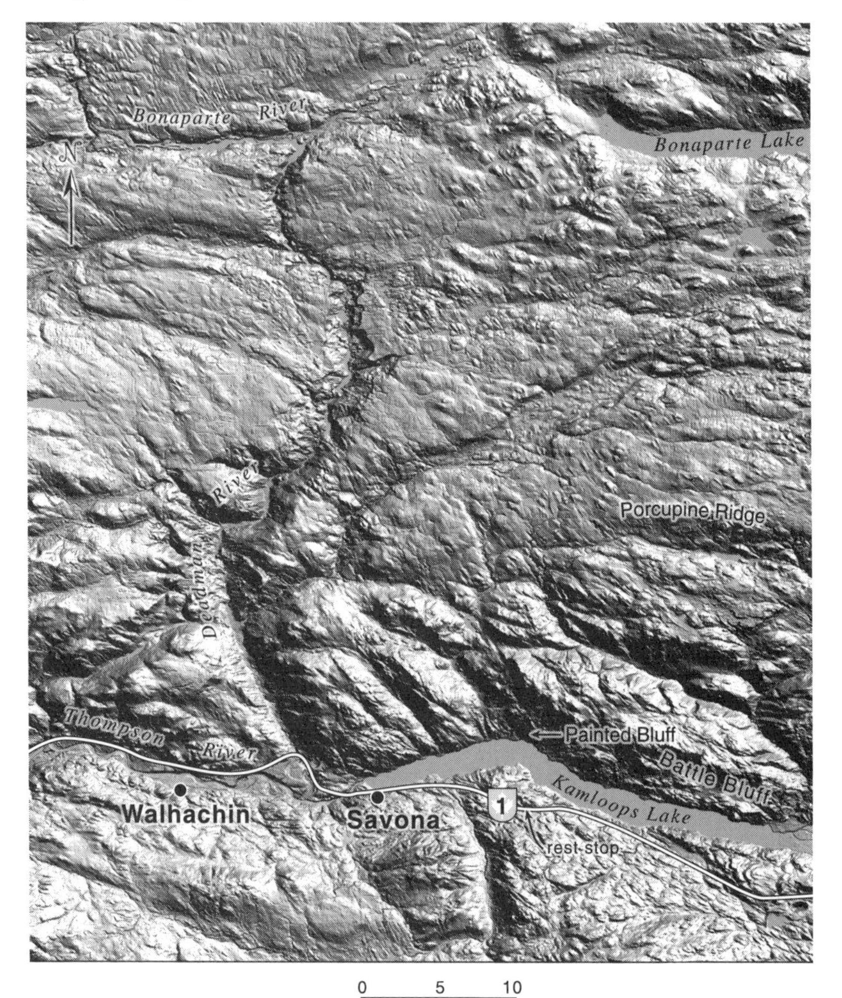

Glacial Lake Deadman and the Late-Glacial Deadman River

The lowermost Deadman River valley, about 30 kilometres east of Cache Creek and 50 kilometres west of Kamloops, contains one of the most voluminous deposits of gravel and sand in southern British Columbia. For a geologically short time, the late-glacial Deadman River was much larger than its modern successor because ice had diverted water into its drainage during the final glacial retreat between about 14,000 and 12,000 years ago, and at that time the river carried and deposited an enormous amount of sand and gravel. In addition, water that collected in ice-dammed lakes upstream was released catastrophically when the ice dams ruptured. These short-lived but gigantic floods, or *jokulhaups*, an Icelandic term meaning "glacier burst," are more effective in moving sediment than the steadier flow associated with running water in nonglacial conditions.

Where Highway 1 starts its descent northeast into the Deadman Valley, it follows a gently curved escarpment, which is a river cutbank, for several hundred metres. The radius of curvature of this cutbank, which is a measure of the size of the stream that carved it, is comparable to those of cutbanks

The view looking south at the lower Deadman River valley shows the cutbank incised in delta deposits by the large, late-glacial Deadman River. Flat-lying surfaces on the opposite hillside mark shorelines of Glacial Lake Deadman. The present small river runs near the foot of the cutbank.

along the present Thompson River and much greater than those of the present Deadman River. The late-glacial Deadman River not only carved the cutbank but also planed off the surface extending out from the foot of the cutbank.

At its lower end, the river discharged into Glacial Lake Deadman, an ice-dammed lake that occupied the lower Deadman River valley and downstream parts of the adjacent Thompson River valley. The discharge into the lake built an enormous delta that effectively trapped most of the gravel and sand and even some of the silt carried to it. We know that glacial ice persisted at the southern end and eastern flank of the lower part of the delta because its surface topography is irregular and contains kettles, depressions that formed at sites where buried ice masses eventually melted.

The highest historic shoreline of the lake lies at the 550-metre contour immediately west of the mouth of the Deadman River valley, where it appears under appropriate lighting as three horizontal bands crossing the hillside. The uppermost band is the steeper wave-cut shore cliff; below this is the flatter beach; and lower again is another moderately steep face that is the front of a deposit built just below the waterline from materials eroded from the shore cliff. On the south side of the Thompson Valley, you can see the corresponding shoreline on the grassy and rocky hillside about 200 metres above the Thompson River opposite the mouth of the Deadman River. The highest level of Glacial Lake Deadman can be traced north for 17 kilometres up the Deadman Valley and eastward for at least 32 kilometres, but no traces occur westward down the Thompson River valley.

The occurrence of well-developed deltas and shorelines implies that the lake level remained constant to within a few metres for a period long enough to permit streams and waves to create their characteristic landforms. This stability requires a bedrock spillway at the outlet of the glacial lake because a spillway over unconsolidated sediment would be quickly eroded and the rapidly falling lake level would leave little or no record of its shores. Several shorelines at different levels have been recognized for Glacial Lake Deadman. The intermittent occurrence of shorelines, one below the other, reflects the repeated opening of new and lower outlets as the ice retreated.

To the casual observer the shorelines appear to be horizontal, but careful measurements show that this is not quite the case. A tilt of about 4 metres per kilometre to the southwest is recorded on a high, old shoreline, although tilts of about 2 metres per kilometre are more common. During and following the retreat of the ice sheet, vertical rebound was greater where the ice was thicker and lasted longer. This explains the tilt.

During the retreat of the last ice sheet, residual ice occupied the valley bottoms, particularly around the margins of the Interior Plateaus, so water drained toward the interior. At one time, Deadman River received runoff from the upper Bonaparte River, runoff from the vicinity of Bridge Lake on the Miocene basalt plateau, and most or all of the water released from the melting ice sheet in the mountains to the west. Until an ice dam near Spences Bridge melted, the Thompson River drained southeastward to the Okanagan and Similkameen Valleys. The water flowed down the Okanogan River in Washington, across land already freed of ice, to those parts of the Columbia River that lay well beyond the limits of the ice sheet.

Volcanic Rocks of the Nicola Group

West of the Deadman River, rocks of the Nicola group of the Quesnel terrane are mainly red and green volcanic breccia of andesitic composition, made up of angular fragments of lava containing small feldspars. East of the outflow of Kamloops Lake, in several roadcuts near Savona, you can see volcanic breccia with fragments up to 50 centimetres across.

About 8 kilometres east of the bridge over the Thompson River at the western end of Kamloops Lake, Highway 1 climbs a hillside, and long roadcuts on the south side expose Nicola rocks. A large pullout on the north side of the highway allows a relatively safe stop, but you must cross the

Coarse-grained volcanic breccia of the late Triassic Nicola group in a roadcut along Highway 1 about 2 kilometres east of the bridge over the Thompson River at the western end of Kamloops Lake and 2 kilometres west of Savona.

highway to examine the rocks. Take care: the hazard from traffic here is even greater than that from rocks falling off the roadcut!

The roadcut is in sedimentary and volcanic rocks, both of which are common in the eastern part of the Nicola group. At the eastern (upper) end of the roadcut are black to grey thinly bedded calcareous argillite and siltstone. Diligent searching of the debris at the base of the rock face reveals small, finely ribbed fossil clams called *Monotis*, relatives of modern scallops. The fossils date the rock as latest Triassic. The western (lower) part of the roadcut is made of massive greenish beds composed of fragments of volcanic rock and also small black crystals of the pyroxene mineral augite. The rocks probably were deposited from slurry that slid down the flanks of a submarine volcano into a fairly deepwater basin that contained the sediments present in the eastern end of the roadcut. In places, the beds with coarser volcanic fragments cut into the finer-grained beds. The roadcut features numerous faults, and in places the bedding is fairly steep.

The volcanic rocks in this part of the Nicola group are darker than those to the west, typically contain either pyroxene or hornblende crystals, and are of basaltic rather than andesitic composition. South and east of Kamloops, volcanic rocks similar to those at the western end of the roadcut grade laterally eastward into shale, siltstone, and tuff that resemble strata at the eastern end of the roadcut, and there almost the entire late Triassic section of Nicola rocks is sedimentary.

Limited Lifespan of Kamloops Lake

The Thompson River flows out of the western end of Kamloops Lake. The lake's eastern end near Kamloops is at the confluence of the clear, west-flowing South Thompson River, from which nearly all sediment has been lost during passage through lakes farther east, and the sediment-rich, south-flowing North Thompson River, which lacks upstream settling basins. A delta being built at the inflow end of the lake just west of Kamloops advanced westward by 320 metres between 1949 and 1973. Although the lake's depth averages 70 metres and in places reaches 150 metres, the continued influx of sediment will completely fill the lake in about 4,000 years at the present rate of deposition. Were the annual supply of sediment to be distributed uniformly over the entire lake bottom, it would form a layer 1.9 centimetres thick. However, because deposition is concentrated at the delta front, the lake is destined to shrink in area faster than in depth.

Kamloops Lake has already completed well over half of its life span; it will disappear when the basin is entirely full of sediment. An extensive, 60-kilometre-long alluvial plain that stretches from the head of the lake to Kamloops and northward up the North Thompson to Fishtrap Rapids records the original extent of the lake basin, which was filled with meltwater during the melting of the last remaining chunks of ice about 10,000 years ago.

Bluffs across Kamloops Lake

The rest stop on the north side of Highway 1 at the top of the hill provides a panoramic view of the north side of Kamloops Lake. To the northeast across the lake, cliffs and bluffs of Eocene volcanic rocks, which weather brown, extend from the skyline down to the lake. The rocks are of the same age as the generally flat-lying lava layers that form the hills northeast of Cache Creek, but here the layers are faulted and tilted and in a few places dip relatively steeply. Some of the dips are original; detailed studies show that the volcanic pile is made up of coalescing cones. The Eocene lava overlies Triassic volcanic rocks and shallow intrusions related to the Iron Mask batholith, which is exposed south of the lake and west of Kamloops.

Low down to the north and across the lake, banded pink and green beds form the Painted Bluffs, consisting of flows and pyroclastic beds rich in the dark minerals pyroxene and olivine. Near lake level on the north side of the lake, the Canadian National Railway passes through a long tunnel cut in a dark red pyroxene-rich rock in Battle Bluff. All these rocks belong to the Triassic parts of the Quesnel island arc. To the northwest, Triassic volcanic rocks are overlain in places by chert-pebble conglomerate and sandstone probably of mid-Cretaceous age.

At the eastern end of the bluff, two sheets of 50-million-year-old Eocene gabbro, presumably sills, wrap up and over it. They are accompanied by tuffaceous mudstone and sandstone that are less resistant to erosion than the gabbro and contain pollen of middle Eocene age. Farther east, the lower south-facing slope is a 200-metre-thick section of yellow tuffaceous mudstone overlain by shattered pillow lava, which is in turn overlain by glassy, granulated lava that cooled quickly by contact with water. All three units probably accumulated in a body of standing water that collected in a down-dropped fault block. The overlying rocks, which were deposited above water, include andesitic flow breccia about 500 metres thick, basaltic andesite flows and breccia about 300 metres, and related mudflows.

Iron Mask Batholith and Associated Copper Mines

Near the junction of Highways 1 and 5, about 6 kilometres west of Kamloops, is the Iron Mask batholith, a complex intrusive body composed of gabbro, diorite, syenite, and some ultramafic rock. From its composition and its age of about 205 million years, we think it is the intrusive equivalent to the pyroxene-rich volcanic rocks of the eastern Nicola group and of the rocks below the Eocene volcanic rocks north of Kamloops Lake.

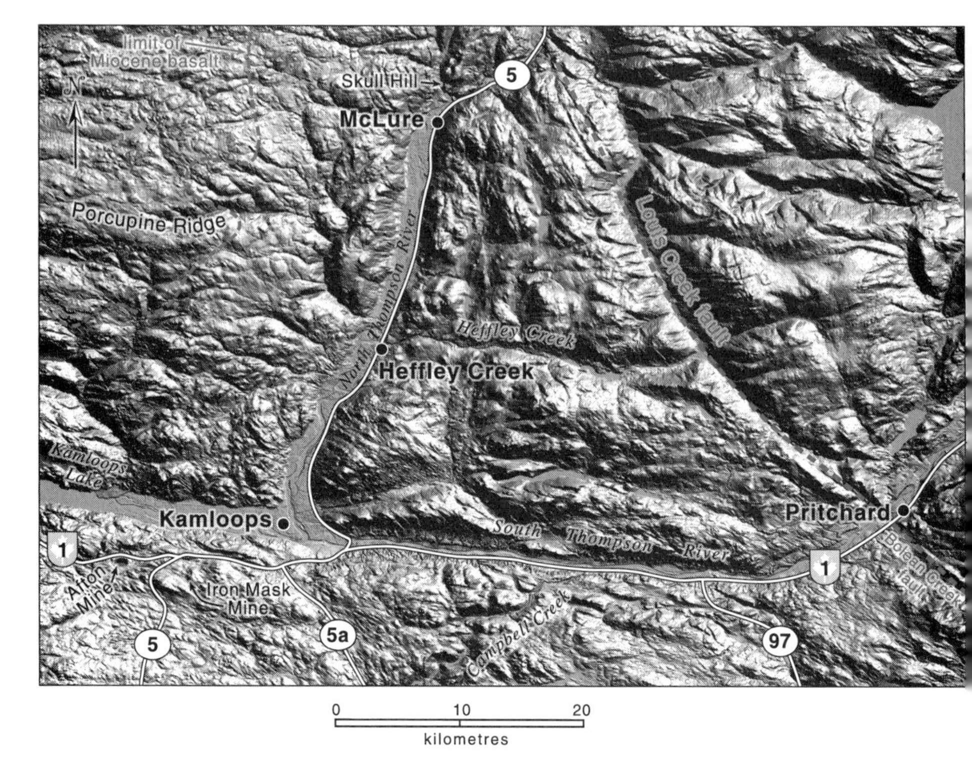

Shaded-relief image of topography along Highway 1 and Highway 5 in the Kamloops area. —Image constructed using Canadian Digital Elevation Data obtained from GeoBase

Like several other similar Triassic intrusive bodies, such as the Guichon Creek batholith, the Iron Mask batholith contains concentrations of copper and gold. Ore in the old Iron Mask Mine, located 1 kilometre southeast of the junction of Highways 1 and 5, was discovered in the late 1800s and was mined intermittently between 1901 and 1928. The Afton Mine, 4 kilometres to the west of the junction, came into production in 1977, after Highway 1 had been relocated to make room for the open-pit mine. Though mining ceased in 1989, the top of the open pit, the flat-topped and grassed-over waste dump, and the smelter building are visible from the highway.

The Afton ore is in shattered diorite and syenite near the western end of the Iron Mask batholith. The ore minerals were originally the copper sulfide minerals bornite, chalcopyrite, and minor chalcocite, but later these were altered by the oxidizing effects of descending groundwater, which dominate from near the surface to depths up to 400 metres. The alteration converted most of the bornite and chalcocite into finely disseminated native copper

The Afton copper and gold open-pit mine in 1982.

and the iron oxide mineral hematite. The native copper contains gold, which was the most valuable product of the mine. The alteration found at the Afton Mine is unique among all of the mineral deposits in the area.

Eocene volcanic and sedimentary rocks, including some coal, have been down-faulted into the batholith and are found within the confines of the open pit. The land surface in Eocene time was probably close enough to the deposit to be the source of the downward-percolating and oxidizing groundwater.

Kamloops White Silt

For about 50 kilometres east of Kamloops, much of the bottom of the South Thompson Valley contains a voluminous deposit of silt that settled out in what has been called Glacial Lake Thompson, which was located behind a mass of ice that dammed the valley now containing Kamloops Lake. The silt forms the buff-coloured, steep slopes and cliffs on the lower valley walls east of Kamloops and has a width of as much as 1.2 kilometres. The silt has been drilled to a depth of 40 metres without reaching its base. The top surface of this deposit lies slightly above the 500-metre contour on the valley wall and dips gently toward the centre of the valley. The South Thompson River has entrenched it to depths of as much as 130 metres. If the 150-metre deepest part of the floor of Kamloops Lake is projected eastward into the South Thompson Valley to give us an idea of that valley's depth, then the original thickness of the silts may have reached 300 metres and the volume of silt probably is measurable in cubic kilometres.

Kamloops white silt forms the terrace at the base of the north side of the South Thompson Valley east of Kamloops. The higher hills are underlain by Paleozoic and Mesozoic rocks.

The silt is light coloured and commonly contains flakes of white muscovite mica that sparkle in the sunlight. Microscopic examination reveals a high proportion of quartz and feldspar and lesser amounts of rock fragments, iron- and magnesium-rich minerals, and reddish mineral garnet. In the upper part of the deposit, 1- to 2-millimetre-thick layers of sediments made of clay-sized grains alternate regularly with silt layers that average at most a few centimetres thick. These layers are comparable in thickness to those accumulating today in Kamloops Lake, and together with the regularity and lateral continuation of the beds, point to their deposition by annual cycles, with the silt layers representing accumulations following high runoff in the late spring and early summer. The silt layers are called *varves.*

Unusual in this silt deposit is the gradual but persistent downward increase in varve thickness, from 1 to 2 centimetres at the top to a phenomenal 10 to 15 metres in the lowest exposures. The thick layers low in the deposit tend to be sandier and display internal coarse layering, which is probably a product of short-lived storms or periods of rapid melting. A count of the varves indicates that the visible part of the deposit accumulated in only a few decades. Accumulation of the lower portion, which is hidden by alluvial fans or lies below river level, might have required another one or two decades. Both the western and northeastern ends of the deposit display irregularities in bedding and surface topography, which may be due to the removal of support when buried ice melted, the grounding of icebergs that floated on the lake, or post-depositional slumping.

The rapid accumulation of this huge silt deposit, possibly within a period spanning less than one century, and the whereabouts of both its source and the absence of really coarse sediment that might be expected to accompany it, demand attention. Perhaps it is the product of some gigantic glacier burst, a mechanism that could account for the short time span of the event, but the original site of the pent-up water is not known. The necessary volume of coarse sediment might be found in the nearly contemporaneous Deadman delta, but the Deadman River of the time was not known to tap a source rich in quartz, feldspar, muscovite, and garnet. To provide these minerals in quantity demands a metamorphic or granitic source such as that in the Omineca Belt to the north, east, or southeast. This problem remains unresolved.

Between Kamloops and Pritchard

Along Highway 1 east of Kamloops, the hillsides north of the South Thompson River above the silt terraces are made of a variety of rocks in the Quesnel terrane. The rocks immediately northeast of Kamloops are volcanic rocks, interbedded shale, and small intrusions of Triassic age, but between 15 and 22 kilometres east of Kamloops (or for 5 kilometres west of the large cement plant and quarry in Permian limestone north of the highway), the rocks are late Paleozoic in age. The Paleozoic strata, known as Harper Ranch group, include cherty argillite, siltstone, volcanic sandstone, and breccia derived mainly from volcanic rocks; conglomerate made of pebbles and cobbles of volcanic rock, chert, and limestone; and bodies of limestone, one of which is quarried above the cement plant. The rocks range in age from latest Devonian to middle Permian and appear to represent remnants of old volcanic arcs and their accompanying sediments.

A variety of fossils, including the large calcareous forminifera called *fusulinids*, are preserved in the limestone bodies. The fossils are similar to others of comparable age in parts of the western United States. They may have lived in reefs that formed around island arcs in the eastern part of the ancestor of the Pacific Ocean, called Panthalassa, and possibly not too far from the margin of the supercontinent Pangea, a fragment of which became North America. A limestone lens about 6 kilometres north of the highway contains middle Permian fusulinids of similar age to, but very different from, those in the Cache Creek limestone along Highway 99. The latter are similar to Permian fusulinids found today in Asia.

View from Highway 1 looking northward across the South Thompson River. At the base of the hill is a lens of Carboniferous limestone within the volcanic sandstone, chert, and argillite that make up most of the hill. The light-coloured bluffs on the right are a remnant of the Kamloops silt beds that were deposited in a late-glacial lake.

Sedimentary and volcanic rocks, which for a long time were undated, unconformably overlie late Carboniferous limestone at the top of the hill northeast of the cement plant quarry. They were once thought to be equivalent to the late Triassic Nicola group to the west, but a recent discovery of an early Jurassic fossil in the gully near the top of the bluff shows they are younger. Early Jurassic volcanic and sedimentary rocks, similar to those in this locality, are known about 100 kilometres to the north, and near Rossland and Salmo on Highway 3 in southeastern British Columbia. They are the younger part of the late Triassic and early Jurassic Quesnel island arc.

A Jurassic age makes sense because due south of here in the uplands south of the South Thompson River, the late Triassic rocks are mostly shale and siltstone, with only minor amounts of volcanic rock. The bluffs of volcanic rock north of the highway descend gradually to the east and are exposed in a roadcut on the south side of Highway 1 near Pritchard, which is about 37 kilometres east of Kamloops. They are the easternmost rocks of the Intermontane Belt exposed along Highway 1.

HIGHWAY 3

East Gate of Manning Provincial Park—Osoyoos

155 kilometres

See the map on page 266 for the east end of the route at Osoyoos.

At the east gate of Manning Provincial Park, you are already well within the Intermontane Belt. The Pasayten fault, which trends in a north-northwest direction for nearly 250 kilometres, is about 4.5 kilometres west of the park gate and separates the Methow terrane of the Coast Belt from the granitic and gneissic rocks of the Eagle plutonic complex of the westernmost Intermontane Belt.

The gneissic and granitic rocks form a continuous tract extending for about 130 kilometres north-northwestward from here. In places the tract is as much as 20 kilometres wide. Southward into Washington these rocks grade into a region with fewer gneissic and more granitic rocks. Although the rocks superficially resemble one another along the entire tract, their age changes in the segment between Highway 5 and Highway 1. The southern part, called the Eagle plutonic complex, consists mostly of latest Jurassic and early Cretaceous rocks. The northern part, called the Mount Lytton complex, is of latest Permian and early Mesozoic age. Together, they form a tract of formerly deep-seated rocks that were uplifted along the southwestern edge of the Intermontane Belt in late Jurassic and earliest Cretaceous time. A less deep-seated expression of the structures associated with these rocks may be the Ashcroft syncline south of Cache Creek.

About 0.6 kilometres west of the east gate, the granitic gneiss of the Eagle plutonic complex gives way to metamorphosed and cleaved volcanic and sedimentary rocks of the Nicola group of Triassic age. As the distance from the gneiss increases, the grade of metamorphism diminishes, although the prevailing westward dip of the foliation in the Nicola rocks persists. This suggests that the gneissic rocks to the west were pushed eastward over the Nicola group.

Similkameen River Valley

Highway 3 follows the Similkameen River for much of the way across the southern Intermontane Belt. From the park headquarters to the east gate of Manning Provincial Park and a few kilometres beyond, alluvium deposited near the margin of an ice age glacier floors the Similkameen Valley. As the river approaches Similkameen Falls, about 5.8 kilometres east of the east gate but not visible from the highway, its channel becomes incised. The

Fragmental Nicola group rocks near Similkameen Falls. The flattened, elongated fragments in the rock give it a crude foliation that dips westward at about the same angle as the granitic gneiss of the Eagle plutonic complex just to the west.

highway climbs slightly and crosses a gravel flat for a short distance and then descends abruptly into a rock-walled canyon. The river passes around the gravel flat and has begun the slow task of cutting into the resistant bedrock. Farther downstream, the modern channel cuts into the walls of the older, gravel-filled channel, and highway travellers follow the position of the old channel.

At the pullout near Copper Creek, about 4 kilometres east of Similkameen Falls, you can look northeast across the river and see an exposure of lake-deposited silts, one of the few tangible signs of an ice-dammed lake that once occupied valleys in the region.

North of the pullout, Highway 3 leaves the Similkameen River and its canyons and climbs onto the upland at Sunday Summit. It then begins the long descent to Whipsaw Creek and the Princeton Basin. Scattered roadcuts along the highway north of Copper Creek are in Eocene lava, volcanic breccia, and tuff that belong to a widespread continental arc.

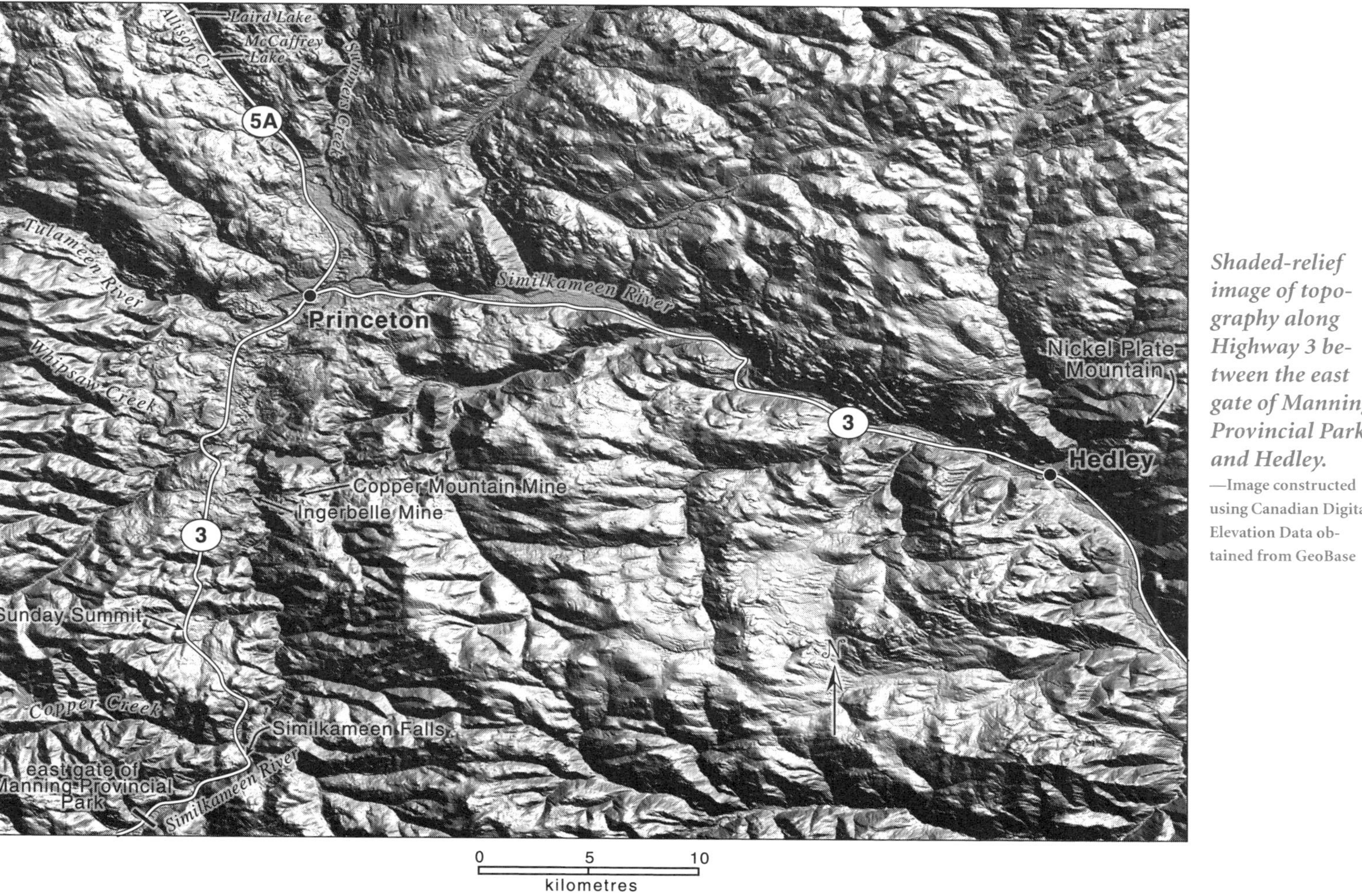

Shaded-relief image of topography along Highway 3 between the east gate of Manning Provincial Park and Hedley.
—Image constructed using Canadian Digital Elevation Data obtained from GeoBase

Copper Mountain Mines

About 10 kilometres north of Sunday Summit and 23 kilometres south of Princeton are the remains of a blocked-off paved road. This was the route of Highway 3 before it was rerouted in the 1960s to permit mining of the Ingerbelle copper deposit, a move matching the rerouting of Highway 1 around the Afton deposit west of Kamloops. Even more coincidental was the involvement of Eocene rocks at both sites, a feature that prompted the tongue-in-cheek prospecting suggestion that the intersection of the base of Eocene rocks with a major highway is the place to find an ore deposit.

Visible to the northeast across the Similkameen River valley from the blocked-off old highway is the mine at Copper Mountain, from which over 100 million tons of copper ore were mined, initially by tunnelling and later by open-pit mining. The deposit was discovered in 1892 by surface prospecting, but the mine was not brought into full production until 1925. It operated from then until 1993, apart from periods of near or complete shutdown in 1931–36 and again in 1957–71.

The Ingerbelle deposit, below and just east of Highway 3 on the west side of the Similkameen River, then was developed and a new and larger mill opened in 1971 to serve both deposits. The Ingerbelle deposit was mined as an open pit during the first eight years of its life, and then open-pit mining was transferred back to Copper Mountain across the river. The ore ran at 0.4 per cent or better in copper and was transported across the canyon of the Similkameen River on a suspended conveyor belt to the Ingerbelle mill for treatment.

As with several other copper deposits in the Canadian Cordillera, the mineral deposits are within lava and fragmental volcanic rock of the late Triassic Nicola group adjacent to slightly younger diorite and syenite intrusions. The copper sulfide minerals chalcopyrite, bornite, and chalcocite are dispersed in scattered particles and films along fractures in the volcanic rocks and localized at intersections of faults and breccia pipes, which are chimneylike conduits below a volcano. The Copper Mountain intrusions probably were emplaced toward the back, or eastern side, of the late Triassic island arc of the Quesnel terrane.

Princeton Basin

The descent on Highway 3 from the upland surface near Sunday Summit to the floor of the Princeton Basin, about 600 metres below the summit, takes the traveller past several points of interest. First, the partly grass-covered debris on the eastern side of the highway is waste from the Ingerbelle Mine, which has been graded and seeded. Below this, the highway curves into the lower end of a glacial meltwater channel that was cut into the Triassic bedrock by an eastward-flowing, glacially diverted stream. The collapse of the channel walls has covered its floor with loose rock. Above the highway at the end of this curve is the Ingerbelle mill, and almost directly ahead across the Similkameen River is the tailings disposal area for the mill, located in another meltwater channel.

North of this, the highway crosses middle Eocene sedimentary rocks containing coal beds. These overlie the Eocene volcanic rocks passed to the west near Sunday Summit. The widely scattered exposures of Eocene sediments include brown bentonite layers formed by alteration of volcanic ash. These layers are very slippery when wet and upon drying shrink to produce a popcornlike texture on the soil surface. Near the bottom of the basin, the highway descends along a narrow ridge that separates the Similkameen River valley to the east from the Tulameen River valley to the north.

Eocene sedimentary rock in the Tulameen River valley, below the level of the highway, was the source of most of the 2 million tons of lignite to subbituminous coal mined in the Princeton Basin between 1909 and 1961. Ancient fires in the underground coal have baked and reddened the enclosing shale in places. Eocene shale baked to brilliant brick red and yellow can be seen in a cliff on the north bank of the Tulameen River, and again in downtown Princeton behind the buildings on the west side of Bridge Street in the block north of Billiter Street. The burned rock was used by First Nations people as pigments and gave rise to the old place-name Vermilion Forks, used by Hudson's Bay traders. The name was changed to Princeton in 1860 to commemorate a visit by the Prince of Wales.

The Princeton Basin is both a topographic and a structural depression that is about 25 kilometres long from north to south and up to 7 kilometres wide. It contains a stratigraphic sequence composed of the Eocene volcanic rocks crossed near Sunday Summit, overlain by the coal-bearing sedimentary rocks passed much lower down on Highway 3. The combined thickness of the two units probably approaches 3 kilometres. The structure of the basin resembles a partly open trapdoor that is hinged on its western side, so that the volcanic rocks exposed near Sunday Summit dive eastward below the sedimentary rocks that comprise the upper part of the basin fill. The eastern side of the basin is bounded by a large normal fault, along which the Eocene rocks have been down-dropped to lie against late Triassic Nicola strata and, locally, Eocene volcanic rocks. Such a structure is called a *half-graben*, a term derived from the German word for "grave" and used by geologists for a structurally depressed area bounded on one side by a normal fault. South, east, and west of the basin, the nearly flat-lying Eocene volcanic rocks are at elevations of 1,500 to 2,000 metres. A rough calculation that combines the topographic elevation of the base of the Eocene rocks surrounding the basin with the probable thickness of the Eocene strata suggests that downward movement on the normal fault along the eastern side of the basin is between 4 and 5 kilometres.

The sedimentary rocks in the upper part of the basin fill range from coarse angular breccia thought to have been laid down by debris flows, through coarse sandstone, shale, and conglomerate, to layers of coal and bentonite, a type of clay derived from weathered volcanic ash. The size of the sand grains tends to decrease toward the south end of the basin, where most of the coal is found. The rock fragments in the conglomerates are mainly granitic and probably were derived from a source to the northeast, where today a large granitic pluton is exposed. The direction of stream flow at the time of deposition, as determined from crossbeds, channels, and the orientation of elongate fossil plant fragments, was mostly to the south or southwest.

You can examine the sedimentary rocks a few blocks south of the Princeton town centre, at the foot of the grade that brings Highway 3 down to the level of the Similkameen floodplain. There, in an excavation behind the bus station, tilted beds of yellow sandstone containing a few coal layers are

Tilted beds of Eocene sandstone, behind the bus station in Princeton, contain thin coaly layers.

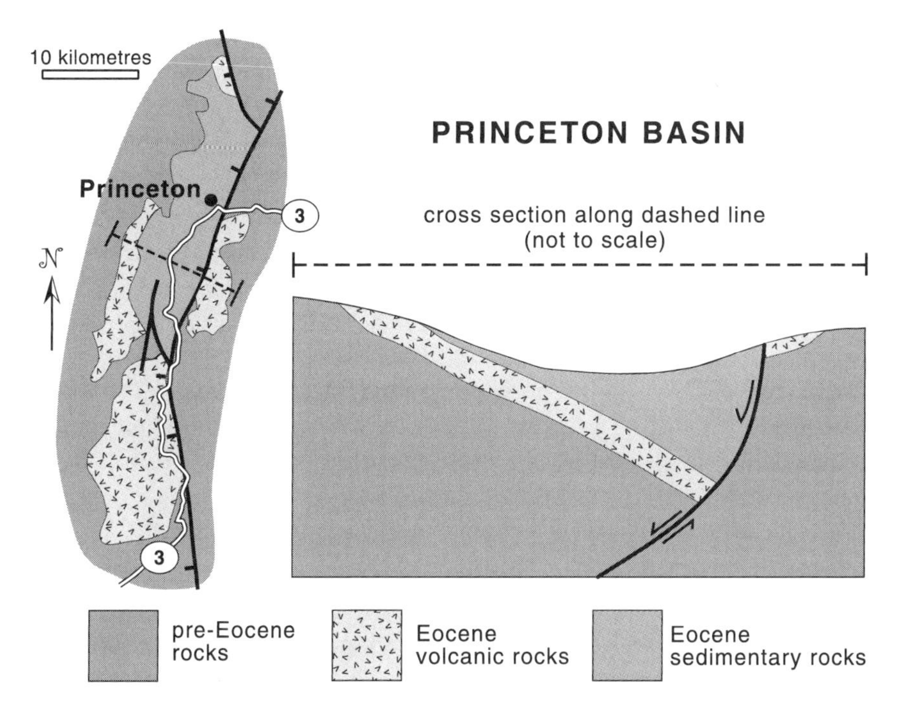

Geological sketch map and cross section of the Princeton Basin and surrounding area.

visible. The sandstone has been planed off along a flat surface, a classic unconformity, by a prehistoric course of the Similkameen River, which then deposited a layer of gravel on top before shifting to a new course.

Fossils from sedimentary rocks and isotopic dates from volcanic rocks in the Princeton Basin provided the first solid information on the age of Tertiary rocks in the region. Prior to 1960 and the widespread use of isotopic dating, these rocks had been assigned ages ranging from Eocene to Miocene on the basis of plant fossils and then-current ideas on climatic and floral changes with time. In 1933, the fossilized tooth of a *tillodont*, a primitive bearlike mammal, was discovered in coal in the Pleasant Valley mine in the basin. Tillodonts are known only from Paleocene and Eocene strata, and the tooth was from a species known only from middle Eocene rocks. Then in 1961, biotite from an ash bed about 40 metres below the main coal seam yielded a date of 49 million years, which corresponds to middle Eocene on the geological time scale.

Incidentally, in the early 1980s a sharp-eyed naturalist discovered the fossilized lower jaw, only 12 millimetres long, of an early form of shrew in the waste dump from the Pleasant Valley mine. By carefully searching the vicinity he found the impression of the other side of the same jaw.

Plant fossils from the Princeton Basin indicate that in Eocene time the climate was distinctly warmer and more humid than it is today. The most abundant fossils are twigs, with needles still attached, of a coniferous tree of the genus *Metasequoia.* Today, *Metasequoia* is native only to central China. The modern flora of the Princeton Basin is significantly different from that on the coast and indicates the present arid conditions in the area. A comparison of Eocene flora from the lower Fraser Valley near Vancouver and that in the Princeton Basin does not show the strong inland increase of aridity reflected by the modern flora. This difference suggests that the coastal mountain barrier, which today creates a rain shadow and is responsible for the dry climate in the interior, was much less prominent in Eocene time.

Between Princeton and Hedley

Tailings, or finely crushed waste rock, from the predecessor of the Ingerbelle Mine mill, line the south side of Highway 3 just east of Princeton. The barren piles of grey silt constitute a major source of dust in dry windy weather. In wet weather, rainwater leaches acid from sulfide minerals, notably pyrite, in the tailings. The leaching of acid could have been alleviated had an ample cover of limestone-rich material been added when the tailings piles were first deposited, and then a vegetative cover established on top. Far greater efforts are made to do this today than formerly, when environmental laws were less strict.

Along Highway 3 about 6 kilometres east of Princeton, granitic dikes that weather to a pink colour intrude greenish volcanic rocks of the Nicola group. The dikes are undated but are probably related to one of three groups of granitic rocks in the region. The early Jurassic Bromley pluton, about 193 million years old, is exposed on the north side of the valley from just east of here almost to the bridge across the Similkameen River. About 5 kilometres northeast of here, and crossed on the road leading northeast from Princeton to the Okanagan Valley, the large middle Jurassic Osprey Lake batholith is isotopically dated at 166 million years and intrudes both the Nicola group and the Bromley pluton. About 10 kilometres south of the highway is the early Cretaceous, 105-million-year-old Verde Creek granitic stock, and still farther south and straddling the international boundary, the large Cathedral Lakes pluton is of the same age. The last two intrusive bodies are probably related to the Spences Bridge volcanic rocks of the same age, which extend more or less continuously north-northwestward, from near Princeton to well beyond Spences Bridge on Highway 1.

Just east of the dikes and north of the highway, the Nicola volcanic rocks are replaced to the east by Nicola siltstone and shale of late Triassic age. The lateral change between Triassic volcanic rocks to the west and sedimentary rocks to the east can be traced northward from here for at least 120 kilometres and is crossed by Highway 1 just east of Kamloops. The sedimentary rocks were deposited in the westernmost part of the basin that lay east of the late Triassic Quesnel island arc and separated it from the Triassic continental margin.

The sedimentary rocks are exposed on Highway 3 about 3 kilometres east of the bridge over the Similkameen River. There, thinly bedded argillites and siltstones contain occasional, small, finely ribbed fossil clams related to modern scallops. Still farther east, near Hedley, there are beds of limestone, and on the south side of the Similkameen River south of Hedley, late Triassic shale contains large boulders and cobbles of limestone of the same age, containing microfossils. The limestone fragments probably slumped off the margins of high-standing, possibly fault-controlled limestone reefs and were deposited in the deeper-water parts of the basin.

Gold Mines and Hedley

The discovery in the mid-1890s of gold-bearing ores near the summit of Nickel Plate Mountain led to development of the Nickel Plate Mine, and in 1900 the settlement of Hedley was established at the base of the mountain. It was not until 1909, however, that rail service linked Hedley to Princeton and to Oroville in Washington. The abandoned grade of a branch of the Great Northern Railway underlies the road on the north bank of the Similkameen River, west of the highway bridge west of Hedley.

Mine development involved installation of a 3-kilometre-long heavy wire rope for an inclined tramway, the scars of which are still visible at the south end of Nickel Plate Mountain. Because it was not practical to splice it in the field, a very long line of packhorses hauled the entire rope up the steep hillside. Each carried a loop or two, and lengths of cable spanned the gaps between the horses. What the wranglers said to get the horses to march in unison is not recorded and is perhaps best left to the imagination!

Production commenced at the Nickel Plate Mine in 1904 and averaged 100 to 200 tons of ore per day until the mine closed in 1931. The increase in the price of gold and a decline in costs during the 1930s depression allowed the mine to reopen in 1935 and continue production until 1955. It reopened for a second time in 1987, using an open pit rather than underground mining, but is now inactive. A second mine, the Hedley Mascot, commenced production in 1936, mining the central part of a major orebody that had already been mined out at both ends from the Nickel Plate Mine. A grudge

The tramway route above Hedley runs up the hill from the lower left and led to mine buildings at the top. The dark bands in the bedded sedimentary rocks exposed in the steep slope in the foreground are diorite sills.

by the owner of the Mascot fractional claim barred extraction of this central part through the Nickel Plate workings. A third operation recovered gold by mining the tailings from the original Nickel Plate concentrator at Hedley. The tailings were transferred to pits with waterproof linings, sprinkled with water, and treated with special bacteria to aid solution of the gold. Meanwhile, Hedley not only survived two mine closures but has become a community housing senior citizens.

The ore responsible for this activity is in *skarn*, a rock formed as an alteration product where diorite sills and dikes intruded late Triassic limestone. The limestone was altered and now contains the skarn minerals garnet, pyroxene, and quartz and also the sulfide ores arsenopyrite, pyrrhotite, chalcopyrite, and some pyrite. Gold, both in native form and as bismuth-gold telluride, tends to be associated with the arsenopyrite. The ore is concentrated along the walls of dikes and a swarm of dioritic sills that followed weak zones between the bedding planes in the limy rocks. The altered area along the sills may have contained open fractures along which mineralizing solutions percolated from nearby *stocks*, small intrusions of diorite that are roughly circular in cross section; the stocks may also have been the sources of the sills and dikes. The hillside east of Hedley provides a magnificent cross section in which the stratified limestone is grey, the sills and dikes are dark and weather to a rusty colour, and the dioritic stocks form blocky, homogeneous, light grey bodies.

Few mining camps have been as strongly and repeatedly influenced by one person as Hedley. Shortly after the Nickel Plate orebody was staked and the claims were surveyed, an independent individual named Duncan Woods somehow learned that a small area had been missed in the staking. Woods staked this area, some 17.2 acres in extent, and called it the Mascot Fraction. Over the next two or three decades, the Hedley Gold Mining Company worked the Nickel Plate orebody downward and to the west until they reached the Mascot Fraction. Hedley Gold tried to buy the fraction, but their offers only offended Woods, who ultimately vowed that the Mascot Fraction would never be sold to a big mining company. Finally, in 1933, Woods met a group of Vancouver businessmen who agreed to his terms. Mining started in 1936 using some of the same workings as Hedley Gold for access and ventilation, but as Woods had decreed, the Mascot ore was processed in a separate mill. Some of the profits from the Mascot Mine were reinvested in other mining properties, notably the Bralorne Gold Mine about 90 kilometres northwest of Lillooet, and the Pacific Nickel Mines near Hope. When the Nickel Plate Mine at Hedley was up for sale in 1977 as a potential open-pit mine, it was bought by Mascot Gold. How would Duncan Woods have felt had he lived to see this event?

Between Hedley and Keremeos

About 6.5 kilometres southeast of Hedley, Highway 3 passes the mouth of Winters Creek, which follows the trace of a fault separating late Triassic beds to the west from Mississippian to Triassic argillite, sandstone, chert that in places contains radiolarians, pillow basalt, and minor limestone. On the valley wall about 2 kilometres to the southeast of the fault contact, blocks of limestone containing Ordovician corals are within a sandstone matrix containing early Carboniferous microfossils. Today, the nearest known fossiliferous Ordovician beds in Canada lie about 150 kilometres to the east, and just how far away they were at the time of deposition is unknown. Near the Keremeos garbage dump, west of town, are slide blocks of pillow basalt and red and green bedded chert containing Carboniferous radiolarians. These rocks appear to be ocean floor deposits that unconformably underlie the Triassic strata. There is a remote possibility that they can be correlated with the Slide Mountain terrane, and they appear to be very different from arc-related rocks of the same age that lie northeast of Kamloops.

These mainly late Paleozoic rocks support high, steep cliffs that are home to a seasonally migrating flock of mountain goats. The cliffs overlook a valley floored with an accumulation of sediment mostly deposited by the Similkameen River; the valley continues eastward for nearly 30 kilometres, as far as Keremeos. Throughout the valley, extensive talus aprons and fans cover the lower slopes and infringe on the valley floor. In several places, such as one about 23 kilometres east of Hedley just east of a big slide block by the road, roadcuts across the toes of the fans expose a bed of white volcanic ash several centimetres thick. The ash came from Mount Mazama, a Cascade volcano that erupted violently about 6,700 years ago and now is the site of Crater Lake in southwestern Oregon. In other places, roadcuts display gravel with well-rounded pebbles and with an irregular upper surface including kettles, depressions that formed when buried ice blocks melted at the end of the last ice age.

The white Mazama ash layer shows up in a roadcut east of Keremeos. —R. Turner photo

Mount Kobau

Mount Kobau, 15 kilometres southeast of Keremeos, rises to an elevation of 1,875 metres and is made of micaceous and graphitic quartzite, and fine-grained mica, chlorite, and hornblende schist. Though a great thickness of rock is present, metamorphism has destroyed any fossils, and the age of the rocks is unknown. Based on the range of rock types present, they could be more-metamorphosed equivalents of the mostly late Paleozoic rocks between Hedley and Keremeos to the west; but alternatively, they could be much older and be related to the early Paleozoic strata near Kootenay Lake in southeastern British Columbia.

Saline Spotted Lakes

Highway 3 leaves the valley floor about 3 kilometres north of the international boundary and climbs and doubles back southward into a broad, treeless valley leading to Richter Pass. Two kilometres east of the pass is the larger of two Spotted Lakes that contain concentrations of magnesium sulfate. A smaller but more intensively explored lake is a few kilometres to the south in Washington. Both lakes were mined during World War I. The deposits were refined in a plant at Oroville, Washington, and sold either as Epsom salts or for dyeing fabrics and tanning leather. The lake surfaces display circular pools of brine separated from one another by bodies of foul black mud whose surfaces become deceptively covered by an efflorescence of powdery white crystals in dry weather. Cattle have been mired in this mud, and mining operations were restricted to periods of dry or cold weather because of the difficulty of getting out in the lakes. The pools of brine are floored with bodies of crystalline epsomite, or hydrous magnesium sulfate, as much as 4.5 metres thick, capable of supporting tunnels that stayed dry under the impervious muds and described as being "rock hard." Drilling revealed a layer of gypsum, or calcium sulfate, under the epsomite and on top of glacial till or bedrock.

Osoyoos

Osoyoos lies just north of the international boundary at the southern end of the big Okanagan Valley (whose continuation in Washington is spelled Okanogan). Here, the valley marks the boundary between the Intermontane and Omineca Belts. Osoyoos, with an elevation of about 275 metres, is at the lowest point in the southern Canadian Cordillera east of the coastal regions or lower Fraser Valley. The climate in the valley bottom is the most desertlike in British Columbia, although in many places irrigation obscures its true nature.

HIGHWAY 3A
Keremeos—Kaleden
32 kilometres

Highway 3A, the shortcut from the Similkameen River valley to the central Okanagan Valley, follows Keremeos Creek northward from Keremeos for its first 12 kilometres. The valley of Keremeos Creek is incised as much as 1,000 metres below the rolling upland surface. The floor of the valley is relatively flat, probably because meltwater streams deposited sediment here at the end of the last ice age. The valley floor deposits have been partly or completely buried by postglacial alluvial fans and talus cones from the valley walls. This alluvium is too coarse to have been effectively washed away by the modern Keremeos Creek. The conical forms of these younger

Shaded-relief image of topography along Highway 3A between Keremeos and Kaleden. —Image constructed using Canadian Digital Elevation Data obtained from GeoBase

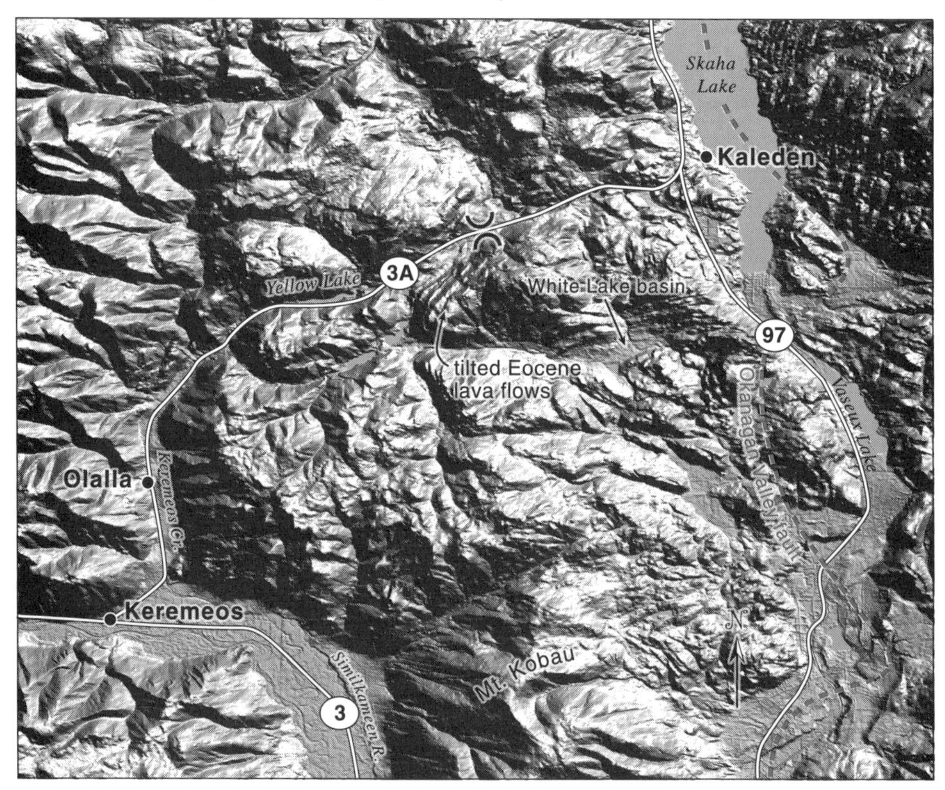

Talus cones are formed by erosion from the rock walls northwest of Keremeos.
—R. Turner photo

deposits are particularly well exhibited here by virtue of the relatively steep slopes and the open vegetation.

Olalla Pyroxenite Stock

About 6.5 kilometres from the junction with Highway 3, Highway 3A passes through the village of Olalla, which lies near the middle of an intrusion of early to middle Jurassic age that is 2 to 4 kilometres in diameter. The intrusion, or stock, is made of very dark pyroxenite, an iron- and magnesium-rich rock composed mainly of the mineral pyroxene. Pale granitic dikes intrude the stock, and all are intruded by dikes of pink quartz-free and potassium-rich syenite. In places, mineralized bedrock has encouraged prospecting, but no producing mines have developed.

Eocene Rocks near Yellow Lake

About 11 kilometres northeast of Olalla is the outlet of Yellow Lake, which occupies a westward-draining glacial meltwater channel that has been blocked by a debris fan from the north. Lava flows near the base of the succession of Eocene volcanic rocks are exposed in roadcuts along Highway 3A by the lake. The lava overlies Eocene conglomerate, sandstone, and shale, hidden in the valley bottom here but exposed on the slopes west and southwest of Yellow Lake.

The Eocene lava and sedimentary rocks dip eastward at angles of up to about 25 degrees and are cut by numerous north-trending normal faults that dip steeply at the surface. Individual flows are visible in bluffs east of the lake. From the lake outlet to the hilltops about 6 kilometres to the east, the volcanic rocks are about 1,500 metres thick and are generally andesitic in composition but are unusually low in silica. They feature distinctive feldspar crystals that range from single rhomb-shaped crystals to small flowerlike rosettes set in a fine-grained matrix. Small cavities in the lavas contain zeolites, hydrous aluminum silicate minerals. The andesitic lava flows are overlain unconformably by a silica-rich rhyolite sequence up to a few hundred metres thick, which is covered in turn by a 1,000-metre-thick

This view from the White Lake basin shows easterly tilted and faulted Eocene lava flows in the area east of Yellow Lake. —R. Turner photo

stack of sedimentary rocks containing plant fossils of Eocene age. Neither the rhyolite nor the overlying sediments are exposed along Highway 3A. The lower sedimentary and volcanic part of the Eocene pile is widespread and can be traced, between areas isolated from one another by older rocks, at least as far east as the Kettle River valley along Highway 3, about 80 kilometres east-southeast of here.

Although the Eocene rocks in the Yellow Lake area are everywhere highly faulted, they extend continuously for nearly 20 kilometres to the east, as far as the Okanagan Valley, where they can be seen to form the upper, or hanging, wall of the Okanagan Valley fault. This large normal fault dips at a shallow angle westward and brings Eocene rocks to the west into contact with the Vaseux gneiss in the lower wall, or footwall of the fault. The gneiss ranges in age from possibly Precambrian to early Tertiary. Along the southern Okanagan Valley the fault marks the boundary between the Intermontane and Omineca Belts.

Meltwater Drainage Patterns

For 4 kilometres east of Yellow Lake, Highway 3A follows a westward-draining glacial meltwater channel up to the divide between the Similkameen and Okanagan drainage basins. About 3 kilometres east of Yellow Lake, a road branches to the south through a golf course, and from Highway 3A you can see the side road climbing a gravel scarp about a half kilometre away. The top of this scarp is a gravelly terrace on which two kettle lakes are located 0.5 to 3 kilometres beyond the scarp. The terrace formed at the end of the last ice age when ice still blocked the Yellow Lake channel and covered the first half kilometre of the channel to the south and the sites of the two lakes beyond, barring escape of water to the east. Meltwater at this stage drained to the west through a pass about 3 kilometres south of Yellow Lake. With continued melting of the ice, the Yellow Lake channel opened and the higher, southern route was abandoned. During this later stage, much of the gravel carried by meltwater from the Okanagan ice tongue accumulated at the head of the northern channel, which forms the pass followed by Highway 3A. Although most of the gravel deposited during this stage has been excavated and removed for highway construction, the remaining patches clearly show it had been built up to a common surface that sloped gently west toward the head of the meltwater channel.

In the final 9 kilometres before its junction with Highway 97, Highway 3A descends northeastward across glaciated terrain with occasional exposures of Eocene volcanic rocks.

HIGHWAY 5
Coquihalla Pass—Little Fort and Beyond
250 kilometres

See the map on page 218 for the Kamloops area.

Between the tollbooth east of Coquihalla Pass and its junction with Highway 1 about 6 kilometres west of Kamloops, Highway 5 is a magnificent road for travelling and scenery but a bad place to stop and look at the rocks exposed in the numerous roadcuts. The traffic is very fast, and stopping is frowned upon for safety reasons. North of Kamloops, Highway 5 is older and narrower and follows the valley of the North Thompson River, but even there the roadcuts are mostly in dangerous places.

Northward Migration of the Coquihalla Drainage Divide
Just west of Coquihalla Pass, northbound travellers drive through a narrow gap, emerging abruptly onto a precipitous slope overlooking the upper part of the valley of the southward-flowing Coquihalla River. Although concrete barriers block the view directly downslope at this place, one can sense the very steep drop, which is more than 300 metres in scarcely 500 metres of horizontal distance. Vegetation is sparse and scattered, and bare rock and talus abound. The highway turns northward and within 100 metres passes a sign announcing Coquihalla Pass at an elevation of 1,244 metres. Within the next few hundred metres, treetops start to show up over the concrete barrier wall on the outside of the road, announcing the presence of a gentler, more stable hillside that is closer to the valley bottom. This narrow shelf widens northeastward to more than 1 kilometre wide near the Falls Lake exit. A matching shelf with a precipitous slope below it can be seen on the opposite side of the Coquihalla River. Near the tollbooth the two shelves converge to form a nearly continuous and fairly flat surface scored by a minor rock-walled canyon containing the headwaters of the Coquihalla River. Still farther northeast along Highway 5, at Coquihalla Lakes, the canyon is no longer present and the highway follows the bottom of a broad, U-shaped, north-trending valley that merges with the valley of the northward-draining Coldwater River. One cannot escape the thought that the broader valley here at one time contained the Coldwater River drainage. Numerous ice age features support the theory that the Coldwater-Coquihalla drainage divide appears to have shifted about 9 kilometres to the north, in the direction of the ice retreat, because of rapid downcutting by the Coquihalla River.

About 1 kilometre northeast of the Falls Lake exit on Highway 5, and about 2 kilometres southwest of the tollbooth, the highway crosses a bridge

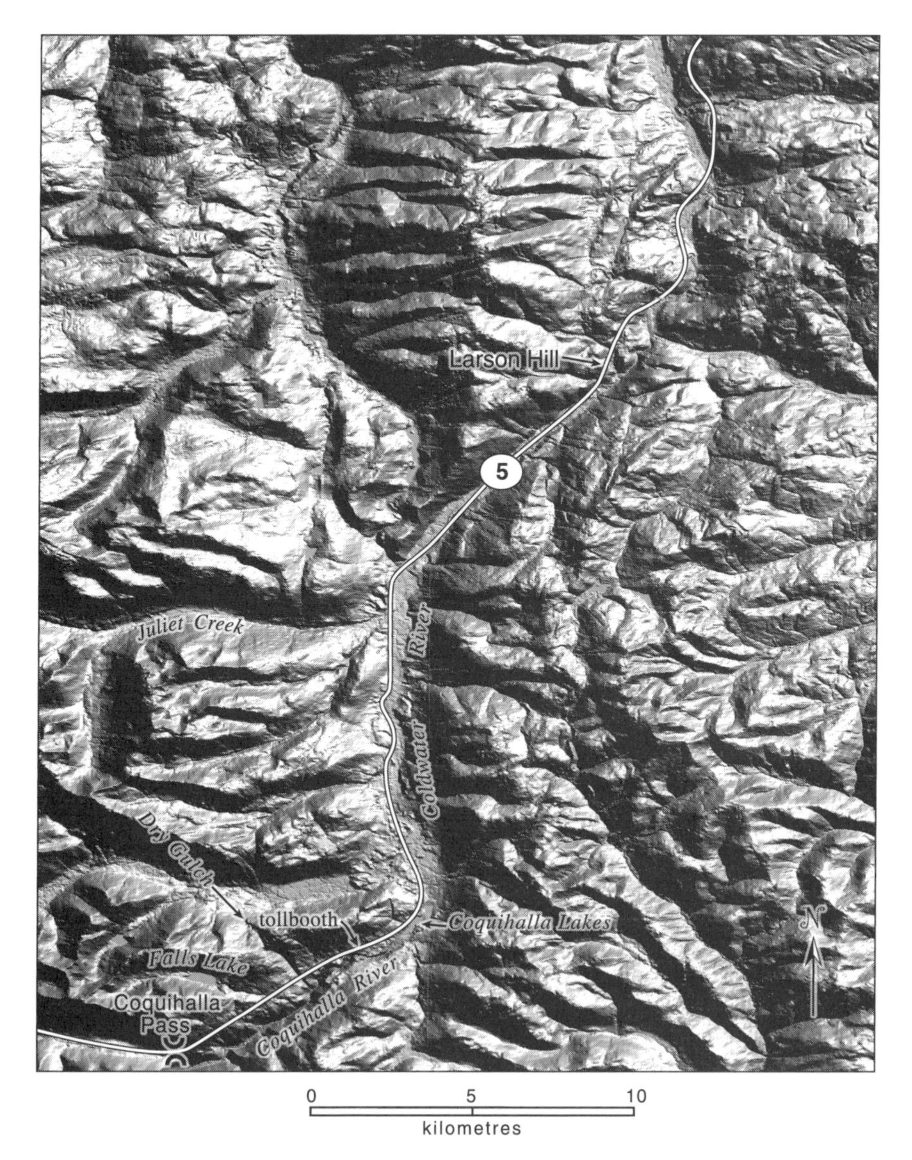

Shaded relief image of topography along Highway 5 in the Coquihalla Pass area and upper Coldwater River valley. —Image constructed using Canadian Digital Elevation Data obtained from GeoBase

over a remarkable waterless canyon called Dry Gulch. The canyon runs directly downslope and for much of its length is over 100 metres deep. Its walls are steep, sparsely vegetated, and clearly unstable. Dry Gulch looks like a water-cut channel in most respects, but the head of Dry Gulch lies near a single notch in the crest of the ridge, instead of dividing into a series of smaller tributary valleys as you'd expect in a normal drainage. The other side of the headwall ridge faces a large eastward-draining valley, which

View northwestwards across the upper Coquihalla River valley from Coquihalla Mountain. Across the valley, which marks the trace of the Coquihalla fault, are 48-million-year-old rocks of the Needle Peak pluton. The highway emerges from Coquihalla Pass high on the northwest side of the valley, and the tollbooth is off the photo to the right.

today contains the headwaters of the Coldwater River. If this valley had been blocked by ice during the last ice age, it is likely that glacial meltwater was ponded until it overflowed the ridge, eroding Dry Gulch in a sudden torrent.

With the shrinking of the ice dam toward the end of the last ice age, water escaped southward along the valley past the Coquihalla Lakes area, carving the canyon into the flat valley floor near the tollbooth. As the ice front melted northward past Larson Hill, which is about 15 kilometres north of the tollgate, a new and lower escape route for glacial meltwater opened to the north, and water stopped spilling into the Coquihalla drainage. Alluvial fans fed by side streams developed at the pass and blocked the water flow, forming the Coquihalla Lakes.

One might expect the northward-draining Coldwater Valley to contain silty deposits from the lake that ponded in front of the ice. However, any such lakebed deposits are concealed by an extensive cover of gravel that the retreating ice spread across the valley. In addition, at least one meltwater channel contributed sediment, and vigorous mountain streams including the upper Coldwater River and Juliet Creek deposited postglacial alluvium in the valley. Gravel from these sources has migrated down the faces of

Dry Gulch, crossed by a bridge on Highway 5, was probably formed by a torrential flood of glacial meltwater.
—R. Turner photo

undercut slopes along the Coldwater River valley, concealing the finer-grained lakebed deposits at lower levels.

Eagle Plutonic Complex

Near Coquihalla Pass, Highway 5 crosses the boundary between the Coast and Intermontane Belts. Eocene granitic rocks of the Needle Peak pluton that intrude rocks of the Methow terrane are part of the Coast Belt, whereas the granitic and gneissic rocks of the Eagle plutonic complex are here the easternmost element of the Intermontane Belt. The boundary between the two, elsewhere marked by the Pasayten fault, here may lie between the dark rocks west of Highway 5 as it emerges from the pass and the granitic rock just to the north that weathers to a tan colour. The latter, dated at 105 million years old, or mid-Cretaceous, intrudes the late Jurassic grey gneissic granodiorite, which is dated at 155 million years, exposed in the roadcuts near the tollbooth. The Jurassic and Cretaceous granitic rocks form the Eagle plutonic complex, which along with Permian and Triassic gneisses near Lytton, form a long belt up to 20 kilometres wide that extends continuously from the international boundary to north of Lytton, a distance of 200 kilometres. The narrow belt, once deeply buried and later uplifted and eroded, is the southwesternmost part of the Intermontane Belt.

About 8 kilometres north of the tollbooth on Highway 5, roadcuts expose dark, foliated metamorphosed volcanic rock. Near the contact with the Eagle plutonic complex, the metamorphosed volcanic rock has an irregular foliation that in places parallels the gneissic layering in the complex. You can see the contact between gneissic and metavolcanic rocks on logging roads east of the tollbooth. There, the parallel foliation and gneissic layering suggest that the foliation and layering formed concurrently during late Jurassic time, about 155 million years ago. From 5 to 15 kilometres east of the contact, the cleavage has disappeared completely, the metamorphism is far less, and the volcanic rocks can be readily identified as belonging to the Nicola group of late Triassic age and part of the early Mesozoic Quesnel island arc.

Spences Bridge Continental Arc

In the 20 kilometres between Coquihalla Pass and the base of Larson Hill, Highway 5 follows the broad, 1-kilometre-wide open valley of the Coldwater River. At Larson Hill, the valley floor becomes abruptly restricted to little more than the width of the Coldwater River. To avoid the canyon and the sharp curves low in the valley, the highway climbs sharply and does not return to the river for another 8 kilometres. In this hillside stretch are numerous roadcuts in reddish brown to greenish volcanic flows and fragmental rocks of the Spences Bridge group, which are the remains of an early Cretaceous continental arc about 105 million years old.

The Spences Bridge rocks are preserved in a long, narrow structural depression, which is locally discontinuous where broken by younger faults. East of the Fraser fault, the depression extends from west of Cache Creek in the north, south-southeastward for a distance of about 200 kilometres, and is nowhere more than about 30 kilometres wide. West of the Fraser fault, similar rocks are in the vicinity of Churn Creek, a tributary of the Fraser River about 90 kilometres northwest of the exposures west of Cache Creek.

About 8 kilometres north of Larson Hill, Highway 5 descends the western side of the valley of the Coldwater River and crosses it. Columnar jointing in streaky, dark red to pink Spences Bridge rhyolite flows is visible in bluffs on the eastern side of the river just south of the bridge. On the eastern side of the Coldwater River valley, the road ascends to about the 1,000-metre contour. About 1 kilometre north of the river crossing, a roadcut in layered volcanic sandstone and shale of the Spences Bridge group contains thin carbonaceous beds. Fossil leaves, pollen, and spores collected mainly from this roadcut show a flora that consisted mainly of ferns and conifers, but with rare flowering plants. Flowering plants appeared about 110 million years ago, in early Cretaceous time, but did not dominate the vegetation until toward the end of Cretaceous time. Accordingly, the age of these sedimentary rocks based on plant fossils agrees with isotopic dates of about 105 million years from elsewhere in the Spences Bridge group.

North of the Spences Bridge beds, the road passes through numerous roadcuts in mostly greenish but locally red volcanic rocks of the late Triassic Nicola group, intruded in a few places by bodies of light-coloured granitic rock. Most of the Nicola rocks are massive, but some thin beds of dark shale and lighter tuff are present. Numerous faults and fractures cut the rocks.

Columnar-jointed rhyolite of the Spences Bridge group above the east side of the Coldwater River, near the Highway 5 bridge across it.

Coaly (dark) *layers are interbedded with tuffaceous sandstone and shale of the Spences Bridge group in roadcuts along Highway 5 about 1 kilometre north of the bridge over the Coldwater River.*

Coldwater Fault System

Well-bedded buff-coloured sandstone of Eocene age is exposed low down in the Coldwater River valley north of the Highway 5 bridge but, because of the width of the highway, can only be seen readily by travellers driving south. The Eocene rocks are bounded by part of a system of north- to north-northeast-trending interconnected normal faults that form the Coldwater fault system. It extends for about 130 kilometres, more or less parallel with Highway 5, from south of the tollbooth, northeast through the Merritt basin, and north of this along Clapperton Creek, to link up with a north-northwest-trending fault that crosses Kamloops Lake. The Coldwater fault was active in Tertiary time.

Views to the northwest show the undulating surface of the Interior Plateaus region; the highest parts here reach about 1,700 metres. Gravel and sand deposits representing an abandoned late-glacial delta lie near the southernmost of the two exits from Highway 5 to Merritt.

Valley Basalt at Merritt

Highway 5 crosses the broad Nicola River valley between the two Merritt off-ramps and remains close to the valley floor for a few kilometres. Just

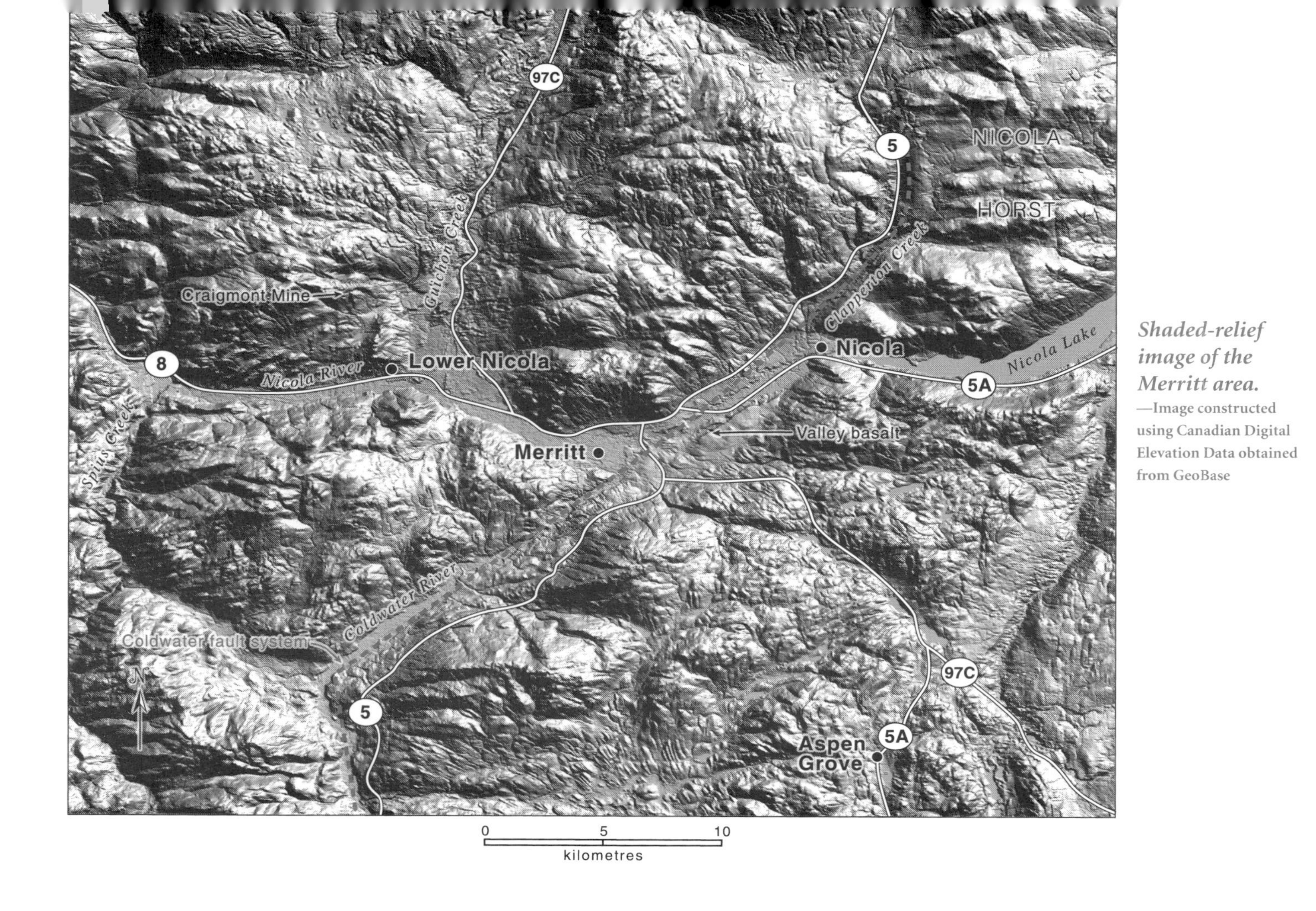

Shaded-relief image of the Merritt area. —Image constructed using Canadian Digital Elevation Data obtained from GeoBase

Valley basalt, geologically young lava flows, forms a treed bluff east of Highway 5 just north of Merritt. —R. Turner photo

north of Merritt, and visible across the grassy fields on the valley floor east of the highway, is a treed bluff below a nearly flat-topped bench. The bench is made of horizontal sheets of olivine-bearing basalt, one of which has provided a mid-Pleistocene age of about 500,000 years. Similar basalt is exposed on the northwestern side of Nicola Lake about 10 kilometres east of Highway 5, and on the eastern side of Courtney Lake about 20 kilometres to the southeast, as well as at several less-accessible sites in between. With one exception, all occurrences are near the present valley floors, indicating the valleys had already been deeply cut by mid-Pleistocene time. These are some of the youngest volcanic rocks in the southern interior of British Columbia, and in older geological reports they are referred to appropriately as "valley basalt," in contrast to the older, mainly Miocene "plateau basalt."

Nicola Horst

North of Merritt, Highway 5 follows the sharply incised valley of Clapperton Creek and climbs steadily up a gradient of about 660 metres in 15 kilometres before reaching a broad, shallow valley at the rolling upland surface at an elevation of about 1,400 metres. The numerous roadcuts are in highly fractured, mainly green and locally red volcanic rocks of the Triassic Nicola group.

Here, Highway 5 runs near the surface trace of the Coldwater fault, or as this segment of the fault is sometimes called, the Clapperton fault. This is a normal fault whose western side is down-dropped relative to the eastern side. West of Highway 5, rocks on the western side of the fault are little metamorphosed, contain fossils in places, and have never been buried deeper than a few kilometres. However, the eastern side of the fault, east of Clapperton Creek, contains granitic rocks of the 65-million-year-old Nicola batholith, which has cooling dates of 60 and 49 million years, indicating that it cooled in Paleocene to Eocene time, presumably when it was uplifted by fault movement and exposed. Bodies of metamorphic rock within the batholith contain mineral assemblages that suggest the rocks were metamorphosed at depths of 12 to 15 kilometres.

Interpretation of east-west-oriented, deep seismic reflection profiles made across this area in 1988 suggest that the west-dipping Coldwater fault continues to depths greater than 20 kilometres but flattens out at depth. The eastern side of the batholith is similarly bounded by an east-dipping normal fault, although the displacement on that fault is probably less. The formerly deeply buried Nicola batholith and its associated metamorphic rocks form a *horst.* This term for an elevated structure bounded on both sides by normal faults is taken from the German word for "high" (as opposed to *graben,* the German word for "grave," used for a fault-bounded structurally depressed area). The structure is named the Nicola horst, and the bounding normal faults appear to be related to Tertiary stretching of the crust, the effects of which can be recognized eastward from this area almost as far as the Alberta border in the southern Foreland Belt.

Interior Plateau Topography

Upon reaching the plateau, Highway 5 follows broad valleys within the upland at altitudes ranging between 1,250 and 1,450 metres and avoids the hilltops, which attain elevations of 1,500 to 1,900 metres. The highway passes through pine and spruce forest that hides all but a few of the many lakes. The roadcuts tend to be shallow, and long stretches of the road have no visible bedrock. The gently sloping ground received and retained a blanket of glacial deposits that cover the bedrock.

Descent to Kamloops

A viewpoint where truckers are invited to stop and test their brakes before the steep descent toward Kamloops is about 10.5 kilometres north of the Logan Lake–Lac Le Jeune off-ramp. In clear weather the crest of Porcupine Ridge, capped with Miocene lava, is visible on the distant skyline to the north. It is about 30 kilometres north of Kamloops Lake and is 1,862 metres above sea level. "Plateau basalt" lava flows of middle Miocene age, about 15.5 million years old, form the gently rolling plateau surface that caps the ridge. The original thickness of basalt here was about 100 metres. However, about 10 kilometres south of Porcupine Ridge, the base and top of the Miocene lava lie about 300 metres lower than their counterparts at the ridge. It is possible that since middle Miocene time, the lava on Porcupine Ridge has been uplifted by 300 metres relative to the lava to the south. If so, the lava should be flexed or faulted somewhere in between, but alas, exposures are so rare in the intervening space that confirmation of this is still lacking.

For the next 13 kilometres, Highway 5 descends about 500 metres. Visible to the east is Sugar Loaf Mountain, a prominent steep-sided knoll made of latest Triassic rocks of the Iron Mask batholith.

North Thompson River Valley

In the 20 kilometres between Kamloops and the settlement of Heffley Creek, Highway 5 runs along the east side of the roughly 2-kilometre-wide floodplain of the North Thompson River. The hillsides bounding the valley are faulted and fractured Triassic sedimentary rocks that mostly belong to the eastern sedimentary part of the Nicola group. North of Heffley Creek, the floodplain narrows, and after another 25 kilometres, the North Thompson River valley narrows at Fishtrap Rapids, where the river flows across bedrock. This is near the probable northern limit of the extensive Glacial Lake Thompson.

At Skull Hill, north of Fishtrap Rapids, a body of black lava superficially resembles a young, late Tertiary or Pleistocene valley basalt in that the top end of the lava is at the mouth of a hanging valley, and the present lava surface appears to bulge outward from there down the steep wall of the North Thompson Valley. However, its isotopic date is about 50 million years, hence the location of its upper end in a valley—which is common in young basalt flows such as those near Merritt—should be regarded as coincidental. The bulge is merely the product of the basalt being more resistant to erosion than adjacent rocks.

North of about 40 kilometres north of Kamloops, and north of McLure and mostly away from the river, are dark sedimentary rocks derived from mainly volcanic sources, limestone lenses up to 4 kilometres long, and lesser amounts of volcanic rock. Small-scale folding and shearing are widespread, and metamorphism is more intense than that near and south of Heffley Creek. It is perhaps for these reasons that few fossils diagnostic of age have been found in the belt of rocks extending northwest between McLure and Louis Creek. However, the range of rock types suggests they are of late

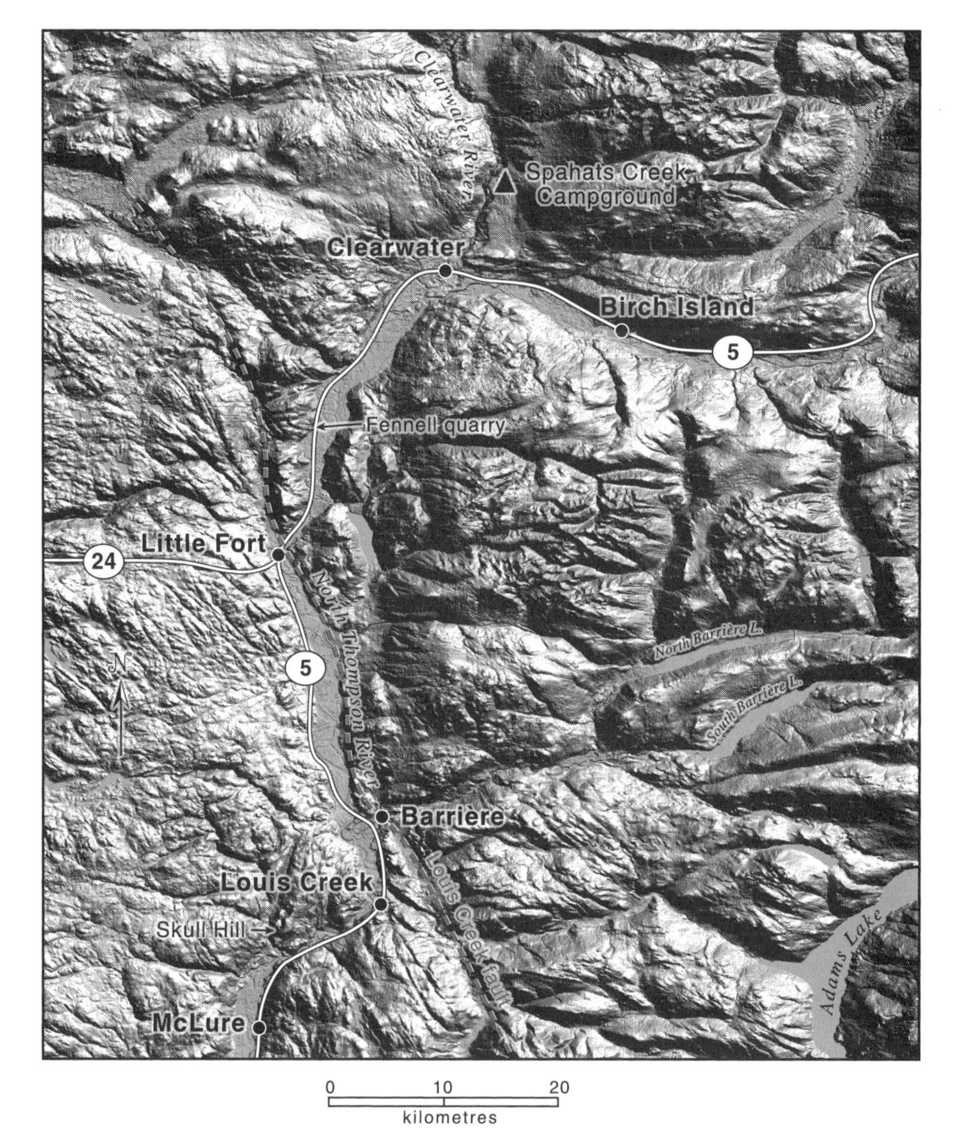

Shaded-relief image of topography along the North Thompson River between McClure and Clearwater. —Image constructed using Canadian Digital Elevation Data obtained from GeoBase

Paleozoic age and related to the Harper Ranch group east of Kamloops and north of Highway 1 or, less likely, to the early Mesozoic Nicola group.

East of Highway 5, and running south-southeastward from near the villages of Louis Creek and Barrière, the large Louis Creek fault extends along the Louis Creek valley southward toward Pritchard on Highway 1. The fault separates generally less-metamorphosed late Paleozoic and younger rocks to the west from more-metamorphosed and generally older rocks to the east. The fault delineates the boundary between the Intermontane Belt and the Omineca Belt in this area.

Highway 5 crosses to the west side of the North Thompson River just northeast of Barrière, and between here and Little Fort, about 30 kilometres north, the North Thompson River valley is wider and alluvium covers its floodplain. The broad alluvium-filled areas in the valley floor mark sites of lakes that formed along the ice margin at the end of the last ice age. Silt deposited in a glacial lake covers the lower slopes alongside the floodplain and in places is deformed, presumably by late movement of adjacent ice. Bedrock exposures along the road are rare and consist mostly of Eocene volcanics.

Slide Mountain Terrane

About 20 kilometres north of Little Fort on Highway 5, a quarry on the west side of the highway exposes black to dark green basalt containing remnants of pillows. You can find the pillows by careful examination of the rocks in the quarry. Look for their curved outlines, most clearly seen on freshly broken surfaces. The rocks belong to the Fennell formation, which is present over an area about 15 kilometres wide and 60 kilometres long, mostly east of the North Thompson River.

The Fennell formation is part of the Slide Mountain terrane. Like other pieces of the terrane to the north, it consists mainly of basalt, bedded chert, gabbro, and minor amounts of serpentinite, all of which are rock types characteristic of ocean basin floors. Chemical analyses suggest that the basalt is similar to that forming today on mid-ocean ridges. Microfossils extracted from interlayered chert beds indicate that the rocks range in age from early Carboniferous to middle Permian.

The rocks differ significantly from the Devonian to Permian rocks of the Harper Ranch group passed on Highway 1 east of Kamloops, which, although of a similar age range, probably formed in an island arc setting. In contrast, rocks of the Slide Mountain terrane may have originated in a back-arc basin that lay between the Harper Ranch island arc and the old continental margin. The Harper Ranch rocks and probably the Slide Mountain rocks were overlapped by the early Mesozoic Quesnel arc.

Repetition of fossils of the same age within the Fennell formation shows that slices of the formation were stacked one upon another by thrust faults that apparently moved the rocks from west to east. The rocks generally have experienced only mild metamorphism, except where intruded by granitic plutons. The time of thrust-faulting, based on evidence from near Kootenay Lake in southeastern British Columbia and from north-central British Columbia, possibly was at the end of Paleozoic time, thus predating formation of the early Mesozoic Quesnel island arc.

Pillows are outlined by curved shapes in basalt of the Fennell formation, part of the Slide Mountain terrane, in a roadside quarry on Highway 5 about 20 kilometres north of Little Fort.

Eocene Coal

Eocene sedimentary rocks, ranging from conglomerate to shale and containing coal, occur in a few places along the North Thompson River and in the lower parts of some of its tributaries. The strata dip rather uniformly at 30 to 50 degrees to the east and southeast, but the occurrences are small and isolated from one another, so the thickness and amount of coal are probably limited. Some of the Eocene exposures are within tightly constrained valleys cut as much as 600 metres deep into older rocks, and like so many of the other early Tertiary strata in southern British Columbia, probably are bounded on one or both sides by faults. The coal seams, some more than 1 metre thick and close to the Canadian National Railway, were investigated in the hope that they could meet the need for locomotive fuel and eliminate the long haul of coal from Alberta, but this idea appears to have been abandoned in the early 1920s.

Clearwater Basin and Wells Gray Provincial Park

Near Clearwater, located at the confluence of the North Thompson and Clearwater Rivers, Highway 5 enters a topographically lower area that extends northward for another 50 to 60 kilometres. This area is home to more than a score of young volcanic vents, cinder cones, and associated valley-filling lavas. The volcanic rocks have disrupted the pre-existing stream drainages and thereby created some spectacular waterfalls. The boundaries of Wells Gray Provincial Park, whose southern tip is about 25 kilometres north of Clearwater, were established to include most of the young volcanic features and the alpine wilderness drained by the Clearwater River.

The young lavas are consistently basaltic, although they do show slight but consistent differences in composition from flow to flow and from vent to vent. The lavas range in age from probably more than 2 million years old to less than 7,500 years old, and span at least three glaciations. Most of the lava was deposited on dry land, flowing down valleys; individual flows are rarely more than 10 metres thick. Pyroclastic vents produced steep-sided cones with or without flows. Some lava flowed into lakes dammed either by older lava piles or by glacial ice, forming pillow lavas, breccias with scattered pillows, or breccias with fine-grained fragments commonly with yellow alteration. Lava surrounded and dammed by glacial ice may have been prevented from flowing downslope; it solidified into flat-topped caps or irregular mounds.

The origin of this volcanic activity is not clear. The rocks are the eastern tip of a roughly east-west-aligned chain of volcanic rocks located near 52 degrees latitude that cross much of British Columbia, with the oldest rocks on the coast and the youngest in this area. A current hypothesis is that, as in Yellowstone, these rocks may reflect the westward movement of the continent across a *hot spot*, a localized upwelling of hot mantle material.

The traveller with an hour or so to spare can drive north from Clearwater and east of Dutch Lake, on a paved but unnumbered road for about 10 kilometres, to one of the southernmost accumulations of basalt and its associated waterfall at Spahats Creek Campground at the southern end of Wells Gray Provincial Park. In addition, a few days spent in the park can be richly rewarding for those with an interest in volcanic landforms. The traveller with little time to spare will have to be content with a distant view of the terrace-capping flows that can be seen 4 to 5 kilometres southwest of Clearwater and about 100 metres above Highway 5.

Uranium Ore at Birch Island

The settlement of Birch Island lies on the south bank of the North Thompson River 11 kilometres east of Clearwater and a few kilometres from the Rexspar uranium deposit. The petroleum crisis of the 1970s generated a need for alternate sources of energy. However, a hue and cry against mining of uranium ores in British Columbia was catalyzed by a proposal to dump mill tailings on the North Thompson River floodplain, where the tailings possibly could be redistributed during high river flows. In addition, the location was close to the Birch Island School, which provided grounds for concern that radon gas would affect the students. The Royal Commission on Uranium Mining in British Columbia was set up in response to these complaints, among others. Before its mission was completed, a seven-year moratorium on mining of uranium ores was proclaimed. By the end of the seven-year period the demand and price for uranium had dropped, and so the incentive for mining this relatively small and low-grade deposit vanished.

For the next 95 kilometres, Highway 5 continues eastward and northward to Blue River across the topographic division called the Shuswap Highlands, to reach the western margin of the Monashee Mountains. For much of this distance, Highway 5 passes roadcuts in mostly middle Jurassic and mid-Cretaceous granitic rocks, but in places there are high-grade metamorphic rocks containing the aluminum silicate minerals sillimanite and kyanite. The metamorphic rocks probably are derived from late Proterozoic and early Paleozoic sedimentary rocks deposited on or near the old continental margin.

HIGHWAY 5A
Princeton—Kamloops
183 kilometres

See the maps on page 225 for the Princeton Basin and on page 247 for the Merritt area.

Princeton sits at the confluence of the Tulameen and Similkameen Rivers and was called Vermilion Forks by Hudson's Bay Company fur traders in the early 1800s. First Nations people came to the banks of the Tulameen River to collect red ochre formed from shale or siltstone baked by fires in adjacent coal seams. The name Vermilion Bluffs refers to a locality in a 60-metre-high cliff on the north bank of the Tulameen River about 5 kilometres northwest of Princeton. Red-baked siltstone also occurs in lower slopes on the south side of the Tulameen River at the western limit of the Similkameen River floodplain. Unfortunately, part of this exposure has been covered with cement to stabilize it, but a part still survives along the lane in the centre of Princeton, a half block west of Bridge Street and a half block north of Billiter Street.

Highway 5A begins just east of the Highway 3 bridge over the present Similkameen River. It heads north and crosses the Tulameen River. After passing kettles, depressions formed by the melting of remnant glacial ice blocks buried in gravel, the highway climbs to the level of a plain of sediments deposited by glacial meltwater. The bedrock here, Eocene sedimentary rock in the upper part of the Princeton Basin, is soft and therefore poorly exposed. Green volcanic rock of the Triassic Nicola group is exposed along Highway 5A about 9 kilometres north of Princeton and about 100 metres north of its junction with the Summers Creek Road. It is overlain unconformably to the north and south by Eocene conglomerate, sandstone, and shale. The Eocene beds are in turn overlain by late Pleistocene gravel, and it is fairly difficult to distinguish weathered Eocene conglomerate from late Pleistocene gravel.

Major Meltwater Channel

For the next 25 kilometres to the north, Highway 5A follows a narrow, steep-sided valley that is incised to a depth of 250 to 500 metres below the upland surface. The valley contains several lakes, each dammed by alluvial fans deposited from side streams. The valley was carved by a large meltwater stream in glacial times, but now the main stream, Allison Creek, is too small to deepen its channel and carry away the alluvium. In places, the dams are tens of metres above the level of the lakes that lie immediately upstream of them, so there is no surface outlet for the water. Such lakes may fluctuate markedly with seasonal changes in water supply and with losses by underground seepage and evaporation. The uppermost 2 kilometres of the valley was the part most actively deepened at the end of the last ice age and is particularly steep and rocky.

North of Aspen Grove, Highway 5A joins Highway 97C, and together they drop to Merritt, which is at the confluence of the west-flowing Nicola River and north-flowing Coldwater River. At the end of the last ice age, water collected in the valley behind a residual tongue of ice that occupied the lower Nicola River valley to the west. The lake, called Glacial Lake Nicola, drained to the northeast along the valley containing the upper Nicola River, and across the divide into Campbell Creek, which flows north-northeastward and joins the South Thompson River about 17 kilometres east of Kamloops.

Type Section of the Nicola Group

In geology, a *type section* is the location where a stratigraphic section is first described in some detail and the place name provides the formal name for a rock unit. Ideally, a type section is as representative as possible of the rock types within a rock unit and their stratigraphic order. Even though later geological mapping may discover more-complete and better-preserved sections, the original description and name generally are retained. In his 1879 description of Triassic rocks in southern British Columbia, G. M. Dawson clearly indicated that his choice of a type section for rocks that he called the Nicola series, and which we now call the Nicola group, was located on the south side of the Nicola River valley between the outlet of Nicola Lake and the mouth of Guichon Creek, which flows into the Nicola River 10 kilometres northwest of Merritt. Dawson noted, however, that apart from a limestone body at the eastern end of the section, the rocks appear to be entirely volcanic. They include volcanic breccia, tuff, and lava, some of which contain *amygdules*, gas bubbles filled with siliceous or calcareous minerals. Elsewhere in the section it is possible to find shale, siltstone, and sandstone made up dominantly of volcanic debris.

Since Dawson's time, the Nicola group rocks have been mapped over a wide area, not least because they and associated intrusions are potential hosts for copper and precious metal deposits. The relative abundance of volcanic and sedimentary rocks varies laterally, and far more fossils have been collected to date the rocks. Geologists now subdivide the Nicola rocks into three volcanic units with different compositions that overlap in time, and to some extent, in space. A mainly sedimentary unit, still part of the Nicola group, lies east of the volcanic rocks, east of a north-south line between Kamloops and just east of Princeton. Dawson's selection more than 120 years ago of a type section near Nicola Lake appears to be a farsighted choice in that it contains all three volcanic units, albeit faulted against one another and somewhat metamorphosed, with widespread green minerals such as chlorite and epidote. Dawson noted one fault but recognized the potential for more and very tentatively suggested a thickness of 2.7 to 3 kilometres, with neither the top nor the bottom of the section exposed, a thickness that is close to most modern "guesstimates."

The bluffs northwest of Nicola Lake are earliest Tertiary granitic rocks at the south end of the 65-million-year-old Nicola batholith, which is exposed in the Nicola horst and whose nature is discussed in **Highway 5: Coquihalla Pass—Little Fort and Beyond.**

Shoreline of Glacial Lake Nicola

For a close-up view of a late-glacial shoreline, turn east up Peter Hope Road, whose junction with Highway 5A is 9.5 kilometres north of the head of Nicola Lake and 3 kilometres southeast of the south end of Stump Lake. Drive up this road for about 0.4 kilometres and look for a gently sloping shelf, which is an old lakeshore shaped by waves in Glacial Lake Nicola,

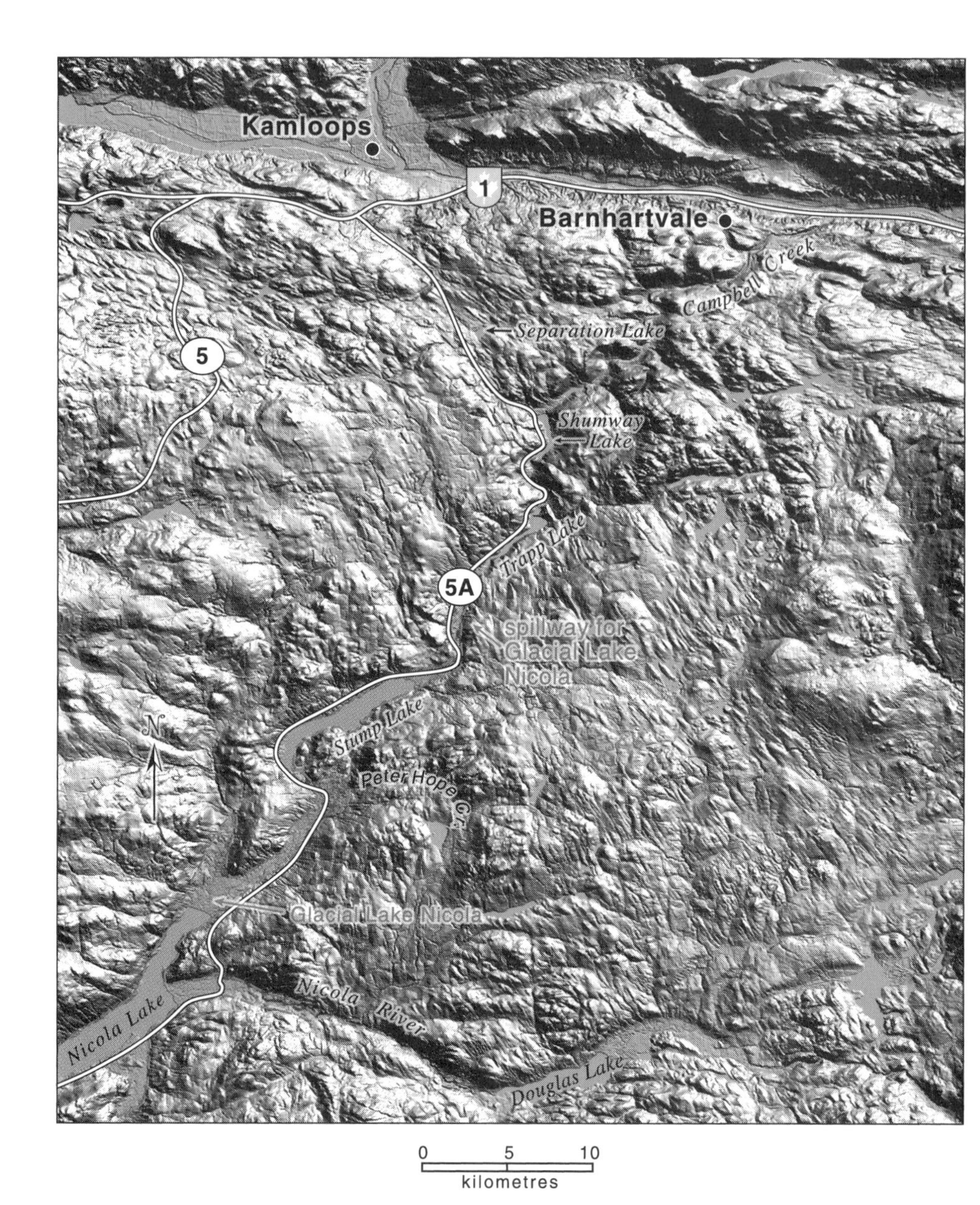

Shaded-relief image of topography along Highway 5A, showing the basin of Glacial Lake Nicola and its spillway. —Image constructed using Canadian Digital Elevation Data obtained from GeoBase

which filled the upper Nicola River valley. Here it is but a few metres wide, but it extends for several kilometres along the contour. Directly above this shelf is a low scarp only a metre or so high that was undercut by waves on the former lake. Below the shoreline shelf, another relatively steep strip was formed by the subaqueous deposition of gravel eroded by undercutting the beach scarp. You can see other examples of this shoreline on the grassy slopes at about this same elevation of approximately 760 metres. The present surface of Nicola Lake and its outflow is at an elevation of about 620 metres, which means that Glacial Lake Nicola was at least 140 metres deep.

Spillway of Glacial Lake Nicola

The low point on the drainage divide that separates Stump Lake from Campbell Creek was the site of the short-lived spillway of the ice-dammed lake that filled the upper Nicola River valley. The spillway briefly carried a stream, much larger than its present flow, capable of cutting the narrow, rock-walled canyon containing Trapp and Shumway Lakes. After the ice dam in the lower Nicola valley melted, the spillway cutting ceased. The tributaries of Campbell Creek, however, continued to move debris onto the valley floor, but the now-shrunken creek could no longer remove it. The debris accumulated and formed the barriers that dammed Campbell Creek into the present series of lakes. In its northern 15 kilometres, Campbell Creek displays continued signs of postglacial ponding by side streams, albeit less pronounced than nearer its head.

An abnormally large delta deposit rests near Barnhartvale, about 15 kilometres east-southeast of Kamloops and just above the confluence of Campbell Creek with the South Thompson River. Its presence indicates that the lake in the upper Nicola River valley was contemporaneous with Glacial Lake Thompson, which occupied the South Thompson River valley and in which was deposited the buff-coloured, terrace-forming Kamloops white silt. The water that spilled down Campbell Creek flowed into Glacial Lake Thompson and deposited sediments into the large delta.

North of where Highway 5A climbs out of the Campbell Creek valley, the road is close to the rolling upland surface, which here is mantled with glacial debris. Northeast of Separation Lake, glacial ice shaped the debris into the streamlined forms called *drumlins*, elongate mounds that form below a glacier. Their streamlined shapes with downstream "tails" show the south-southeasterly direction of ice flow.

A good view to the north is available on the northward route for much of the last 7 kilometres to Highway 1. Porcupine Ridge, visible on the skyline to the northwest on a clear day, is capped with Miocene lava.

HIGHWAY 8
Merritt—Spences Bridge
65 kilometres

See the maps on page 247 for the Merritt area and page 197 for the Spences Bridge area.

Merritt sits in a basin at the confluence of the Nicola and Coldwater Rivers. Though the town was known to be located on Tertiary coal deposits, it was not linked to the main line of the Canadian Pacific Railway until 1907, and then only because Robert Dunsmuir had shut down the coal mines of Vancouver Island, an action that denied the Canadian Pacific Railway its main source of fuel in southwestern British Columbia. Most of the 2.7 million tonnes of coal produced from the Merritt coalfield between 1906 and 1945 were from the Middlesboro Mine, located 1 kilometre south of Merritt's town centre. The deposits are in Eocene sedimentary rocks that mostly lie beneath the valley floor. The sedimentary rocks overlie Eocene volcanic rocks, exposed in the hills southwest of town. This stratigraphy is directly comparable to that at Princeton, although here the main fault, buried beneath the valley floor, is the Coldwater fault.

With diminished demand for coal in the 1940s and 1950s and the termination of mining here, Merritt became dependant on cattle ranching and lumbering until discovery of the Craigmont copper deposit, 14 kilometres northwest of town, revived its economy in the 1950s. The closing of the Craigmont open-pit mine in 1982 again hurt Merritt, but construction of Highway 5 in 1986 and its connections in the following few years brought another renewal as a retirement and recreational community.

Craigmont Mine

The Craigmont Mine lies midway up the eastern slope of the Promontory Hills, on the west side of lower Guichon Creek about 14 kilometres northwest of Merritt. The mine dumps are visible on the east-facing hillside about 6 kilometres north of Lower Nicola, a village on Highway 8 about 9 kilometres west of Merritt.

The Craigmont orebody was unknown until 1954, although an earlier prospect on the hillside 500 metres below it had been staked and trenched before 1939. Because the orebody contained a substantial amount of the iron mineral magnetite, it produced an anomalous high on aeromagnetic maps that had been made from measurements from an airborne magnetometer. This prompted a surface magnetometer survey and a program of soil sampling. Exploration drilling that followed in 1957 found ore-grade copper in four of the seven deep holes. By 1959, 15,000 metres of diamond drilling, 2.5 kilometres of underground workings, and 75,000 cubic metres of stripping had been completed. With the ore reserves thus assured by exploration, the design of an open pit and the planning and construction of a mill followed. Mining started in 1961, initially in an open-pit mine; underground mining continued until 1982. Since then the tailings from the mill have been treated to reclaim magnetite.

Flows, breccias, shale, and sandstone of the late Triassic Nicola group host the orebody. The copper ore is in skarn, a rock formed by alteration of limestone and limy tuff in the Nicola group during intrusion of the southern tip of the 210-million-year-old Guichon Creek batholith, which is exposed a few hundred metres to the northwest. Mineralization consists of magnetite, hematite, and the copper-iron sulfide mineral chalcopyrite, with the principal ore values in the last mineral. The ore is found over a length of 900 metres and a vertical range of 600 metres. It is limited on the southwest and northwest sides by faults, and on the north by the Guichon Creek batholith.

Rock Hoodoos

Two kilometres west of the bridge over Guichon Creek at Lower Nicola, and about 20 metres above the north side of the highway, is an outcrop of pale grey to purplish Eocene lava containing small, elongate black crystals, or "needles," of hornblende. The lava is weathered and eroded into hoodoos, a series of steep-sided pyramids. Normally the term *hoodoo* is applied to pillars developed in weakly cemented Quaternary sediment containing scattered boulders that have sheltered lower parts of the pillar from erosion. Here, west of Lower Nicola, weathering and erosion have apparently proceeded faster along more closely fractured zones, leaving the less-fractured rock standing in relief.

Spences Bridge Group in the Lower Nicola River Valley

For most of the distance between Guichon Creek and Spences Bridge, the lower Nicola River valley is eroded in volcanic rock and lesser amounts of sedimentary rock of the early Cretaceous Spences Bridge group. The rocks were part of a continental volcanic arc that developed on what was then the western edge of the North American continent about 105 million years ago, in mid-Cretaceous time. The Spences Bridge group overlaps both the Quesnel and Cache Creek terranes.

The lower part of the Spences Bridge group here is dominated by pale yellow to green and brown volcanic rocks of andesitic composition ranging from flows to pyroclastic rocks, although minor red to pink rhyolite is also present. The upper part of the group, seen in roadcuts opposite the mouth of Spius Creek, which flows into the Nicola River from the south, consists of much more uniform, dark reddish brown, basaltic andesite flows that characteristically contain amygdules, mineral fillings in holes left by gas bubbles in the flows. The mineral fillings generally are soft, pink or white zeolites, hydrous aluminum silicates.

In places, such as about 40 kilometres northwest of Merritt and visible across the Nicola River from the highway, well-bedded tuffaceous sandstone and shale with some coal beds separate the upper and lower volcanic units. Microscopic pollen and spores from these rocks indicate that the beds were deposited toward the end of early Cretaceous time, an age confirmed by an isotopic date of 105 million years from the lava. The lower

A swarm of dikes intrude andestic tuff and breccia of the Spences Bridge group, on the north side of the lower Nicola River valley.

lavas and breccias erupted from coalescent *stratovolcanoes*, volcanic edifices built of alternating layers of lava and pyroclastic debris that exploded from the vent so that the thickest deposits lie closest to the vent. The more-fluid basaltic andesite flows in the upper part of the Spences Bridge group formed *shield volcanoes*, broad, low domes built by runny lava flows. At several places, but best seen in the lower end of the valley near Spences Bridge, dike swarms within the volcanic rocks fed lava flows higher in the volcanic pile.

The V-shaped cross section of the lower Nicola River valley downstream of Spius Creek is unexpected in view of its trend nearly parallel to the direction of ice flow during at least the later part of the last glaciation. Perhaps the valley was shaped by rapid outflow of water from Glacial Lake Nicola after the ice dam at its western end broke. The volume of unconsolidated deposits here also appears to be low in comparison to many other valleys of the western Interior Plateaus, although they are prominent in places in the lower part of the valley.

HIGHWAY 24

100 Mile House—Little Fort

97 kilometres

Between 100 Mile House and Little Fort, Highway 24 crosses an upland that is part of the Interior Plateaus, here about 1,100 metres high. For the western half of the route, the bedrock, or what little of it is exposed, consists of flows of Miocene basalt. Similar flows located 10 to 40 kilometres to the west are about 8 million years old and those to the southeast or east are about 15 million years old. Stocks of 6- to 8-million-year-old gabbro, a wholly crystalline, slowly cooled, intrusive equivalent of the basalt, form Lone Butte and Huckleberry Butte, two distinctive hills north of the highway. The gabbro stocks filled volcanic vents through which the lavas first reached the ground surface.

The plateau in the western part of Highway 24 extends to the northwest nearly as far as the Fraser River. It is very flat and perhaps is the best example that we see of the topography on the Interior Plateaus. The two gabbro buttes rise less than 100 metres above their immediate surroundings yet stand out clearly from a distance of 20 kilometres. The western half of the highway rarely rises above the 1,150-metre contour and nowhere drops below 1,100 metres. It is close to the height of the drainage divide between Bridge Lake and Lac des Roches, east of which water flows into the North

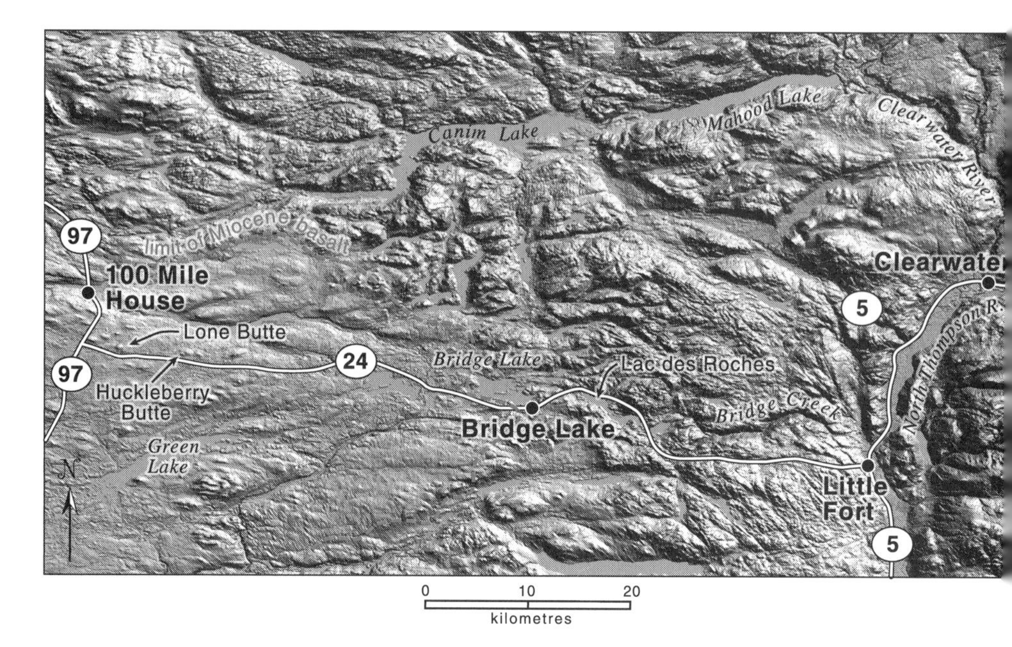

Shaded-relief image of topography along Highway 24 between 100 Mile House and Little Fort. —Image constructed using Canadian Digital Elevation Data obtained from GeoBase

Thompson River at Little Fort, and southwest of which water flows into the Bonaparte River to eventually join the Thompson River at Ashcroft. These characteristics support the idea that the topography of the region is due largely to outpouring of lava in Miocene time, and that the effects of subsequent erosion are minimal, consisting primarily of scattered basalt boulders from the underlying bedrock and glacial landforms that have a relief of probably only 10 to 20 metres.

Bridge Lake, at an elevation of 1,133 metres, marks the eastern limit of the Miocene basalt, and the highway east of here descends through a relatively thick sequence of lava exhibiting well-defined columnar jointing. Farther east, older rocks are exposed and include Eocene lava, early Mesozoic granitic rocks, and Triassic volcanic and sedimentary rocks of the Nicola group.

About 11 kilometres east of the easternmost exposure of Miocene basalt and 10 kilometres east of Bridge Lake Provincial Park is a viewpoint from which you can look southward across a grassy field to Lac des Roches and beyond. The plateau surface rises from about 1,350 metres above sea level near the north end of Lac des Roches to 1,600 metres about 20 kilometres to the southeast. For most of this distance, the plateau is made of Eocene

The knoll capped by the tower is Lone Butte, a volcanic vent filled with 6- to 8-million-year-old gabbro.

volcanic rocks that in places rest on early Mesozoic granitic rocks but elsewhere are faulted against them. The absence of the 6- to 8-million-year-old Miocene basalt here may be because the land stood too high at that time to be overtopped by the lava flows.

HIGHWAY 97
Osoyoos—Vernon
265 kilometres

Highway 97 runs along the Okanagan Valley (spelled Okanogan in its southward continuation in Washington). Its southern end in Canada is at an elevation of 275 metres, the lowest point in the southern Canadian Cordillera east of the lower Fraser River drainage. The natural, non-irrigated valley floor is semidesert. The valley, from south to north, contains Osoyoos, Vaseux, Skaha, and Okanagan Lakes, and the availability of water, together with the sunny and generally mild climate, has provided a great incentive for fruit growers. Vineyards have flourished in the last few years.

Vaseux Gneiss
About 30 kilometres north of Osoyoos, near Vaseux Lake, the Okanagan Valley narrows where it is constricted by bluffs of hard Vaseux gneiss. The layering in the gneiss is mainly flat-lying or shallowly dipping. A variety of

Shaded-relief image of the Okanagan Valley between Osoyoos and Vernon. —Image constructed using Canadian Digital Elevation Data obtained from GeoBase

Vaseux gneiss, exposed at the base of a cliff above a side road east of Highway 97 and just north of Vaseux Lake, contains lighter-coloured granitic layers and complex folds.

generally lighter-coloured and variably foliated granitic bodies intrude the gneiss, some lying in the plane of the gneissic foliation and others cutting acutely across it.

The gneiss has apparently undergone a long and complex evolution. Isotopic dates of different parts of the gneiss provide a variety of ages that range from late Mesozoic through early Tertiary, with hints that the oldest part may be Precambrian. Regardless of the age of the rocks, they seem to have remained buried to a great depth until early Tertiary time, about 50 million years ago.

Okanagan Valley Fault

The Okanagan Valley mostly follows the trace of an enormous system of west-side-down, interlinking normal faults, collectively known as the Okanagan Valley fault, whose southern part marks the boundary between the Intermontane and Omineca Belts. The fault system begins about 80 kilometres south of the international boundary and can be traced north along the valley to a point near Peachland west-southwest of Kelowna. It then takes a jog to the east to near Kelowna and then continues just east of the valley as far as Vernon, at the north end of Okanagan Lake. From there, it heads north-northeastward, within the Omineca Belt, to cross Highway 1 at Sicamous, where its continuation is called the Eagle River fault. The total length of the fault system is about 250 kilometres. Erosion by ice and water has exploited weaknesses in the rocks caused by faulting and has created the valley.

In early Tertiary time, the rocks on the western side of the fault were dropped down on a normal fault relative to those rocks on the eastern side. Regionally, rocks west of the valley are mainly little-metamorphosed late Paleozoic through Tertiary strata and granite. To the east are gneiss and granitic rocks of mainly Mesozoic and early Tertiary age and less-metamorphosed late Paleozoic, Mesozoic, and early Tertiary sedimentary and volcanic rocks and granitic intrusions, juxtaposed on other Tertiary normal faults with the high-grade metamorphic rocks.

In the southern Okanagan Valley, the normal fault is unusual in that the fault surface dips westward at a shallow angle of about 10 degrees. Here, the fault puts unmetamorphosed Eocene volcanic and sedimentary rocks and underlying weakly metamorphosed Mesozoic and Paleozoic strata on top of the gneiss.

To view the gentle westward dip of the fault surface, take the Oliver Ranch Road off the north side of Highway 97 about 2 kilometres north of Vaseux Lake. Drive up onto the vineyard-covered hilltop and look south. The surface above the bluffs of flat-lying gneiss on both sides of Vaseux

The Okanagan Valley fault viewed southward from the crest of Oliver Ranch Road. The land surface that dips from east to west (left to right) *into the saddle is close to the fault surface. Below it, in the footwall of the fault, is the Vaseux gneiss; above it, in the hanging wall of the fault and underlying the hill to the west* (right) *are low-grade metamorphic rocks capped by Eocene sedimentary and volcanic rocks.*

Lake dips westward at an angle of roughly 10 degrees into a saddle, above which are Paleozoic and Mesozoic low-grade metamorphic rocks and unmetamorphosed Eocene volcanic rocks. Directly west of here, the entire western side of the Okanagan Valley is made of Eocene volcanic rocks, and just north of here and east of the village of Okanagan Falls, Eocene rocks on the eastern side of the valley form the pink to brown bluff. The fault surface below here has been penetrated by holes drilled for mineral exploration.

Interpretation of deep seismic reflection sections made farther north across the Okanagan Valley suggests that the fault surface descends westward at a relatively shallow angle to depths of at least 20 kilometres. Based on its metamorphic grade and the youngest isotopic dates, the gneiss was as much as 20 kilometres below the surface during the early Eocene, about 50 million years ago, when the volcanic rocks were being extruded and deposited on the surface. Furthermore, the dip of the fault surface averages about 15 degrees. A simple calculation shows that the amount of down-dip movement on the fault may be as much as 80 kilometres! This kind of fault may also be called a *detachment fault*. Above it, in the hanging wall of the fault, and mapped in detail within the Eocene strata that lie between here and Yellow Lake on Highway 3A, are numerous small, steeply dipping normal faults that merge at depth into the detachment fault surface.

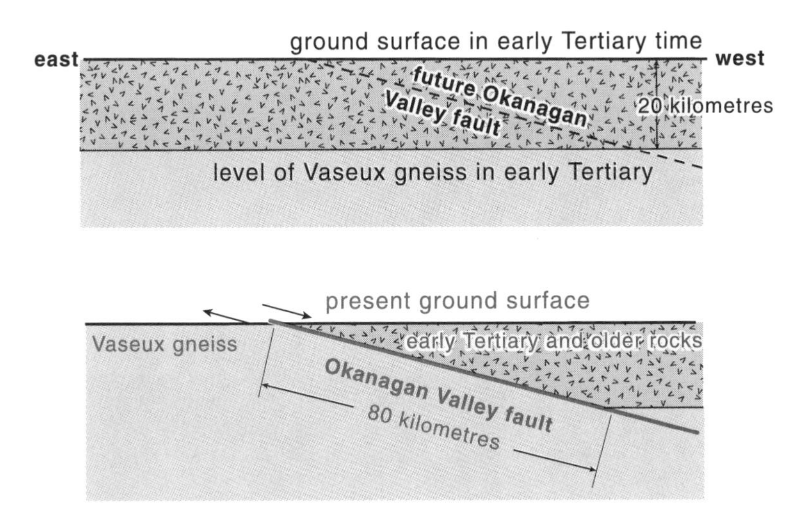

The amount of displacement on the Okanagan Valley fault near Okanagan Lake may be as much as 80 kilometres.

The Okanagan Valley fault is the westernmost of several very large normal faults that lie between here and the Rocky Mountains, about 200 kilometres to the east.

At Okanagan Falls, the road crosses to the western side of the Okanagan Valley. The many roadcuts along Highway 97 north of Okanagan Falls and north of the junction with Highway 3A are in rubbly, fragmental early Tertiary volcanic rocks that are bedded in places.

Roches Moutonnées

Just south of the town of Penticton, and visible on the eastern side of the north end of Skaha Lake, are a series of rounded, elongate hummocks that are smooth at their northern ends and on top, and irregular with bluffs at their southern ends. These distinctive forms, called *roches moutonnées*, French for "sheep rocks," show the erosional power of ice. Glaciers advancing from north to south down the valley smoothed off the upstream sides of the hummocks but plucked rocks away from their downstream sides.

Roches moutonnées lie on the eastern side of the northern end of Skaha Lake, just south of Penticton. —R. Turner photo

View of the eastern side of southern Okanagan Lake; the bluffs directly above the lake are silt and clay deposited in Glacial Lake Okanagan.

Glacial Lake Okanagan

Highway 97 crosses then recrosses the canal connecting Skaha and Okanagan Lakes on the western side of Penticton. Immediately northwest of town, the highway is nearly at lake level. Above the road and visible to the northeast across the lake are prominent, buff-coloured terraces formed of silt and clay deposited in Glacial Lake Okanagan.

Northward, the highway ascends these terraces. The greenish to buff bedrock here includes thick beds of boulders and other debris that slumped and flowed down the sides of Eocene volcanoes, as well as thinly bedded, finer-grained rocks with black coal layers. Where the road descends again to lake level, the roadcuts are in highly fractured, altered greenish granitic rock probably of early Jurassic age. The reach of Okanagan Lake between Peachland and Kelowna roughly follows the east-northeast trend of a fault that crosses the main Okanagan Valley fault, offsetting its surface trace for about 25 kilometres to the east. As a result, north of Peachland, near the junction with Highway 97C at Westbank, the road crosses buff-coloured, greenish, and locally red early Tertiary volcanic rocks down-dropped in the hanging wall of the Okanagan Valley fault. Southeast of here, across the lake, the higher glacially smoothed bluffs and hills are mainly gneiss in the footwall of the Okanagan Valley fault.

The tan, pink, and grey bluffs of rhyolitic to andesitic lava on Boucherie Mountain, rising over 400 metres west of Okanagan Lake, are the remnants of an Eocene volcanic complex.

Highway 97 crosses to the east side of Okanagan Lake at Kelowna and continues northward to Vernon. The bedrock for a distance of about 25 kilometres south of Vernon is in Jurassic and early Tertiary granitic rocks that underlie the Eocene volcanic rocks, close to the Okanagan Valley fault that here lies just east of the highway.

Northwest of Vernon, Highway 97 heads northwest toward Kamloops. The north-northwest-trending valley containing Falkland is located near the trace of another major fault system, which intersects the Okanagan Valley fault near Vernon and links up with the Louis Creek fault north of Highway 1. This fault system, here called the Bolean Creek fault, marks the boundary between the Intermontane and Omineca Belts. East of the fault are mostly metamorphic rocks, derived from early Precambrian and Paleozoic strata, and west of the fault are little-metamorphosed late Paleozoic and Triassic strata belonging to the Harper Ranch and Nicola groups, capped by Eocene volcanics. The nature of the fault system is not certain. It may be a Jurassic thrust- or reverse-fault system, with rocks to the east raised up and moved over those to the west, but this is obscured by a small amount of Tertiary movement on the fault system as well. Tertiary volcanic rocks are abundant west of the fault but scarce east of it. Permian limestone of the Harper Ranch group is exposed in a roadcut along the valley about 10 kilometres east of Falkland. There, the upper walls of the valley are massive reddish brown cliffs of flat-lying Eocene volcanic rock, which are crossed on the northernmost part of the highway.

HIGHWAY 97C
Okanagan Valley—Merritt
105 kilometres

See the maps on page 266 for the eastern part of the route and page 247 for the western part of the route near Merritt.

Highway 97C is a high-speed road with confining concrete barriers and few safe stopping places. The summit elevation at about 1,700 metres is one of the highest on any major highway in southern British Columbia—somewhat surprising as this is not a mountainous region, but rather an undulating upland surface eroded in rocks mostly older than Eocene age. The surface is well above the 1,100-metre elevation of the erosion surface that lies beneath the Miocene basalt to the north along Highways 24 and 97. This surface originally may have approximated the level of the pre-late Miocene surface, but along with other parts of the southernmost Intermontane Belt, may have been differentially warped upward in post-Miocene time.

West of the Okanagan Valley, Highway 97C ascends Trepanier Creek. For the first 5 kilometres or so the roadcuts are buff-coloured, white, pale green, and locally red volcanic rocks of Eocene age with some interbedded sediments. For about the next 20 kilometres, roadcuts are in fractured, greenish granitic rock of early Jurassic age. Near the top of the ascent, the smooth, flat skyline to the south is grass-covered waste dumps from the Brenda Mine, a large open-pit mine active from the 1960s to the 1980s. Molybdenum ore was extracted from fractures in the granitic country rock.

From the highest part of the road, there are good views northward over high, rolling, treed topography with small lakes that is typical of this part of the Interior Plateaus. The roadcuts here are mostly in dark green volcanic rocks, weathering to a rusty colour, with interbedded sedimentary layers that belong to the late Triassic Nicola group. Also along this part of the highway, Nicola strata are intruded by the northernmost tip of the middle Jurassic, 165-million-year-old Osprey Lake batholith, a large granitic body that extends south almost to Highway 3 near Hedley. Studies show an

View northward from near the summit on Highway 97C west of the Okanagan Valley. The bedrock of the high, undulating southern Interior Plateaus here is early Jurassic granitic rock intruding Triassic strata and plastered with glacially deposited sediments.

unusually wide variation in the isotope geochemistry of the batholith. This variation hints that the batholith may have originated by melting in the deep crust of both mantle-sourced volcanic island arc rocks in the Quesnel terrane and rocks derived from the westernmost attenuated parts of the old North American continent. In turn, this supports other lines of evidence that these terranes had been accreted to the old continental margin by middle Jurassic time.

From the summit, the highway begins a long descent and the vegetation changes from forests to the open grasslands typical of the lower elevations in this part of the Intermontane Belt. The roadcuts are mainly in Triassic volcanic rocks of the Nicola group, including banded, red and green, fine-grained volcanic sandstone and shale. Highway 97C joins Highway 5A at Aspen Grove and descends to Merritt, which lies in the basin near the confluence of the Nicola and Coldwater Rivers. Triassic rocks underlie the higher country east of the basin, and rocks of the early Cretaceous Spences Bridge group form the higher ridges west of the basin. Eocene volcanic rocks overlain by Eocene coal-bearing sandstone and shale underlie the lower hillsides and the valley floor of the topographic and structural basin containing Merritt. Young, 500,000-year-old basalt flows form a bench on the valley floor (**see "Valley Basalt at Merritt" in Highway 5: Coquihalla Pass—Little Fort and Beyond**).

HIGHWAY 97

Cache Creek—100 Mile House

113 kilometres

North of Cache Creek, Highway 97 follows the valley of the Bonaparte River. To the west, the hillsides are littered with rocks, which are not glacial erratics but blocks of harder rock in the Cache Creek mélange, weathered free of their softer matrix (**see "Cache Creek Mélange" in Highway 1: Lytton—Cache Creek**). The cliffs to the east are Eocene volcanic rocks, and the boulders on the eastern valley floor below the cliffs have fallen off them.

A brilliant yellow zone, visible on the west side of the valley north of the junction with Highway 99, is an area of Cache Creek rocks altered by former hot springs activity. A body of granitic rock, hidden beneath the floodplain of the Bonaparte River and discovered by drilling in a search for ore minerals, probably was responsible for the alteration. Highway 97 trends northwestward, climbing uphill past foliated sandstone and shale

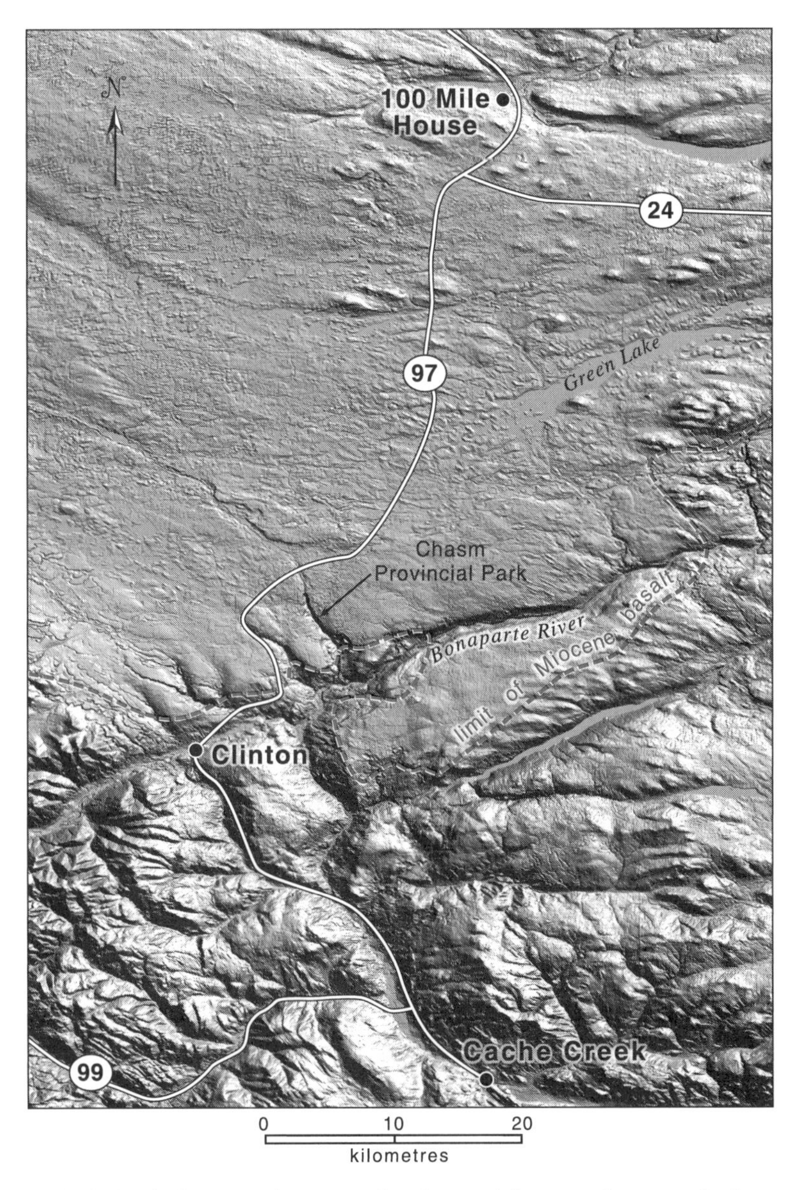

Shaded-relief image of topography along Highway 97 between Cache Creek and 100 Mile House. Note the subdued topography of the plateau of Miocene basalt north of Clinton. —Image constructed using Canadian Digital Elevation Data obtained from GeoBase

View south from the head of the Chasm, cut in Miocene basalt flows.

probably deposited in Jurassic time at the same time as the Ashcroft formation. North of Clinton, the road climbs on top of the surface of Miocene basalt flows that is the most typical landform of the Interior Plateaus.

The Chasm: A Late-Glacial Meltwater Channel

About 16 kilometres north of Clinton, a side road to the east leads to Chasm Provincial Park, where an enormous flood of glacial meltwater carved a valley into the southern edge of the plateau about 10,000 years ago. Leave your car in the parking area, and take the trail to the east that follows the fence above the 100-metre-deep valley.

On the other side of the fence, and visible from the head of the valley, are cliffs of nearly flat-lying Miocene lava flows. The valley ends abruptly in a steep slope and cliff at the north end below the trail, but its south end opens up into the lower country northeast of Clinton, at the southern limit of the Miocene lava flows. The valley gives a good idea of the rocks underlying the surface of the plateau.

Atop the Miocene Basalt

Once on the top of the flat plateau, exposures of the basalt are few and far between. Most of the basalt seen occurs as glacially disturbed boulders, weathering to a rusty colour, although roadcuts do expose a few flows.

A basaltic lava flow exposed in a roadcut along Highway 97 south of the junction with Highway 24.

Highway 97 follows the old wagon road to the Cariboo goldfields near Barkerville, east of Quesnel. Before the route along the Fraser Canyon was opened in the 1860s, one of the stopping places was the original 100 Mile House, located 100 miles from Lillooet, the beginning of the old overland route.

HIGHWAY 99
Highway 97—Fraser River
55 kilometres

Just west of its junction with Highway 97, Highway 99 crosses a farm road on the western edge of the Bonaparte River floodplain. In the 1860s this farm road was the Cariboo Wagon Road to the Cariboo goldfields, which lay about 300 kilometres to the north, east of the town of Quesnel. The large structure among the ranch buildings on the south side of Highway 99 was once a hotel on the wagon road and is worth visiting.

Highway 99 continues through the belt of Cache Creek mélange and passes roadcuts in locally reddish sandstone, shale, and conglomerate of mid-Cretaceous age that is made of pebbles and grains eroded mostly

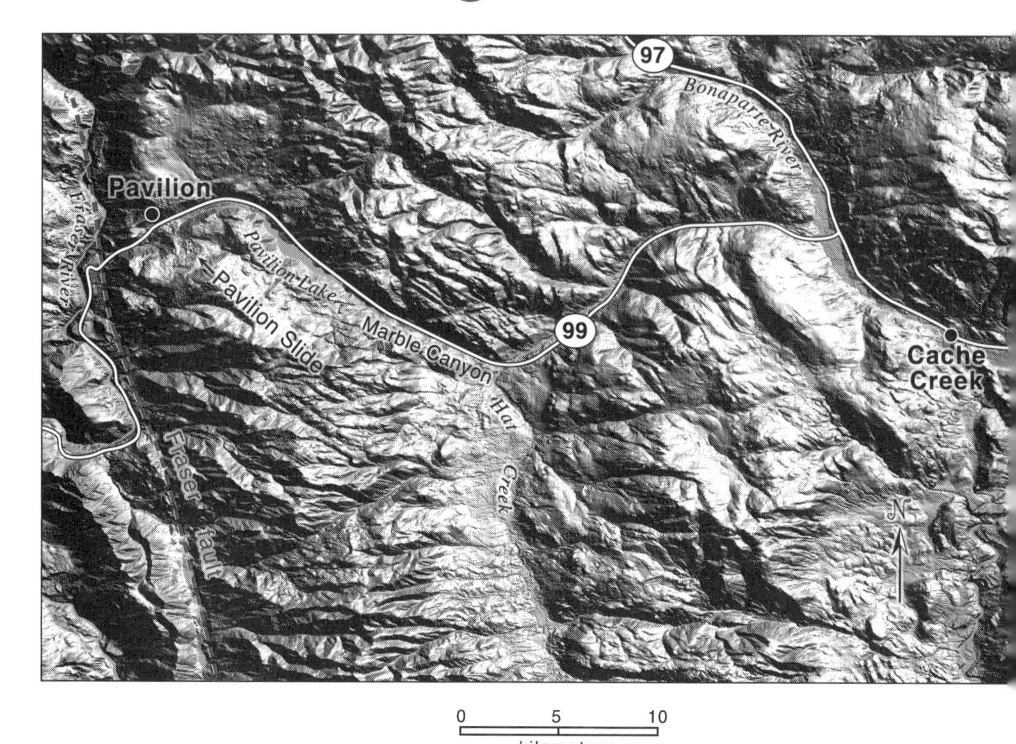

Shaded-relief image of topography along Highway 99 between Cache Creek and the Fraser River. —Image constructed using Canadian Digital Elevation Data obtained from GeoBase

from the Cache Creek terrane and deposited by rivers and in alluvial fans. These beds, which weather to a buff colour, are discussed in more depth in **"Faults at the Belt Boundary" in Highway 12: Lytton—Lillooet.**

Hat Creek Coal and Old Underground Fires

About 20 kilometres west of the junction with Highway 97, a road to the south follows Hat Creek. Two enormous deposits of *lignite*, a soft, brownish coal intermediate between peat and hard subbituminous coal, have been found beneath the floor of upper Hat Creek valley. The lignite and its accompanying beds are susceptible to erosion and are mostly buried beneath the valley floor. Most of the information on the size, shape, and quality of the lignite deposit comes from the nearly five hundred holes drilled by British Columbia Hydro in the 1970s to assess the feasibility of fuelling a large thermal power plant in this area. There are no immediate plans to exploit the coal deposit.

The coal beds consist of about 300 metres of middle to late Eocene strata, much of it lignite, with interbedded carbonaceous shale and lesser amounts of siltstone and sandstone. The strata are gently folded and cut by many steep faults. The structural depression that contains the Hat Creek beds probably formed at the same time as movement on the Fraser fault, between 47 and 35 million years ago.

The northwestern part of the lignite deposit burned at some time in the past and produced brick red to yellowish *clinker*. Spontaneous combustion, particularly where disturbances created openings admitting air, probably started the underground fires. As long as there was access to air and fuel, such fires could burn for decades. Where enough heat was generated, the adjacent rock melted and became brightly coloured clinker, which looks like lava but has a very different origin.

To see the clinker in a burnt area, drive 2.4 kilometres south along the creek, park, and walk southwest. Wade across Hat Creek and continue southwest for another 300 metres to a curious depression with a gently sloping floor that is almost entirely devoid of vegetation. The underground combustion of the coal and the collapse of its roof rocks probably formed the depression, known locally as "Dry Lake." You can find fused clinker on the hillside to the north.

Marble Canyon

At its junction with the Hat Creek road, and for about 10 kilometres to the west, Highway 99 runs through Marble Canyon, carved in an enormous mass of limestone, with steep walls that rise up to 500 metres above the valley floor. The limestone, which forms a body about 90 kilometres long in a north-south direction and up to 20 kilometres wide, is a part of the Cache Creek terrane. Similar limestone bodies are found in all other parts of the Cache Creek terrane as far north as southern Yukon. Such limestone bodies also occur in several places, such as Japan, around the margin of the Pacific Ocean, where they are, like the Cache Creek limestones, associated with such oceanic rocks as pillow basalt and radiolarian chert in accretionary complexes. Here, the limestone is middle Permian to late Triassic in age, but elsewhere it ranges from early Carboniferous to middle Permian age.

How did such large masses of limestone, which formed in shallow water, come to be associated with rocks such as pillow basalt, radiolarian chert, and serpentinite—rock types characteristic of the deep ocean basin floors? Based on observations made in the Cache Creek terrane in northern British Columbia, the answer seems to be that the limestone bodies are the remains of enormous atolls or reefs that grew at or near sea level upon seamounts (such as parts of the Hawaiian chain) or on volcanic ocean

Massive limestone of the Cache Creek terrane forms the north wall of Marble Canyon.

plateaus surrounded by the deep ocean floor. The chemistry of most Cache Creek volcanic rocks, which is similar to that of modern oceanic island basalts erupted within plates, also supports this suggestion.

You can find fossils in the Marble Canyon limestone on the north side of Highway 99 about 250 metres east of its junction with the Hat Creek road. Much of the limestone here is finely recrystallized, but in places it consists entirely of calcareous shells about the size and shape of a small fingernail and slightly lighter-coloured than the matrix. You can see the complex, internal details of these shells with a hand lens if the rock is wet. The fossils are shells of fusulinids, single-celled animals that lived in shallow water in late Carboniferous and Permian time.

These particular fossils belong to a family of middle Permian fusulinids widespread in Asia and as far west as the Mediterranean region, but on the eastern side of the Pacific Ocean they are known only in rocks like those in the Cache Creek terrane. A different kind of fusulinid with a similar age range is common in the continental interior in the central and southwestern United States, elsewhere in the Canadian Cordillera, and in parts of the Andes. In British Columbia, such fusulinids are in limestone of the Quesnel terrane northeast of Kamloops, and in the Slide Mountain and

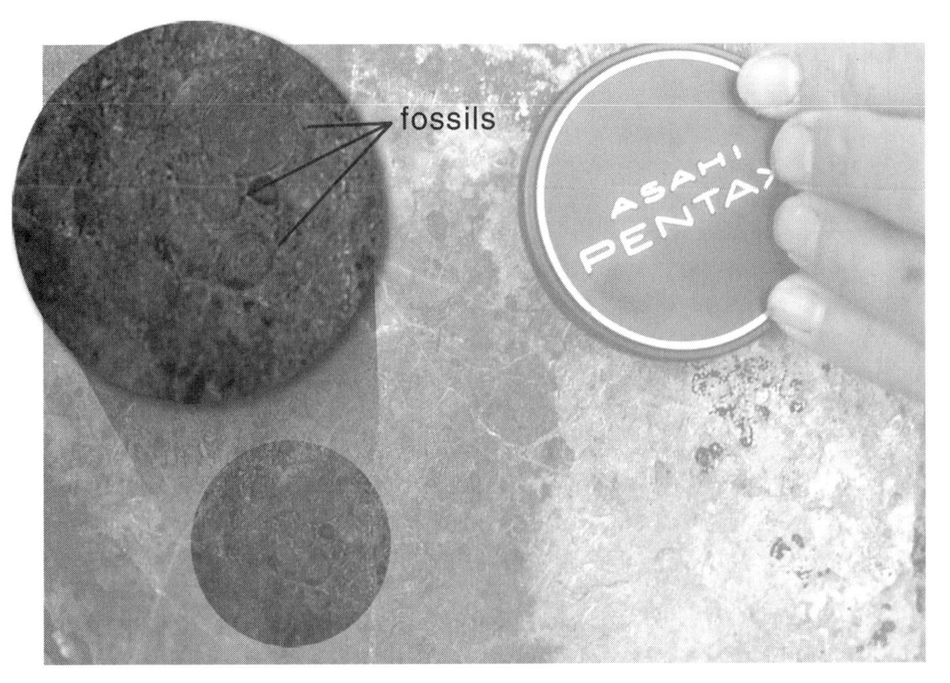

In this outcrop, almost the entire rock is made of the shells of fusulinids, which are oval or round bodies with fine internal structures that can be seen with a hand lens; the largest are little-finger-nail sized.

Detailed internal structure of a middle Permian Cache Creek fusulinid, magnified to about ten times its actual size. The fusulinid has been mounted on a glass slide and ground so thinly that it is translucent, showing its internal structure. —Photo courtesy of Lin Rui and Walter Nassichuk

Stikine terranes. Today, a distance of just over 100 kilometres separates the "Asiatic" fusulinids in Marble Canyon from "American" fusulinids of the same age found northeast of Kamloops. In Permian time they may have been several thousands of kilometres apart. The Asiatic fossils probably were carried on an oceanic "conveyor belt" into the early Mesozoic subduction zone and accreted to the Quesnel terrane.

Strange Life Forms on the Floor of Pavilion Lake

Pavilion Lake lies at the western end of Marble Canyon, and much of its floor is coated with a white layer of carbonate, which is not surprising considering the surrounding limestone. What is surprising are the fields of coral-like communities on the lake floor. These evidently began forming about 11,000 years ago, shortly after glaciers retreated from the area, by the photosynthetic activities of microscopic single-celled plants. They resemble life forms that created large belts of carbonate in early Cambrian time, about 540 million years ago.

West of Marble Canyon, Highway 99 traverses mainly rocks of the western Cache Creek terrane. These are of early Triassic to early Jurassic age and include chert, argillite, and sandstone derived from volcanic rock. In places, they are intruded by granitic rocks of late Jurassic age.

Pavilion Slide

A major earth flow descends the southeast side of the valley across from the village of Pavilion, and Highway 99 crosses its toe. The flow is monitored, and although the upper part appears to be moving very slowly, the bottom part is stable. The flow appears to originate on a surface that features red clay, which probably was a soil that formed by weathering just before the overlying mid-Cretaceous non-marine sandstones and shales were deposited. You can see these rocks in place, with red and buff-coloured flat-lying beds, at the top of the northwest side of the valley, from a point on the highway about 2 kilometres east of the earth flow. Related rocks, also red in places, were crossed east of Hat Creek and along Highway 12 near the Fraser River north of Lytton.

Boundary of the Intermontane and Coast Belts

West of Pavilion the road enters the steep-walled, terraced valley of the Fraser River, whose course here follows the surface trace of the Fraser fault system. Near where the road enters the valley, red and white rocks of the late Cretaceous Spences Bridge group are visible low down near the river. The high western valley wall, which is the eastern end of the Camelsfoot Range, is underlain by sedimentary rocks of the Jackass Mountain group, the youngest part of Methow terrane. The boundary between the Intermontane Belt and Coast Belts, here delineated by a strand of the Fraser fault system, lies between the two. The boundary is crossed at the big bend in the Fraser River to the south, described at the end of **Highway 99: Vancouver—Lillooet and Beyond.**

South of here and west of the road, about halfway between where Highway 99 enters the Fraser Valley and the big river bend, is an isolated knoll of Eocene volcanic rock that weathers to brown. This, like the Eocene block at the east end of the big bend in the Fraser River to the south, is a fault-bound block caught in the Fraser fault system.

Omineca and Foreland Belts

Columbia and Rocky Mountains

The Omineca and Foreland Belts are names applied by geologists to the bedrock of the mountainous eastern part of the Canadian Cordillera, which extends southward for about 2,500 kilometres from the Arctic Ocean near 70 degrees latitude to the international boundary at 49 degrees latitude. In the south, this mountainous region is up to 400 kilometres wide, with the easternmost section continuing for 50 to 100 kilometres into the province of Alberta. The great longitudinal valley called the Rocky Mountain Trench separates the two belts in British Columbia. The name "Omineca" is taken from the Omineca Mountains, which lie west of the northern Rocky Mountain Trench between 54 and 56 degrees latitude, about halfway along the belt, in north-central British Columbia. The name "Foreland" refers to the position of that belt in the easternmost Canadian Cordillera and to its structural style. A *foreland* is a stable area marginal to a mountain belt toward which rocks were thrust-faulted and folded.

In this book, we've combined the two belts in one section because their geological evolution is so closely linked. However, there are major differences between them. First, and perhaps most distinctively, the Omineca Belt contains extensive areas of high-grade metamorphic rock, whereas most rocks in the Foreland Belt are little metamorphosed. Second, granitic and in places volcanic rocks are widespread in the Omineca Belt, whereas the Foreland Belt is made largely of sedimentary rock. Third, the Omineca Belt is the region where the terranes of the Intermontane Belt tectonically overlapped rocks that were deposited *near* the old continental margin of Laurentia (part of which eventually became the North American continent), whereas the Foreland Belt contains rocks deposited *on* the old continental margin.

Many names are applied to the topographic features in southeastern British Columbia. On the western side of the southern Omineca Belt, the Shuswap and Okanagan Highlands have relatively subdued topography but grade eastward into the rugged Monashee, Selkirk, and Purcell Mountains. These mountains, which are traversed by the upper reaches of the Columbia River, together with the Cariboo Mountains farther north, collectively are called the Columbia Mountains. Elevations increase gradually from the 1,200- to 1,800-metre Interior Plateaus up to summit heights of about 3,500 metres in the mountains. Along with this, annual precipitation increases from about 70 centimetres in the west to over 250 centimetres in the higher ranges west of the Rocky Mountain Trench.

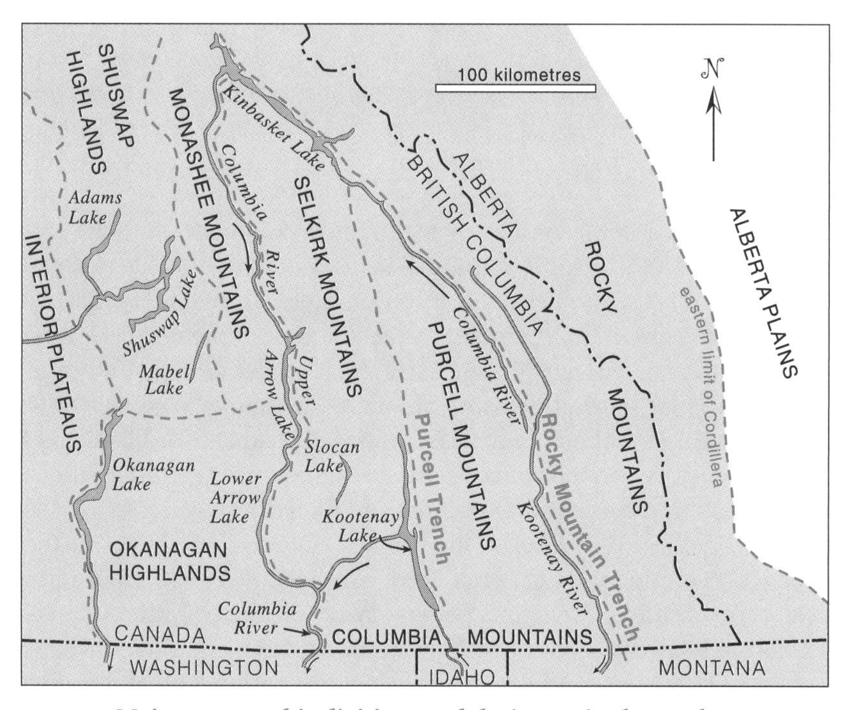

Major topographic divisions and drainages in the southern Omineca and Foreland Belts.

East of the Columbia Mountains, and forming the boundary between the Omineca and Foreland Belts, is the Rocky Mountain Trench, an enormous linear system of interconnected valleys. It is probably the single most obvious topographic feature within the entire Canadian Cordillera when viewed from space. The trench extends north-northwestward from northwestern Montana to the boundary between British Columbia and the Yukon at 60 degrees latitude. From there its continuation, called the Tintina Trench, runs northwestward into east-central Alaska.

East of the trench, the Rocky Mountains are the easternmost ranges in British Columbia. Several summits are around 3,500 metres high, and 3,954-metre Mount Robson in the western Rockies northwest of Jasper is only just exceeded in elevation in British Columbia by Mount Waddington in the Coast Mountains. The spectacular Rocky Mountains display the nature of their bedrock and structure far more clearly than any other part of the Cordillera. The different layers of sedimentary rocks forming the Rockies erode in distinctive ways, with massive limestone and quartzite units making cliffs, and interlayered shale forming gentler slopes or underlying valley bottoms. The Continental Divide, separating west- from east-flowing streams, marks the boundary between southeastern British Columbia and southwestern Alberta. At the Rocky Mountain Front in Alberta, where the easternmost cliff-forming limestone is exposed, the mountains drop dramatically to the rolling foothills, which give way within 20 to 30 kilometres to the Alberta plains, here about 1,000 metres high.

Most of this part of southern British Columbia is drained by the Columbia River and its tributaries, with the exception of the Shuswap Lake and Adams Lake watersheds, near and north of Highway 1, which drain westward via the South Thompson River into the Fraser River. The Columbia River and its tributaries exhibit a somewhat surprising drainage pattern. The Columbia begins in the southern Rocky Mountain Trench at Canal Flats and flows northward along the trench for about 300 kilometres before circling around the north end of the Selkirk Mountains and then flowing south between the Selkirk and Monashee Mountains. It eventually crosses into the United States south of the town of Trail. Its major tributary in Canada, the Kootenay River, flows southward out of the Rocky Mountains into the Rocky Mountain Trench near Canal Flats, where only a few kilometres of swampy flatlands separate it from the source headwaters of the north-flowing Columbia River. Despite passing so close to the headwaters of the Columbia River, the Kootenay River flows for over 400 kilometres before reaching its confluence with the Columbia. It flows southward along the trench, crosses into Montana, where its name is spelled "Kootenai," and after about 70 kilometres turns northwestward into Idaho. From there, it flows northward along the Purcell Trench back into British Columbia and into the south end of Kootenay Lake, which drains westward to meet the Columbia at Castlegar. Hydroelectric dams built in several places across both rivers impede their flow and form long, narrow lakes. A big and unresolved scientific challenge is sorting out the roles played by glacial events versus late Tertiary to Holocene differential vertical movements associated with tectonic activity in creating these tortuous drainage patterns.

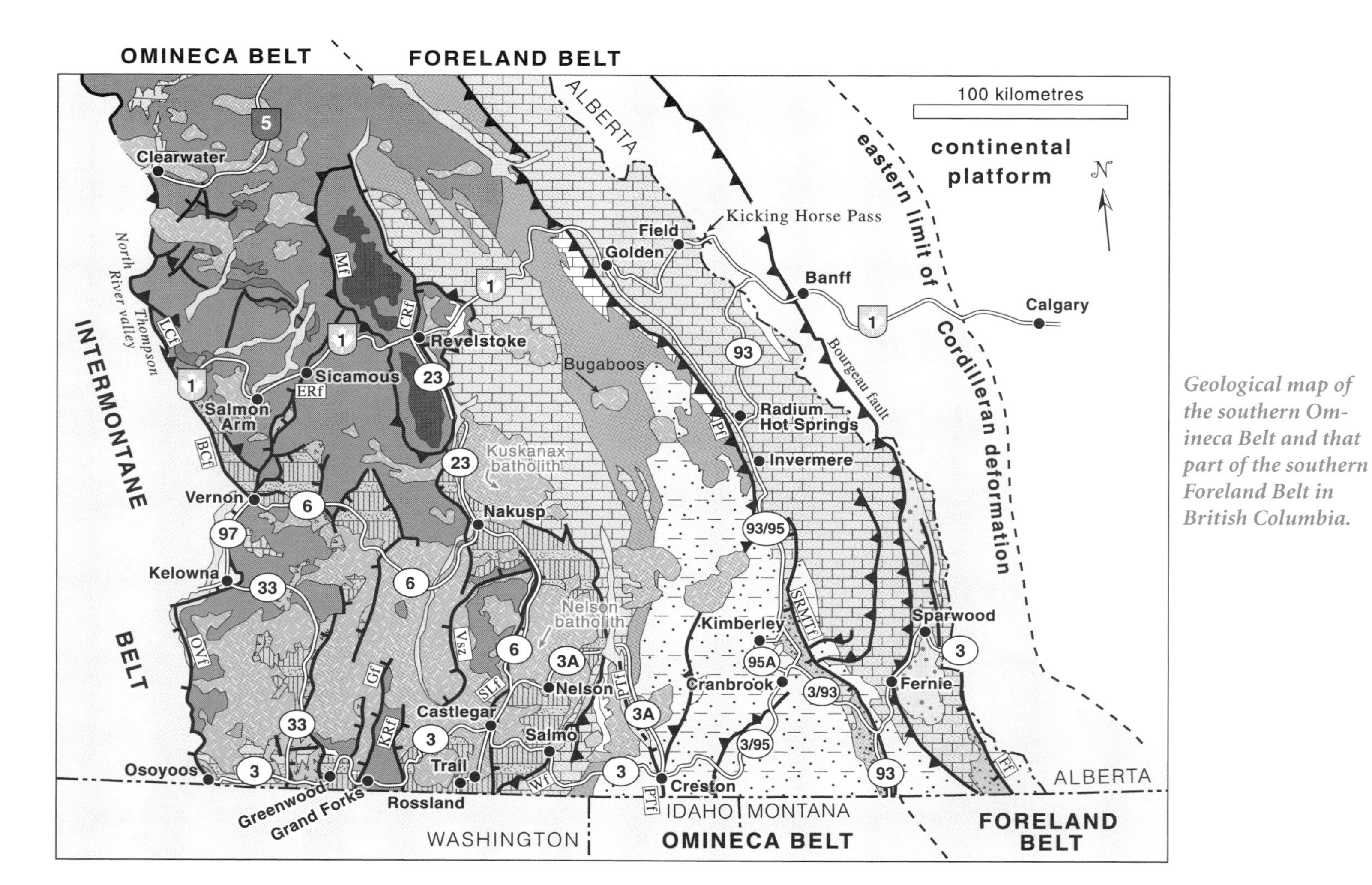

Geological map of the southern Omineca Belt and that part of the southern Foreland Belt in British Columbia.

BEDROCK GEOLOGY AND EVOLUTION

Much of the geologic history of the Omineca and Foreland Belts is closely linked to formation and destruction of the continental margin of Laurentia, the ancestor of the present North American continent. About 750 million years ago, the supercontinent Rodinia began to rift apart into continent-sized fragments, one of which became Laurentia. By the beginning of Cambrian time, about 540 million years ago, the fragments were drifting apart, and the ocean basin that was the far distant ancestor of the present Pacific Ocean started to form offshore of Laurentia. Today, the rocks deposited on and near this Paleozoic continental margin lie in the Omineca and Foreland Belts. In middle Devonian time, about 390 million years ago, the passive margin changed to a convergent margin. In late Paleozoic to early Mesozoic time, between about 360 and 185 million years ago, a time interval corresponding to the amalgamation of all the continents into the new supercontinent Pangea, the convergent plate margin lay offshore, recorded mostly by the island arc rocks in the Quesnel terrane that were separated from the continental margin by a basin partly floored by

Rock units on the geological map.

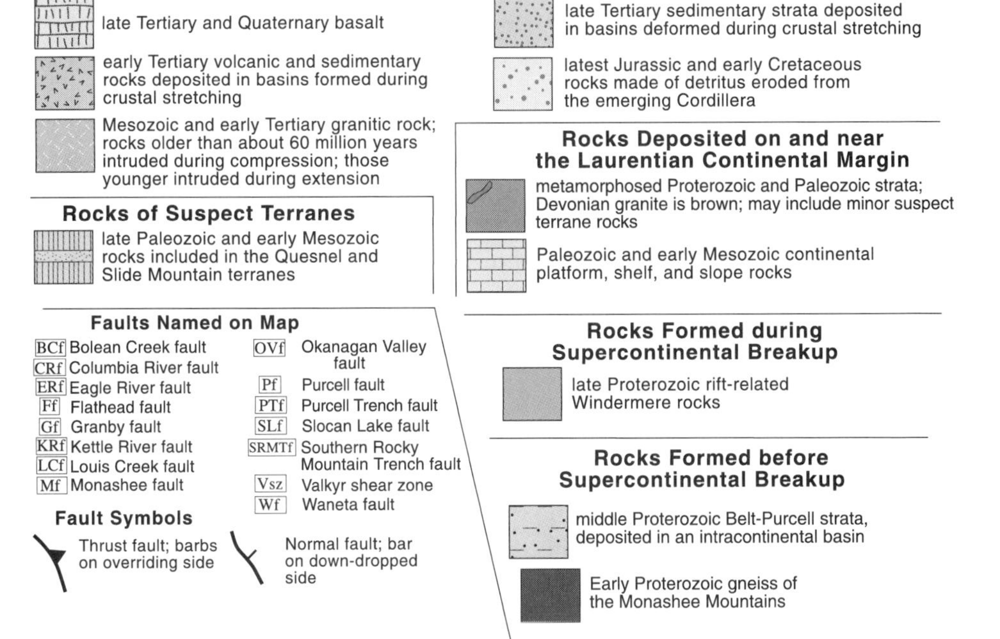

the Slide Mountain terrane. Starting about 180 million years ago, at the end of early Jurassic time, Pangea started to break apart into continental fragments, one of which is the North American continent. As the Atlantic Ocean opened on its eastern side, North America moved toward the convergent plate margin on its western side, accreting the offshore island arc and back-arc basin rocks to the Laurentian continental margin and in the process building mountains. By late Jurassic time, the first Cordilleran mountains emerged above sea level on the site of the future Omineca Belt. Between early Cretaceous and earliest Tertiary time, 110 to 60 million years ago, the old continental margin rocks were thrust back onto and over the edge of the continent, forming the Foreland Belt, and the entire region became elevated above sea level. Subsequently, in early Tertiary time, between about 55 and 45 million years ago, the crust of the region was stretched.

Below, we discuss rocks and structures involved in these events in order of their decreasing age as background for the complex geology covered by the roadside geology descriptions.

Early Proterozoic and late Archean Canadian Shield rocks. Old metamorphic and granitic rocks of the Canadian Shield exposed in eastern Manitoba, northern Saskatchewan, and northeastern Alberta can be traced westward, using oil and gas boreholes and seismic reflection profiles, beneath the Paleozoic through Tertiary sedimentary rocks that blanket the plains. Below southwestern Alberta, Canadian Shield rocks are of late Archean and early Proterozoic age, between 2.6 and 1.8 billion years old. The top of the buried shield can be traced from deep seismic reflection studies westward beneath the Foreland Belt as far as the Rocky Mountain Trench. Beneath the Foreland Belt, or Rocky Mountains, the western edge of the shield, along with its sedimentary cover, is bent downward beneath the enormous load of rocks that have been thrust onto it. Near Calgary, which is just east of the eastern limit of the Canadian Cordillera, the top of the shield is about 4 kilometres below the surface. Near Field, 150 kilometres west of Calgary on the British Columbia–Alberta boundary, it is about 9 kilometres deep; and near Golden, in the Rocky Mountain Trench, it is about 17 kilometres deep. West of the trench, the continuity of the buried shield rocks is not known, but gneiss about 2 billion years old in the Monashee Mountains west of Revelstoke once may have been part of the Canadian Shield.

Middle Proterozoic Belt-Purcell strata. The oldest sedimentary rock in British Columbia, a stack of quartz-rich sandstone, siltstone, and shale with lesser amounts of dolomite and limestone and numerous dark mafic sills, was deposited between about 1.5 and 1.4 billion years ago. The stack is up to 11 kilometres thick in Canada and filled an enormous basin within

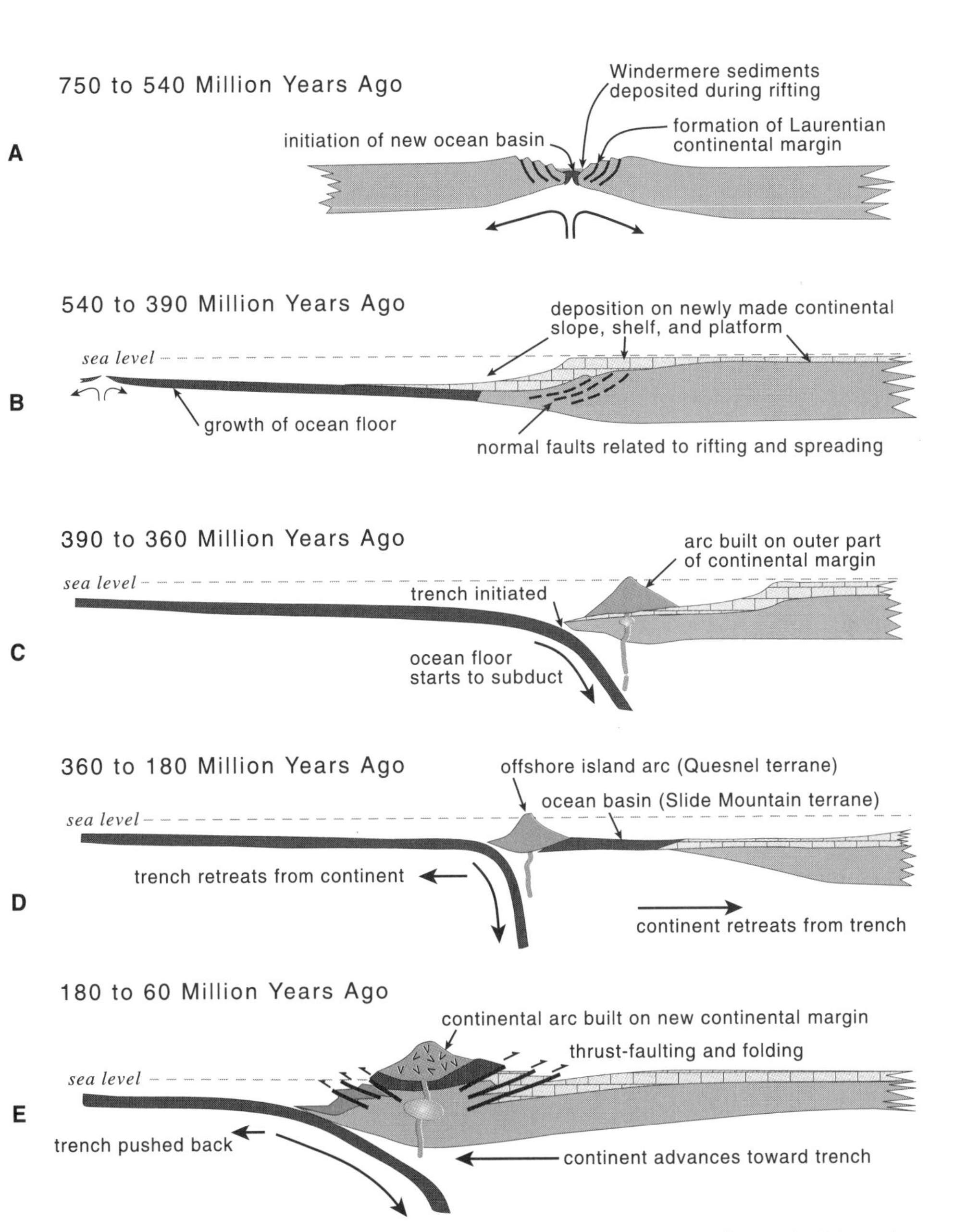

Evolution of the southeastern Canadian Cordillera from the time of initial rifting of the supercontinent Rodinia to later stages of crustal thickening and mountain building. A. *In late Proterozoic time, the supercontinent Rodinia rifted apart.* B. *From late Proterozoic time to middle Paleozoic time, sediments were deposited on the Laurentian continental margin.* C. *In Devonian time, a convergent margin developed and a magmatic arc formed.* D. *In late Paleozoic to early Mesozoic time, the Quesnel island arc was separated from the continental margin by a deep ocean basin as the continents amalgamated into the supercontinent Pangea.* E. *From middle Jurassic to early Tertiary time, the North American continent advanced toward the trench, causing the Cordilleran mountain building.*

an early Proterozoic continent. The rocks, called the Purcell supergroup in Canada, form most of the Purcell Mountains and parts of the Rockies near the international boundary. To the south, in Idaho, western Montana, and northeastern Washington, they are known as the Belt supergroup, and we use the name "Belt-Purcell" in our book.

Late Proterozoic Windermere strata. The Windermere supergroup, named from Lake Windermere in the Rocky Mountain Trench, consists of shale, distinctive coarse-grained sandstone with angular fragments of quartz and feldspar called "grit," local conglomerate, ancient till, carbonate, and minor basalt and evaporite deposits. These rocks, deposited between about 750 and 540 million years ago, extend from northern Alaska to Nevada and are found in Canada in the Omineca and Foreland Belts. They truncate the northeast-southwest structural grain of the buried Canadian Shield rocks

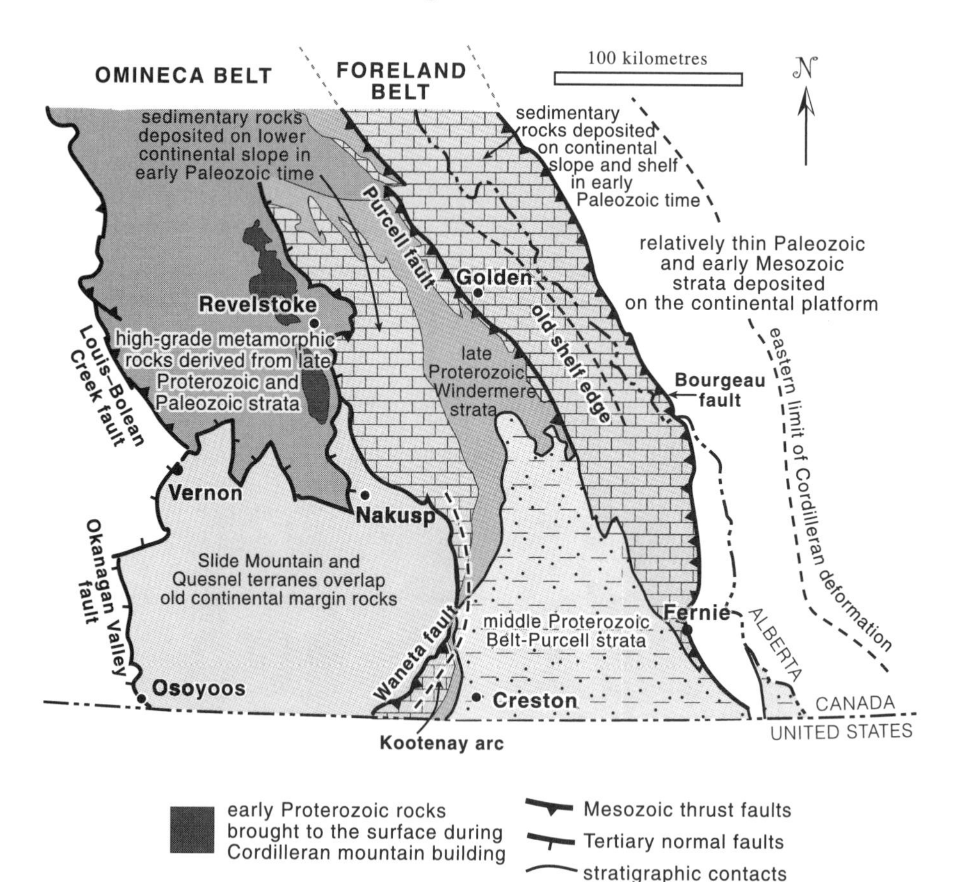

Major sedimentary rock packages deposited on or near the Laurentian continental margin, accreted terranes, and some major geological structures in southeastern British Columbia.

and are the oldest rocks to conform to the general north-northwest trend of the Canadian Cordillera. Windermere strata were deposited during the prolonged episode of rifting that eventually split apart the supercontinent Rodinia into continent-sized fragments.

Paleozoic through mid-Mesozoic continental margin sediments. Limestone and shale with lesser amounts of sandstone were deposited along the Laurentian continental margin that had formed in latest Proterozoic to earliest Cambrian time. The rocks are mostly Paleozoic, but some are as young as Jurassic, and there is a plethora of formal formation and group names, mostly not given herein. These strata make up much of the southern Foreland Belt, and the older Paleozoic rocks, those that were deposited well offshore and are now mostly metamorphosed, extend across the entire Omineca Belt.

The rock types and thickness change from east to west, reflecting the original transition from the high-standing but just submerged continent to the low-standing ocean floor. Relatively thin sequences of mainly limestone deposited in shallow water on the edge of the Laurentian continent are now preserved in the eastern Rockies, east of the Bourgeau thrust fault. Westward in the central Rockies, they change laterally to very thick limestone sequences, deposited on the old continental shelf and upper continental slope, that are interbedded with shale in the westernmost Rockies. Still farther west, in the Omineca Belt, deposits of mainly shale with minor sandstone, basalt, and limestone, now metamorphosed, were laid down mostly in deep water, probably on the lower continental slope.

Devonian arc rocks. An arc was emplaced in and on the mostly deepwater deposits of the westernmost continental margin between 390 and 360 million years ago, in middle and late Devonian time. The arc is represented by small bodies of granite and volcanic rock present in the western Omineca Belt mainly north and west of Sicamous on Highway 1, and thin layers of ash of the same age in Devonian strata in the eastern Foreland Belt in Alberta. Scattered remnants of an arc of the same age are found from northern Alaska to northern California. The emergence of this arc is important because it marks a major plate tectonic change from the early Paleozoic passive continental margin to the convergent plate margin that has persisted to a greater or lesser extent until the present.

Late Paleozoic and early Mesozoic terranes. The Slide Mountain and Quesnel terranes straddle the boundary between the Intermontane and Omineca Belts and underlie a large area in the southwestern Omineca Belt. Scattered and small isolated areas of Carboniferous and Permian volcanic and sedimentary rocks of the Slide Mountain terrane probably formed in a back-arc basin floored by oceanic crust and are unconformably overlain

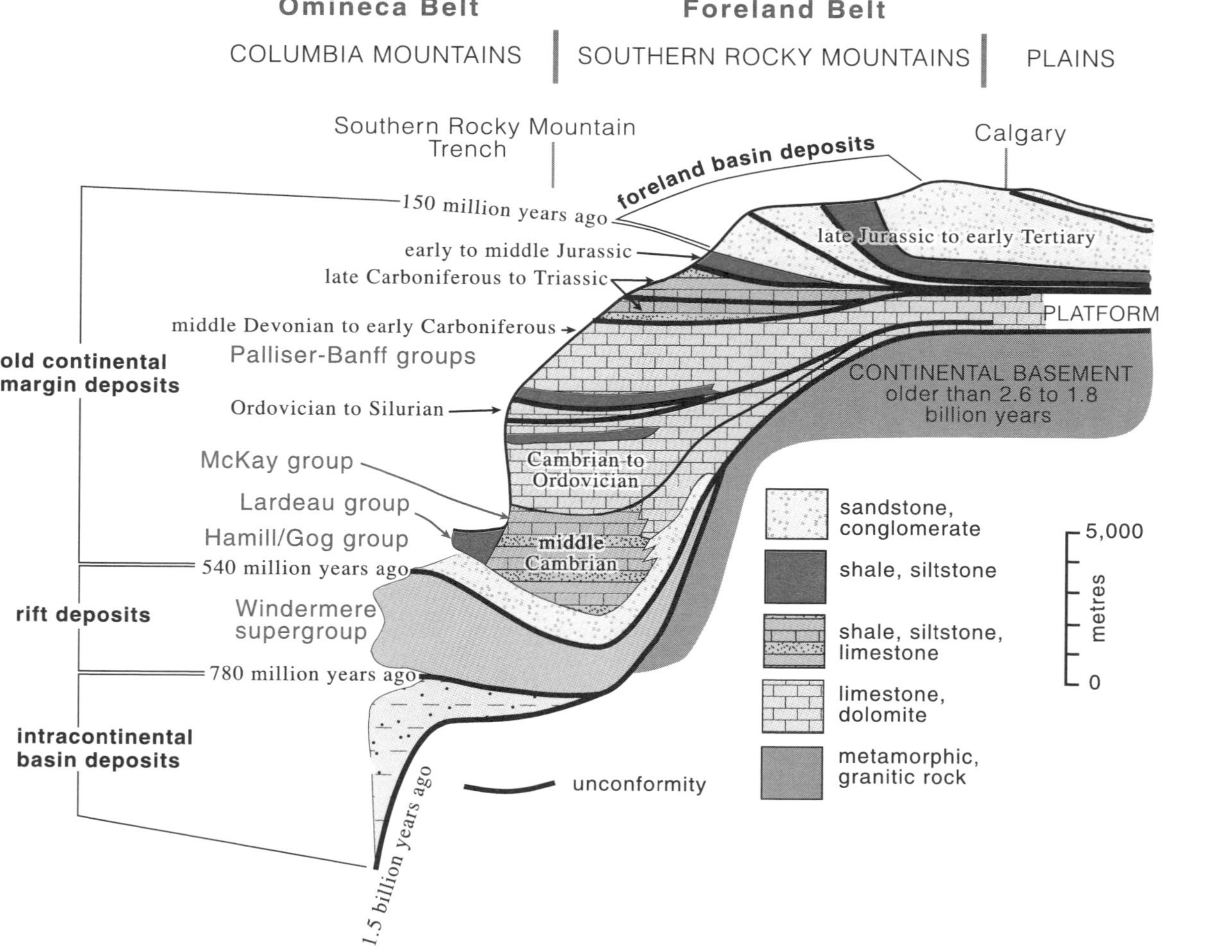

Major rock packages, sedimentary rock types, and thickness changes from west to east in the continental margin deposits of Precambrian through Jurassic time.

in places by middle Triassic to early Jurassic arc-related volcanic and sedimentary rocks of the Quesnel terrane. Permian to early Jurassic fossils in rocks of the Slide Mountain and Quesnel terranes are more similar to those 1,000 to 2,000 kilometres to the south in northern California and Nevada than they are to fossils of the same age and present latitude in the Rockies, which indicates that the terranes may have been displaced northwards from their original locations. The contacts between the terranes and rocks deposited on the old continental margin appear to be faults.

Cordilleran Mountain Building

About 180 million years ago, a big change initiated the transformation of the region of offshore island arcs, back-arc basins, and the old continental margin to the mountainous region that today extends for over 600 kilometres west of the old margin. What caused this change? It seems to us that the best suggestion, and one that has been around for at least eighty years, is that it is related to the Jurassic breakup of Pangea. When the Atlantic Ocean basin opened within Pangea following a prolonged period of rifting on the eastern side of what is now the North American Plate, the plate advanced toward the trench on its western margin, plowed into the mush of offshore basins and island arcs, and collided with the subducting oceanic plate, which could not retreat or subduct rapidly enough to get out of the way. (Something like this is happening today in the Andes.) Warm and weak arc and back-arc basin lithosphere at the western edge of the newly formed North American Plate was caught in a “tectonic vise” made of strong continental lithosphere to the east and stronger oceanic lithosphere to the west. The old continental margin was squeezed, folded, and thrust eastward and westward. The crust became thickened, and some rocks were buried to great depths and metamorphosed and others were uplifted above sea level and eroded. By late Jurassic time, about 150 million years ago, the embryonic Cordillera had emerged above sea level in what is now the Omineca Belt.

We know the change began about 180 million years ago because the youngest marine fossils that lived in the basin between arc and continent are about 185 million years old, and the sedimentary rocks containing them were folded, faulted, and then intruded by granitic rocks that are only slightly younger. In addition, the isotope geochemistry of the middle Jurassic and younger granitic rocks—those less than about 180 million years old that intrude the Slide Mountain and Quesnel terranes, their bounding faults, and old continental rocks—show a continental influence.

Much of the Shuswap Highlands and Monashee Mountains are made of schist and gneiss. Early geologists thought that these rocks were of Archean age, largely because they resemble rocks in the Canadian Shield. However, isotopic dating carried out in the last thirty years shows that most of the metamorphism, with one exception, took place between early Jurassic and early Tertiary time. That exception is the 2-billion-year-old rock just west of Revelstoke that may be a part of the Canadian Shield that was detached during rifting and reattached during mountain building. In the Omineca Belt, late Proterozoic and early Paleozoic deposits, as well as some terrane rocks, were buried to depths of as much as 25 kilometres in places, where they were metamorphosed and started to melt to form gneiss and, ultimately, granitic rock.

By about 110 million years ago, in early Cretaceous time, rocks deposited on the old continental shelf and upper slope, now located in the western Foreland Belt, were themselves being folded and thrust eastward. By earliest Tertiary time, rocks deposited on the old continental platform had been folded and thrust eastward almost as far as Calgary. The folding and thrust-faulting caused by lateral compression finally ceased across the Omineca and Foreland Belts in earliest Tertiary time, about 60 million years ago.

The amount of compression, or shortening, across the southern Foreland Belt can be calculated with some precision because the rocks there are well layered. Once-continuous strata were cut and displaced eastward on numerous mostly flat-lying thrust faults and folds, and the fault slices were stacked one upon the other. The amount of overlap caused by faulting and folding can be removed (graphically!) by pulling the fault slices back sequentially so that the rocks in them are restored to something like their original positions. This restores the rocks west of the Bourgeau thrust fault near Banff, Alberta, in the eastern Rockies, to an original location near Revelstoke, about 150 kilometres to the west. Similarly, the boundary between limestone deposited in shallow water in the east, and deeper-water deposits of calcareous shale to the west, exposed at Field near the British Columbia–Alberta boundary, can be restored to an original position near Vernon at the north end of the Okanagan Valley and about 200 kilometres west of Field. These restorations indicate that at least 300 kilometres of horizontal shortening occurred in just the Foreland Belt during Cordilleran mountain building.

Latest Jurassic through Tertiary Foreland basin rocks. As the Cordilleran mountains emerged above sea level, first on the site of the Omineca Belt and later on that of the Foreland Belt, they were eroded. The products of this erosion are nonmarine and locally marine sandstone, shale, and conglomerate containing important coal deposits. The latest Jurassic and early Cretaceous detritus was deposited in basins now in the Foreland Belt. Continuing convergence, uplift, and erosion resulted in late Cretaceous deposits that are preserved in the easternmost Foreland Belt east of the mountains, and Tertiary deposits that spread eastward across the plains for as much as 700 kilometres.

Tertiary Normal Faulting

A dramatic reversal of the compression that built the first Cordilleran mountains took place in Tertiary time, mostly between about 55 and 45 million years ago. The crust was stretched laterally, rather than compressed, and normal faults that accompanied the stretching have brought together rocks that at one time were at very different depths in the crust.

We don't know what caused the rapid change from compression to extension that started between 60 and 55 million years ago. We can surmise that a change of convergence directions and/or a change in the rates of movement between the North American Plate and various Pacific Ocean plates may have relieved the compression that supported thick crust beneath the Canadian Cordillera, allowing it to stretch and thin. Before 60 million years ago, in earliest Tertiary time, at the end of the long period of compression, the crust in southeastern British Columbia may have been very thick. The modern central Andes, which are being squeezed between the advancing South American Plate and oceanic plates to the west, have a crust up to about 70 kilometres thick in places. Today, the crust in eastern British Columbia varies in thickness from nearly 50 kilometres under the westernmost Foreland Belt to only 30 to 35 kilometres below the Intermontane Belt.

We can see the effects of this pervasive normal faulting most clearly along the east-west route followed by Highway 3 across the Omineca and Foreland Belts. There, unmetamorphosed rocks can be seen side-by-side, across faults, with high-grade metamorphic rocks. Most of the big, roughly north-south-oriented valleys follow Tertiary normal faults. The westernmost big normal fault is the Okanagan Valley fault, marking the western boundary of the southern Omineca Belt. East of this, the valleys that follow big normal faults include, from west to east, parts of the Kettle River, Granby River, Christina Lake, and Slocan Lake valleys; the southern Purcell and southern Rocky Mountain Trenches; and the Flathead Valley. Many of the faults can be traced southward into Washington, Idaho, and Montana, and some disappear beneath the late Tertiary Columbia River basalt. The total amount of extension or crustal stretching across the entire region is uncertain, although guesstimates range upwards to as much as 200 kilometres.

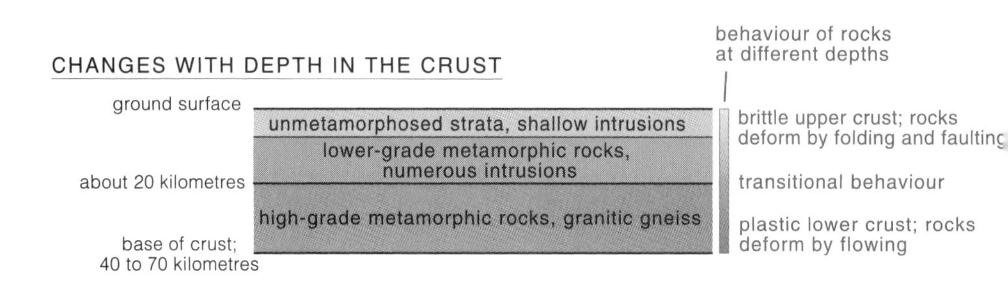

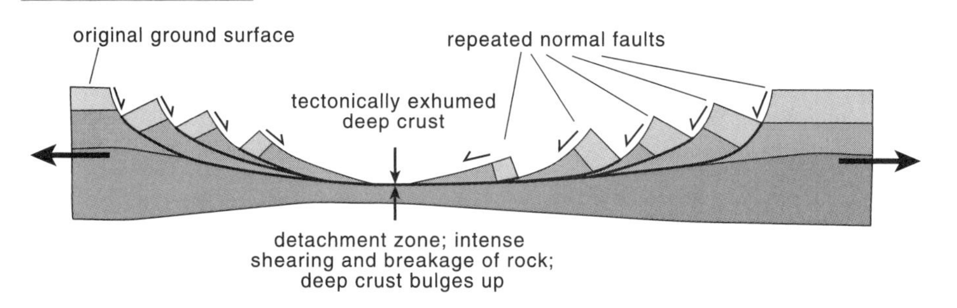

During crustal extension in Tertiary time, plastic stretching occurred in deeper parts of the crust and repetitive normal faulting in higher parts.

In the southern Omineca and Foreland Belts, extension took place mostly from about 55 to 45 million years ago but continued into late Tertiary time on the Flathead and Southern Rocky Mountain Trench normal faults.

The dips of the normal faults range from steep to shallow. Those faults that involve only near-surface rocks tend to have steep dips, whereas those that place near-surface rocks against once-deeply-buried rocks tend to have shallow dips, although exceptions can be found. Shallowly dipping normal faults, called *detachment faults*, mimic landslides on an enormous scale; the Okanagan Valley fault is a well-exposed example. There, you can see gneiss that once was buried to depths of more than 20 kilometres in contact across a shallowly westward-dipping fault surface with Eocene volcanic and sedimentary rocks that were never buried deeper than a few kilometres at most.

Understanding how normal faults may bring deeply buried rocks to the surface resolved a question first raised about forty years ago, when geologists first isotopically dated rocks using the potassium-argon method, which records the time when a rock cools between 500 and 300 degrees

Celsius. They noticed that many intrusions in low-grade metamorphic rocks gave Mesozoic dates, whereas those in high-grade metamorphic rocks in the same region yielded Tertiary dates. The explanation for this is that the high-grade rocks and their intrusions remained deeply buried in the crust and therefore stayed hot. When their 20-kilometre-thick cover was suddenly stripped off—a process called *tectonic exhumation*—by movement on detachment faults, they were suddenly chilled. The gneiss below the Okanagan Valley fault gives cooling ages that are the same as the ages from the Eocene volcanics above the fault.

HIGHWAY 1
Pritchard—Kicking Horse Pass
392 kilometres

The topographic boundary between the Intermontane and Omineca Belts along Highway 1 is singularly unimpressive. East of Kamloops, Highway 1 follows the valley of the west-flowing South Thompson River, and the landscape changes gradually from Interior Plateaus to Shuswap Highlands to Monashee Mountains. From the highway, the valleys become deeper with steeper sides until high, snowcapped mountains become visible, weather permitting, in the wider, wetter Columbia River valley near Revelstoke. Somewhat more obvious from the highway is the vegetation change from the dry Interior Plateaus, with open grasslands, sagebrush, and ponderosa pine, to the dense fir, hemlock, and cedar forest on the wet, western side of the Monashee Mountains.

Near Pritchard, a small place about 40 kilometres east of Kamloops, are the north-northwest-trending, aligned valleys of Louis Creek, north of Highway 1, and Bolean Creek south of the highway. These valleys follow the trace of a system of east-side-up, probable Jurassic-age reverse (or thrust) faults, called the Louis Creek fault north of Highway 1 and the Bolean Creek fault to the south of it. The faults mark the western boundary of the Omineca Belt near Highway 1. West of the boundary, the rocks are at most only mildly metamorphosed, with nonexistent or weak cleavage, whereas east of it the nongranitic rocks are metamorphosed to greenschist and amphibolite, have strong cleavage, and were metamorphosed from rocks that are mostly older, late Precambrian to middle Paleozoic, than those to the west. The base of Eocene volcanic rocks is lower west of the fault system than to the east, suggesting there was also some east-side-up

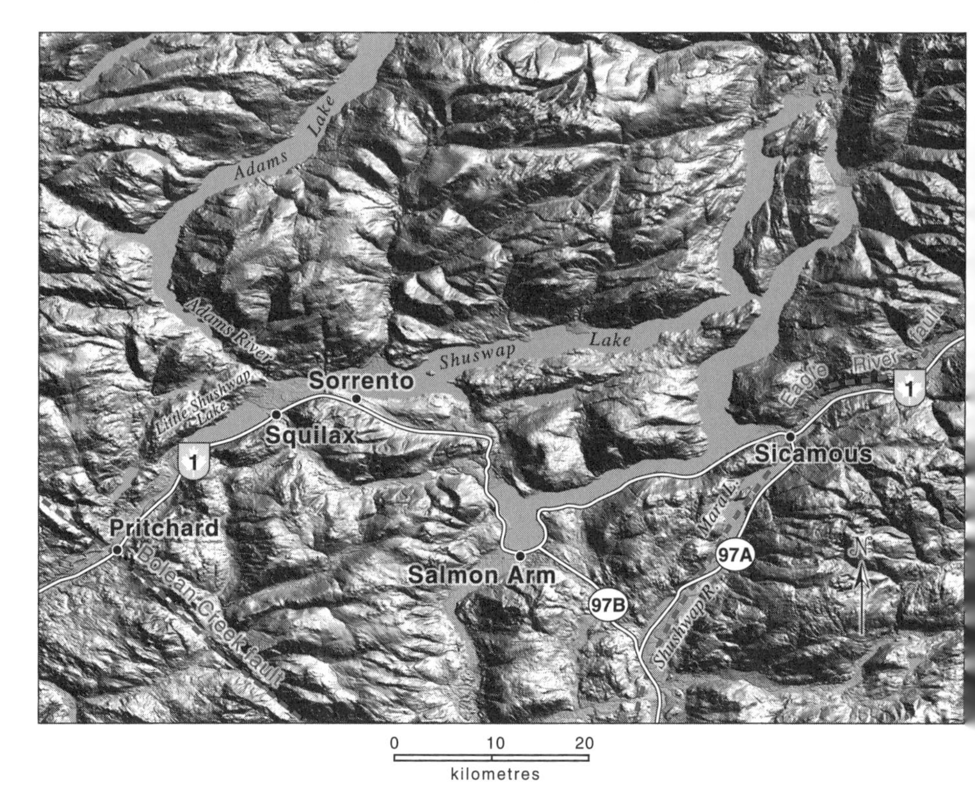

Shaded-relief image of topography along Highway 1 between Pritchard and Sicamous. —Image constructed using Canadian Digital Elevation Data obtained from GeoBase

movement on the fault in Tertiary time, but not nearly enough to account for the difference in metamorphic grade across the fault.

Metamorphic Rocks between Pritchard and Revelstoke

East of Pritchard, metamorphic rocks predominate. About 8 kilometres east of Pritchard the roadcuts are in buff-weathering Cretaceous granitic rock of the Niskonlith pluton, which, about 7 kilometres farther east along Highway 1, intrudes schist, quartzite, and calcareous quartzite, former sedimentary rocks deposited in early Paleozoic time. This is the westernmost of many exposures of metamorphic rock for the next 200 kilometres along the highway.

The metamorphism and accompanying deformation here mask the original nature of the rocks and the order in which they were laid down. They also destroy most fossils that may have been present, although one early Cambrian fossil has been found in limestone near Adams Lake, about

20 kilometres north of Highway 1, and there, fossils of Carboniferous age also are known in less-metamorphosed rocks. Most of the metamorphic rocks probably were derived from late Proterozoic and early Paleozoic deposits, although a few may be younger.

During metamorphism, the combined heat and pressure changes clay-rich rock, such as shale, to phyllite, a rock with a lustrous sheen produced by the growth of minute mica flakes on the cleavage surfaces. With further increases in metamorphic grade, new minerals grow and become visible. Most obvious are dark brown biotite and white muscovite, micas that largely comprise biotite and muscovite schists, which often contain reddish, granular crystals of garnet. Calcium carbonate that once made beds of limestone is recrystallized to coarse-grained marble. Depending on their original compositions, volcanic rocks are metamorphosed to greenish grey chlorite schist, silvery quartz muscovite schist, or black amphibolite.

Other minerals also derived from clay-rich rock may be present in mica schist. From laboratory studies, we know that the specific minerals present can tell us the temperature, pressure, and probable depth at which a metamorphic rock formed. Staurolite, visible as stubby black crystals, generally forms at temperatures between 400 and 500 degrees Celsius. The fibrous clear mineral sillimanite requires temperatures of nearly 600 degrees Celsius. With the normal thermal gradient in the upper part of Earth's crust of about 30 degrees Celsius per kilometre, the temperatures needed to create these minerals would be reached at depths of about 13 kilometres for staurolite and 20 kilometres or more for sillimanite. Kyanite, small bluish lath-shaped crystals that can be scratched with a needle or the point of a penknife drawn along the length of the crystal but not across it, requires similar temperatures to sillimanite but higher pressures; its presence indicates that the rocks containing it once were well over 20 kilometres below the ground surface.

Along Highway 1, rocks once buried to depths of around 15 kilometres are found between Pritchard and Sicamous. The metamorphic grade jumps up at Sicamous, where Highway 1 crosses the east-side-up Eagle River normal fault, which is probably the northern extension of the big Okanagan Valley fault. East of the fault, kyanite- and sillimanite-bearing rocks are exposed along the highway for a distance of about 100 kilometres, to about 25 kilometres east of Revelstoke. These are the highest-grade metamorphic rocks exposed along this section of Highway 1. To the east, the metamorphic grade is lower again and comparable to that between Pritchard and Sicamous.

The presence and abundance of white quartz and feldspar veins and layers provides a rough guide to the metamorphic grade in this region. When the metamorphic grade is relatively low, the white layers typically are small, crosscutting veins and layers parallel with the foliation. You can see such rocks immediately west of the Highway 1 bridge over the arm of Shuswap Lake at Sicamous, where a convenient side road on the south side of Highway 1 permits safe parking. These rocks probably once were buried to depths around 15 kilometres. In the first roadcuts east of the Eagle River fault at Sicamous, the white quartz and feldspar layers are coarser grained, far more abundant, and make up much of the rock. The metamorphic rocks here formed at depths of 20 to 25 kilometres. The layers and veins indicate the partial melting of quartz and feldspar, minerals that are less resistant to heat. At Victor Lake a parking area on the south side of Highway 1 provides a convenient place to stop to examine the gneiss. Be careful when crossing the busy highway. Careful examination of the rock using a hand lens reveals the presence of the colourless fibrous mineral sillimanite.

Isotopic dating tells us that most of the metamorphism in this region took place between about 180 and 45 million years ago. However, higher in the Monashee Mountains between Victor Lake and Revelstoke, isotopic dating has identified gneiss that is 2 billion years old! This date is a bit scary for geologists because the gneiss looks like the other, much younger gneiss

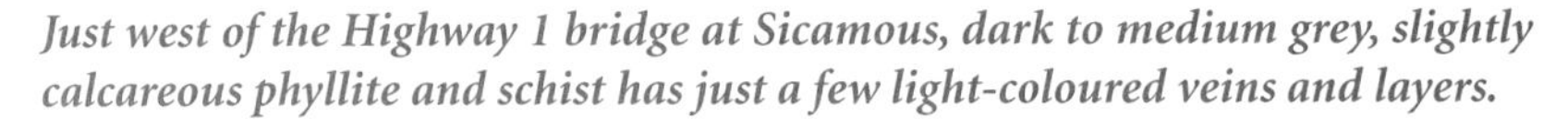
Just west of the Highway 1 bridge at Sicamous, dark to medium grey, slightly calcareous phyllite and schist has just a few light-coloured veins and layers.

Light-coloured granitic gneiss at Victor Lake, about 14 kilometres west of Revelstoke. Here most of the rock is light-coloured quartz and feldspar, and you can see the fibrous mineral sillimanite with a hand lens.

just to the west of it and probably would not have been picked out without careful isotopic dating.

Metamorphosed Devonian Arc Rocks

Orthogneiss is a white, layered rock made from a pre-existing granitic intrusion that was metamorphosed. A body of orthogneiss is exposed in roadcuts south of Little Shuswap Lake and there intrudes schist derived from late Proterozoic and early Paleozoic strata. Isotopic dating of this rock suggests that it crystallized as a granitic rock in Devonian time. Several other bodies of metamorphosed granitic rock of middle to late Devonian age, about 390 to 360 million years old, as well as volcanic rocks with magmatic arc chemistry, are known in the region north of Highway 1. Another metamorphosed Devonian intrusion can be seen on Highway 1 about 30 kilometres east of Revelstoke. In the big picture, these rocks belong to the oldest Cordilleran magmatic arc, which marks the change from the passive continental margin of early Paleozoic time to the convergent plate margin that has persisted to a greater or lesser extent until the present.

Salmon Cycle on Adams River

A short-lived spectacle recurring once every four years, including October 2002, is the great run of sockeye salmon to spawning beds on the lowermost 3 kilometres of the Adams River, where it flows through its delta into the western end of Shuswap Lake. To reach the bridge over the Adams River, turn off Highway 1 at the junction with the road to Squilax, cross the overpass, and head north. Red-bodied, green-headed spawning salmon floor the margins of the river when the run is active.

This is the biggest salmon run in British Columbia, involving about 2 million fish that have escaped the fishing fleets in coastal waters and the mouth of the Fraser River; aboriginal fishermen along the river; and the constriction of the river at Hells Gate. Once at the spawning bed, the female salmon dig nests and lay their eggs in the bottom of the stream in which they themselves were hatched. The males, after fighting off rivals, then fertilize the eggs. The 10-kilometre-long spawning beds can accommodate this large concentration of salmon, but greater numbers would involve excessive losses because late breeders dig out the eggs laid by their predecessors.

The original choice of spawning beds was made in early Holocene time by fish who strayed into terrain from which glaciers had just retreated, seeking gravel coarse enough to allow percolation of oxygen-bearing waters, but not so coarse as to prevent females from preparing the nest and then burying their eggs after fertilization. The first sockeye salmon to have adopted such a spawning bed would have to be from the small percentage of wanderers, willing to migrate to some stream other than the one in which they hatched.

Another requirement for the sockeye is a downstream lake—here Little Shuswap Lake—in which the hatchlings can spend their first year before migrating to the open Pacific to live another three years, then returning to the spawning site to complete their life cycle. The sockeye salmon run on the Adams River peaks consistently between October 10 and 16, with first arrivals about a week earlier and stragglers coming for a couple of weeks afterward.

A sharp decline in the number of spawning salmon has been noted in the past few years in the Adams River, as well as in most salmon streams in the Pacific Northwest. Overfishing is a major cause for the decline.

Possible Slide Mountain Terrane at Shuswap Lake

At the western end of Shuswap Lake between the Adams River and Sorrento, and just west of where Highway 1 leaves the lake, pullouts along the lakeshore permit access to roadcuts in dark green to black metamorphosed basalt intruded by Tertiary rocks that weather to a buff colour. The geochemistry of the basalt is like that of late Paleozoic basalt of the Fennell formation in the Slide Mountain terrane, which is exposed 60 kilometres to the north-northwest along Highway 5. Although the age of the metamorphic rocks here is unknown, their composition hints that they may belong to the Slide Mountain terrane.

Craigellachie: Where Canada Was Linked by Rail

One of the terms under which the colony of British Columbia entered the Canadian Confederation in 1871 was the construction of a railway linking British Columbia to the remainder of Canada. The route of the railway, through what was then known as the Eagle Pass between Sicamous and Revelstoke, had been identified in 1865. The Canadian Pacific Railway was built from the east and from Vancouver on the West Coast, and the two ends were finally linked in a ceremony held at Craigellachie on November 7, 1885. A small park and picnic site on the south side of Highway 1 contains a cairn commemorating this event, and the gilded "Last Spike" is

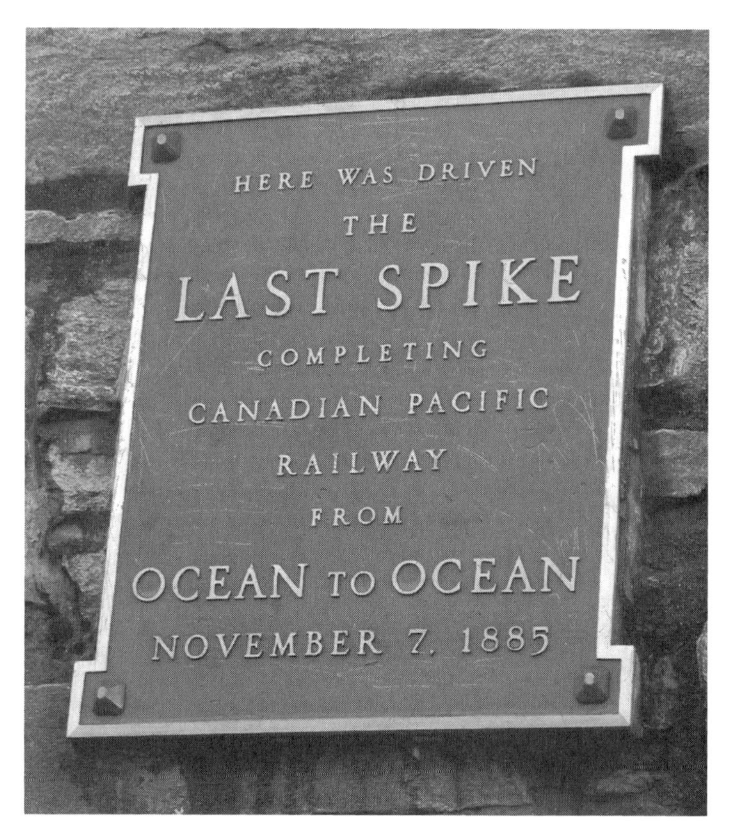

Plaque on the cairn commemorating completion of the railway line across Canada in 1885.
—R. Turner photo

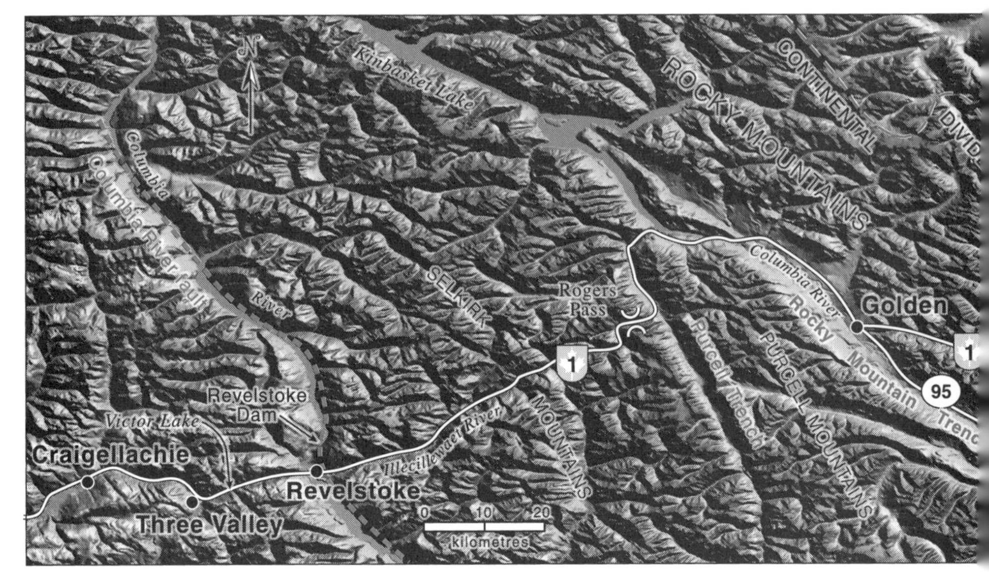

Shaded-relief image of topography along Highway 1 between Craigellachie and Golden. —Image constructed using Canadian Digital Elevation Data obtained from GeoBase

visible on the track behind the cairn. Geological aficionados will appreciate the base of the cairn because it contains representative rocks from all of the provinces and territories of Canada.

Craigellachie also lies near the somewhat arbitrary topographic boundary between the more subdued landforms of the Shuswap Highlands to the west and the more rugged Monashee Mountains to the east, which are the westernmost part of the Columbia Mountains.

Boudins at Three Valley Lake

At Three Valley Lake, roadcuts in the cliffs on the south side of the lake expose dark schist that weathers to a rusty colour. Walk along the lake side of the road, both for safety and to best see the big structures across the road in the schist. You can see that the fairly flat-lying foliation in the schist pinches and swells to make the structures called *boudins*, a geological name taken from the French word for "sausage" because of their supposed resemblance to a string of sausages. These structures in high-grade metamorphic rocks probably formed in the plastic lower part of the crust during stretching in early Tertiary time. Additional evidence for stretching is provided by the white fracture-filling quartz veins where the foliation pinches together, and by local vertical dikes, although these features probably formed as the

Curved foliation in metamorphic rocks at Three Valley Lake outlines the pinching and swelling that forms boudins. Note the Tertiary dikes cutting across the foliation.

schist was getting closer to the surface and cooling, and thus was more brittle. To see these structures, walk along the lakeside just west of the hotel complex and view them from across the highway.

Columbia River Valley at Revelstoke

The valley of the Columbia River separates the Monashee Mountains on the west from the Selkirk Mountains on the east. Near Revelstoke it follows the trace of the Columbia River fault, a large, east-side-down, east-dipping early Tertiary normal fault that extends in a north-south direction for about 200 kilometres. The offset of rock units across the fault surface is probably as much as 30 kilometres. For much of its length, the fault juxtaposes high-grade metamorphic rocks to the west with lower-grade rocks to the east. However, this juxtaposition is not obvious along Highway 1 near Revelstoke, where a structural slice of high-grade metamorphic rocks intruded by Devonian and Mesozoic granitic rocks, and related to the rocks west of Victor Lake, lies immediately east of the fault. The boundary with the lower-grade metamorphic rock to the east is another fault located about 30 kilometres east of Revelstoke.

Just east of the Highway 1 bridge over the Columbia River is the junction with a short road that leads north to Revelstoke Dam, one of the many

The Columbia River flows south from Revelstoke Dam.

dams built on the Columbia River to control water flow and generate electricity. Guided tours at Revelstoke Dam culminate in a visit to a model of the Columbia River Basin showing its many dams and their generating capacities. The reservoir behind Revelstoke Dam stretches northward to near Mica Dam at the north end of the Selkirk Mountains. Kinbasket Lake, the reservoir behind that dam, is 200 kilometres long, and you can glimpse its upstream tip, at high water, by looking north on the descent of Highway 1 into the Rocky Mountain Trench.

During construction of Revelstoke Dam, highly fractured bedrock along the Columbia River fault was exposed when the bedrock was stripped of its cover of unconsolidated Quaternary deposits. Extensive deep grouting had to be applied to the fractured bedrock of the dam foundation. Numerous rock bolts and drainage pipes maintain the slope stability of the fractured high-grade metamorphic rocks forming the eastern valley wall above the dam.

Rogers Pass and the Selkirk Mountains

For a road distance of 124 kilometres, between Revelstoke, on a southward-flowing segment of the Columbia River, to the bridge over the Columbia River as it flows northward along the great valley called the Rocky

Mountain Trench, Highway 1 traverses the Selkirk Mountains. Rogers Pass, about 70 kilometres east of Revelstoke, is at an elevation of 1,330 metres, and surrounding summits reach nearly 3,400 metres. The mountains here are part of Glacier National Park, which features more than four hundred glaciers. East of Rogers Pass, Highway 1 follows the valley of the north-flowing Beaver River along the northernmost part of the Purcell Trench, a big valley that separates the Selkirk Mountains from the northern tip of the Purcell Mountains.

In the Rogers Pass area, snowfall is copious and avalanches are common—numerous avalanche tracks are visible on the valley walls. To protect travellers, snow sheds route avalanches across the highway, and circular areas along the highway are sites for howitzers or mortars to shoot down threatening avalanches. The Canadian Pacific Railway originally ran along the floor of the pass, which separates the westward-flowing Illecillewaet River from the eastward-flowing Beaver River, but partly because of the avalanche hazard and partly to reduce the gradient, it is now routed beneath the mountains through the 14-kilometre-long Connaught Tunnel. Remnants of the abandoned railway track and its bridges are visible in places from the highway, and photos of the early, hazardous days of railroading in the region are on display in the visitors centre at Rogers Pass.

From about 30 kilometres east of Revelstoke almost to the Rocky Mountain Trench, Highway 1 crosses a stratigraphic sequence of mostly low-grade metamorphic sedimentary rocks that were deposited near the continental margin of Laurentia in latest Proterozoic and early Paleozoic time. These rocks can be divided into three big units. The youngest unit, black graphitic phyllite and foliated siltstone called the Lardeau group, with a thin early Cambrian limestone at the base, is exposed in roadcuts and rockslides along the highway from about 30 kilometres east of Revelstoke to 15 kilometres west of the summit of Rogers Pass. It probably was deposited in deep water on the lower part of the old continental slope.

The second unit is buff-coloured to pink massive Hamill quartzite deposited in fairly shallow water during latest Proterozoic to earliest Cambrian time. It forms the steep cliffs and upper parts of the rugged mountains above and just east of Rogers Pass and is exposed in roadcuts and rockslides along Highway 1 as it descends the western side of the Rocky Mountain Trench just above the Purcell fault.

The third and oldest unit, which weathers to a buff to greenish colour, is phyllite, sandstone, and, in a few places, fine-grained garnet-biotite-muscovite schist belonging to the Windermere supergroup of late Proterozoic age, but here called the Horsethief Creek group. It is exposed along Highway 1 east of Rogers Pass, in the Beaver River valley, and in parts of the western valley wall of the Rocky Mountain Trench. All three units have equivalents in the Foreland Belt, although rocks of the same age as the Lardeau group are much more limy there.

These three units in the Selkirk Mountains are folded, thrust-faulted, and reverse-faulted into an enormous structure called the Selkirk fan. The term *fan* comes from the fact that most rocks west of Rogers Pass are folded and thrust westward with east-dipping fault surfaces, whereas those east of the pass are folded and thrust eastward with west-dipping fault surfaces, forming a fanlike array of structures in cross section. The Purcell thrust fault carried the late Proterozoic and younger rocks in the Selkirk fan eastward over Cambrian and Ordovician calcareous shale and phyllite exposed in the bottom of the Rocky Mountain Trench. The surface trace of the Purcell fault is crossed low on the slope of the western side of the trench and about 6 kilometres west of the bridge over the Columbia River. In addition,

Black graphitic phyllite and foliated siltstone of the Lardeau group west of Rogers Pass. Note the tight folds outlined by lighter-coloured siltstone layers and quartz-rich lenses. These early Paleozoic rocks were probably deposited in deep water at about the same time the limestone at Field, near the British Columbia–Alberta boundary, was deposited in shallow water.

View to the east from the summit of Rogers Pass. The massive cliff-forming rocks of the peaks are Hamill quartzite, sandstone that was deposited in shallow water at the edge of the Laurentian continent in latest Precambrian to earliest Cambrian time.

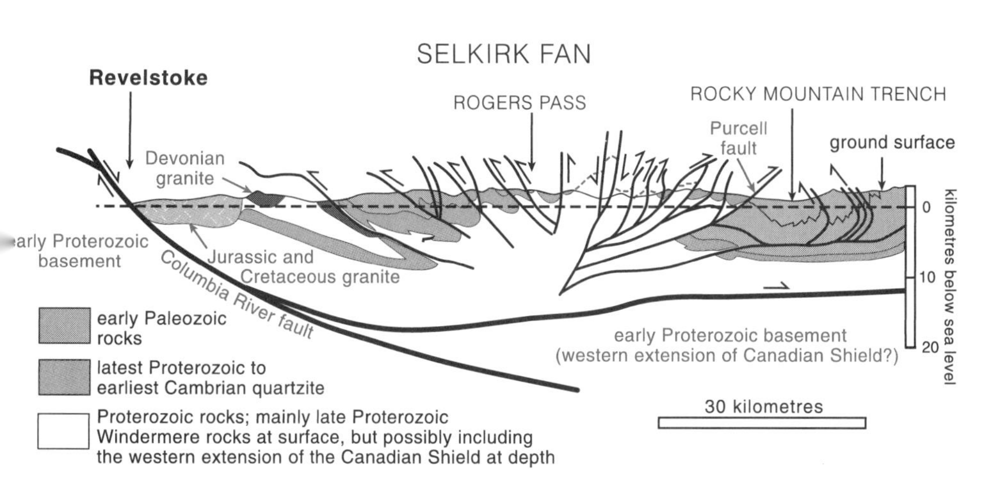

This cross section of the Selkirk fan shows the rock units and structures near the ground surface and their more-speculative distribution at depth.

all rocks in this part of the Selkirk Mountains above a depth of about 15 kilometres are thought to have been carried eastward on a concealed thrust fault that here probably lies at or just above the level of the top surface of the buried Canadian Shield. That thrust fault may be exposed on the elevated west side of the Columbia River fault in the Monashee Mountains. To the east, it is the lowest fault beneath the Foreland Belt, where it absorbs the displacements on the many folds and thrust faults there and eventually ends at the eastern limit of the Cordillera.

Golden Area

Between the bridge over the northward-flowing Columbia River and Golden, Highway 1 follows the Rocky Mountain Trench. It is worth stopping at the Highway 1 viewpoint at the top of the hill just east of Golden for a view westward across the trench. The lower part of the west side of the trench, up to and including the broad bench, is in soft, easily eroded Cambrian and Ordovician shale, calcareous shale, and thinly bedded limestone of the McKay group. Above the bench, the slope becomes steeper. The trace of the Purcell fault follows the break in slope, and above it are the more-resistant late Proterozoic and earliest Paleozoic rocks crossed by Highway 1 on the descent from Rogers Pass.

View westward across the Rocky Mountain Trench from above the east side of Golden. The northernmost Purcell Mountains rise in the distance.

Golden is at the confluence of the Kicking Horse and Columbia Rivers. East of Golden, Highway 1 climbs up over gravel deposits of the delta that formed where the Kicking Horse River flowed into Glacial Lake Invermere at the end of the last ice age. (**see "Glacial Lake Invermere" in Highway 95 and 93/95: Golden—Cranbrook**).

Kicking Horse Canyon

Where Highway 1 enters the narrow canyon of the Kicking Horse River is clearly a bad place to stop. Steel mesh draped over crumbling roadcuts and rock bolt anchors attempt to alleviate the problem of rocks falling on the traveller, but the highly faulted and complexly folded, soft calcareous shale and thinly bedded argillaceous limestone of the Cambrian and Ordovician McKay group is prone to collapse.

To more safely view rocks and structures along this part of Highway 1, we suggest you pull out on the wide shoulder near the south end of the

Roadcut in the westernmost part of the Kicking Horse Canyon contains folded, calcareous shale containing thin, light-coloured limestone beds of the McKay group of Cambrian age. These rocks were deposited in deeper water to the west of the more massive carbonate beds that crop out east of Field.

second bridge over the Kicking Horse River, about 14 kilometres east of Golden. Rocks exposed in the southeast-facing cliff across the river here are thinly bedded limy mudstone and shale of the upper McKay group of latest Cambrian to early Ordovician age, about 490 million years old. The bedding surfaces here are nearly vertical, and a steeply dipping fault cuts the beds. Beds to the southwest (left) of the fault are parallel with it; those to the northeast (right) are parallel with the fault in the lower part of the cliff but truncated by it higher up. How did this structure form? The relationships between the fault and the bedding, when combined with the bigger picture obtained from regional geological mapping, suggest that the fault originated as a relatively flat-lying thrust fault on which rocks above the fault were carried to the northeast. Continuing compression has rotated the beds and the fault into their present orientation. These structures are shown at the eastern (right) end of the Selkirk fan cross section.

Faulted, thinly bedded limestone and shale in a cliff across the Kicking Horse River, viewed from the south end of the Highway 1 bridge 14 miles east of Golden.

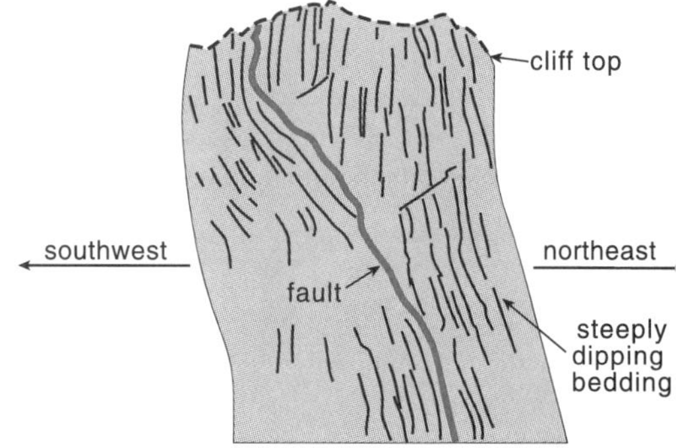

Sketch of the cliff face across Kicking Horse River showing the relationship of the fault and bedding planes.

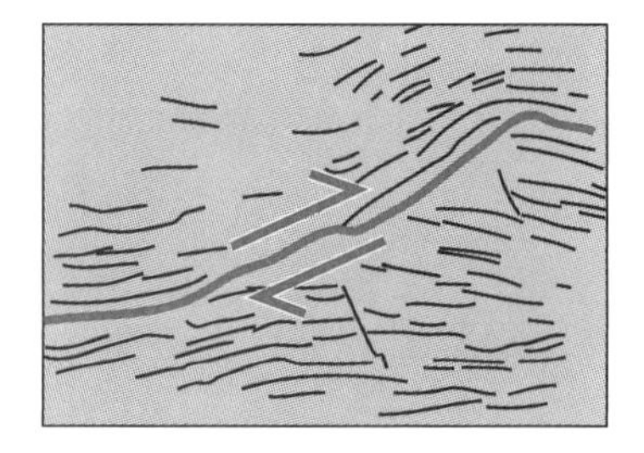

Probable original orientation of the fault and beds before compression rotated them.

The complexly folded and faulted early Paleozoic calcareous shale, thinly bedded limy mudstone, and minor quartzite passed on this part of Highway 1 extend almost as far east as Field, about 40 kilometres east of here near the British Columbia–Alberta boundary. The rocks were deposited at water depths intermediate between those deposited in shallow water, such as the thick deposits of Cambrian limestone encountered near Field, and those deposited in deep water, such as the noncalcareous black graphitic shale and siltstone of the Lardeau group west of Rogers Pass.

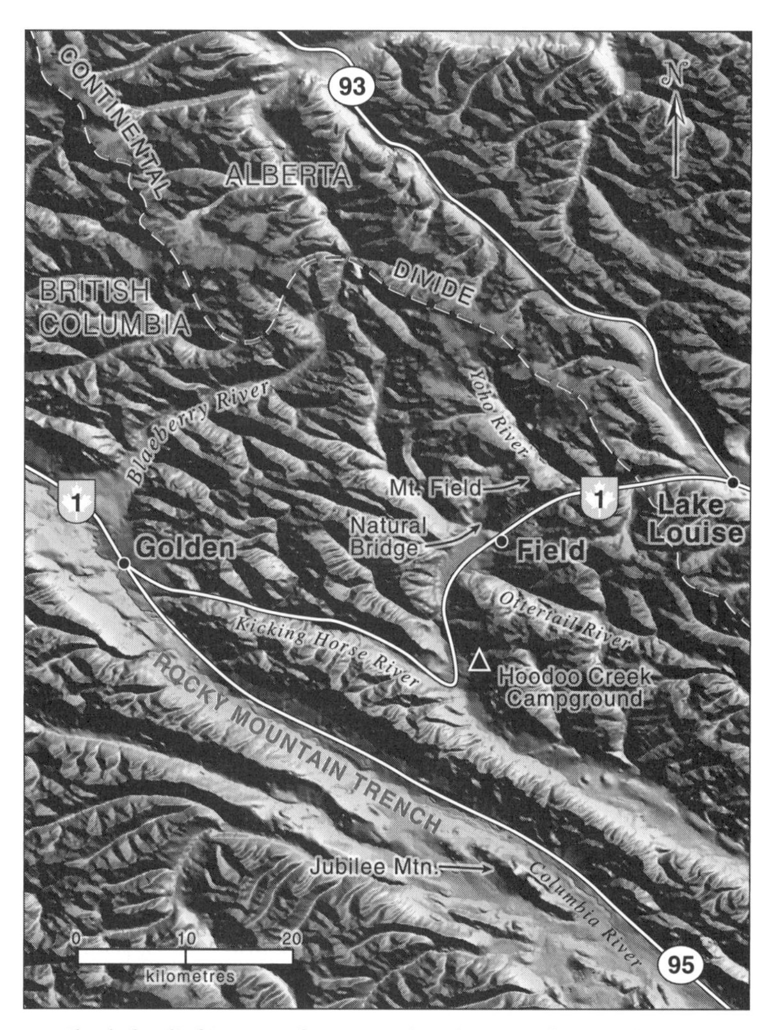

Shaded-relief image of topography along Highway 1 between Golden and Lake Louise. —Image constructed using Canadian Digital Elevation Data obtained from GeoBase

Yoho National Park

About 26 kilometres east of Golden, Highway 1 crosses the western boundary of Yoho National Park; park headquarters are at Field. The park embraces some of Canada's most spectacular scenery, mostly carved in cliff-forming early Paleozoic limestone. Takakkaw Falls in the Yoho River valley spills 254 metres over one of these cliffs. It is the third-highest waterfall in Canada. There are many campsites and hiking trails, but be aware that permits are needed for most activities in the park. Stow away rock hammers—no collecting of rocks, fossils, or plants allowed!

At Hoodoo Creek Campground, on the south side of Kicking Horse Canyon just east of the park's west entrance, a short hike will take you past hoodoos, tall pillars of glacial till. Boulders, the remnants of a harder cap of rock, are balanced on the top of the eroding pillars. Farther upstream, the Kicking Horse River flows through a hole in the rock at Natural Bridge. Turn north on Emerald Lake Road to see this erosional feature.

Kicking Horse Rim and the Burgess Shale

Just east of Field, the softer, thin-bedded, limy shale to the west changes dramatically to the massive, cliff-forming middle Cambrian limestone and dolomite of the Cathedral formation and other units to the east. The massive limestone contains the remains of algae, which need sunlight to flourish, and was laid down in very shallow water as an enormous bank or reef. The bank edge, called the Kicking Horse Rim, was the edge of the Laurentian continental shelf about 500 million years ago. At the time the rocks were being laid down, the rim was a submarine cliff. Oceanward of the cliff, shale was deposited in deeper water.

The Burgess shale is world famous for its beautifully preserved fossils. They are mined from small research quarries located at an elevation of about 2,300 metres above Burgess Pass on the western slope of Mount Field, which is northeast of and on the opposite side of the valley from Field. Fossils of more than 120 different species of hard- and soft-bodied animals, together with the remains of seaweed and other algal material, have been found. The preserved animals apparently lived on the seafloor at the foot of the submarine cliff, and from time to time they were buried with mud swept off or slumped from the cliff. The oxygen content of the sediments containing the fossils was probably very low, so there was relatively little organic decay. The floor of Saanich Inlet, north of Victoria, offers a modern example of the conditions permitting this type of preservation.

An American, Charles Walcott, found the fossils preserved in the Burgess shale in 1909. Walcott, a leading expert in Cambrian paleontology, immediately recognized their importance. In the late 1960s, the Burgess shale

The Burgess shale on Mount Field, home of remarkable fossils, forms the smooth slopes above Burgess Pass.

was resampled as part of a project by the Geological Survey of Canada, and since then research on these fossils has been almost continuous and has resulted in many publications. The exquisite preservation of the fossils in the Burgess shale offers many insights into the "Cambrian explosion," a time when rapid evolution produced a diversity of life-forms, many of which developed hard parts such as shells or bones, which in turn allowed their ready preservation as fossils. Probable fossils of simple organisms such as bacteria are known in rocks as old as 3 billion years, and impressions of soft-bodied animals such as jellyfish are very rare but widely distributed in late Proterozoic rocks, but most major groups of animals first appeared during Cambrian time.

The preservation of these fossils is miraculous. At the quarries, the Burgess shale can be readily split along bedding surfaces, revealing the fossils laid down on them. However, elsewhere in the area, as you can see in roadcuts along Highway 1 just west of Field, the comparable limy mudstone has a strong cleavage imposed during folding. The rocks split on the cleavage, which mostly cuts across bedding, destroying any fossils that may have been present. The cleavage may be undeveloped at the Burgess shale quarries because the adjacent massive carbonate shielded it from most of the effects of deformation.

Visitors are prohibited from collecting fossils in Yoho National Park, but there are several fossil quarries in the park, and guided tours to the quarries can be arranged. Check at the information building in the parking lot by Highway 1 at Field, which has a small display of Burgess shale fossils.

East of Field, in and below the cliffs of Mount Field on the north side of Highway 1, are abandoned tunnels and waste dumps of mines. These were

This fossil crustacean from the Burgess shale was collected by the Royal Ontario Museum in 1991.

in operation until 1946 and produced about 800,000 tons of lead-zinc ore. East of the mine dumps, Highway 1 climbs the hill past the Spiral Tunnels viewpoint. The tunnels were constructed by the Canadian Pacific Railway to reduce the steep gradient between Field and Kicking Horse Pass. One end of the tunnel is above the viewpoint and the other is below.

Roadcuts higher up the hill are in Cambrian carbonate, latest Proterozoic to earliest Cambrian crossbedded quartzite, and grey, green, and red shale, here called the Gog group but equivalent to the cliff-forming Hamill quartzite near Rogers Pass.

Continental Divide

Kicking Horse Pass, at an elevation of 1,627 metres, is on the Continental Divide, which delineates the boundary between British Columbia and Alberta. Rivers west of this boundary flow into the Pacific Ocean. Rivers to the east flow into the Arctic Ocean, Hudson Bay, or the Gulf of Mexico, depending on their latitude. The Rocky Mountain Front, a wall of Paleozoic limestone cliffs, marks the eastern boundary of the mountains about 120 kilometres by road east of the divide. However, the folded and faulted rocks continue still farther east, below the Rocky Mountain foothills, where the Paleozoic rocks are buried beneath Cretaceous sandstone and shale eroded off the emerging mountains. Cordilleran deformation ceases just west of Calgary.

View of the Rocky Mountain Front, looking north from Highway 1 about 75 kilometres by road west of Calgary, Alberta. Cliff-forming Cambrian limestone is thrust eastward, on the McConnell thrust fault, over softer Cretaceous shale and sandstone. The thrust fault, which is gently folded, is at the change in slope at the base of the cliffs.

HIGHWAY 3

Osoyoos—Crowsnest Pass

602 kilometres

Anarchist Mountain Lookout

East of Osoyoos, Highway 3 ascends the steep eastern side of the Okanagan Valley, and near the top of the sparsely vegetated hillside is a lookout on the north side of the highway that provides a splendid view northward along the Okanagan Valley in British Columbia and southward along the same valley, spelled *Okanogan*, in Washington. The lookout is barely 1 kilometre north of the international boundary. Osoyoos is at the south end of Osoyoos Lake, at an elevation of just 275 metres. To the west, the easternmost Cascade Mountains, here made of Intermontane Belt geology, have summits of nearly 2,600 metres.

From this viewpoint you can gain a sense of the enormous thickness of the Cordilleran ice sheets. During the last major glaciation, which culminated

Shaded-relief image of topography along Highway 3 between Osoyoos and Rock Creek and along Highway 33 between Rock Creek and Kelowna.
—Image constructed using Canadian Digital Elevation Data obtained from GeoBase

View northward along the Okanagan Valley from Anarchist Mountain. Osoyoos is below by Osoyoos Lake.

in this area about 15,000 years ago, ice covered the region up to an elevation of about 2,400 metres, which is higher than the rounded mountains flanking the Okanagan Valley. However, rare glacial striae, scratches on the surface of resistant bedrock caused by rocks embedded in the ice, show that the ice overtopped the highest summits in the region during an earlier glaciation.

The bedrock at the lookout consists of granitic rock with a northeast-dipping foliation. Zircons from the granite dated by the uranium-lead method record its time of crystallization at 170 to 160 million years ago, within middle Jurassic time. The rocks belong to an extensive tract of granitic and gneissic rocks that lies almost entirely on the east side of the Okanagan Valley as far north as Vernon, in the uplifted block on the east side of the Okanagan Valley normal fault.

Between Anarchist Mountain and Rock Creek

Granitic rocks in scattered exposures along Highway 3 for 8 kilometres east of Anarchist Mountain intrude poorly exposed, mildly metamorphosed volcanic and sedimentary rocks of late Paleozoic age that most closely resemble rocks in the Slide Mountain terrane, but it is not certain what they are here.

The foliated bedrock at Anarchist Mountain is Jurassic granitic rock.

The last 8 kilometres of the descent to the Kettle River at Rock Creek is almost entirely in Eocene sedimentary, volcanic, and intrusive rocks that weather to a buff colour. Here, sedimentary rocks consisting of tuffaceous sandstone, shale, and some conglomerate and breccia are overlain by volcanic rocks. The volcanic rocks have an unusual composition, like those near Yellow Lake on Highway 3A, with relatively abundant potassium and sodium.

Near Rock Creek, late Paleozoic and early Mesozoic rocks intruded by granitic plutons abut the Eocene rocks. A small area of older rock surrounded by Eocene rock is exposed about 7 kilometres east of Rock Creek.

Rock Creek Placer Gold Deposit

Rock Creek was the focus of an early gold rush. The staking of a placer claim in 1859 attracted hundreds of prospectors the following summer, some of whom came overland from the south to pan for gold in the stream gravels. Rewards were limited and the rush was short-lived, but it alerted Governor Douglas in Victoria to the invasion of commercial interests from the United States. He ordered the extension of the Dewdney Trail from the Pacific Coast to facilitate access to this region. The trail, with additions and modifications, has become Highway 3. Traces of the original road built by the British Royal Engineers are still preserved along Highway 3 between Hope and Manning Provincial Park.

Repetitive Eocene Normal Faults

Between Rock Creek and a point about 5 kilometres east of Midway, most of the rocks are Eocene sedimentary and volcanic rock. The east-dipping sedimentary rocks contain sills and dikes that fed overlying lava flows. The area containing the Eocene rocks is a structural depression, but it is not a simple down-dropped graben bounded on each side by normal faults. The structure involves about fifteen large north- to north-northeast-trending normal faults, as well as innumerable minor ones. The fault traces are mainly, but not entirely, concealed in the creeks and gullies that descend the north side of the valley. The maximum total thickness of the Eocene rocks is only about 3 kilometres, and in most places the thickness is much less, but it superficially appears to be much greater because the faults repeat the same sequence of rock across a horizontal distance of 20 kilometres.

The deposits and accompanying faults formed during the period of regional stretching, about 55 to 45 million years ago, at the same time as the big Okanagan Valley fault was active. Collectively, all the faults exposed in the walls of the Rock Creek–Midway stretch of the Kettle River valley may account for 10 to 20 kilometres of extension. At least some of the faults

View of the north side of the Kettle River valley about 5 kilometres east of Rock Creek. Eocene sedimentary and volcanic rocks are tilted down to the east (right) *at angles between 20 and 50 degrees on repeated normal faults, one of which is shown here.*

were moving at the same time as the sedimentary rocks were being deposited and the overlying volcanic rocks were being extruded. We know this because in places sedimentary beds thicken and coarsen toward the faults, showing that detritus was being eroded off rising fault scarps. In addition, numerous dikes, which are feeders to the volcanic flows, parallel nearby faults, indicating that there was extension during extrusion of the volcanic rocks. A few faults, such as one near the crest of the hill about 5 kilometres east of Midway, are accompanied by large breccia deposits made of large angular blocks of chert, slate, and amphibolite derived from the underlying Paleozoic rocks and from Eocene sandstone. These were formed by enormous landslides originating from the rising fault scarps.

Most faults dip steeply at the surface but a few others dip gently. Some faults that are steep at the surface, curve and become flatter at depth. Not only are they curved vertically, but some are also concave to the west and are described as *listric*, meaning scoop- or shovel-shaped. Downward, the faults may merge with one another into a single big detachment fault like the Okanagan Valley fault. Many fault surfaces are sites of shearing, brecciation, polishing, and deposition of quartz.

This style of faulting resembles *slumping*, a type of landsliding (**see figure on page 298**). In both, rocks break loose along steeply dipping headwalls and their failure or fault surfaces flatten at depth. Rotation of the slump blocks causes formerly horizontal bedding to dip toward the headwall at angles steeper than the underlying failure plane, and the fault displacements typically carry younger strata onto older—the reverse of most thrust faults.

Boundary Mining Camp

A little more than 1 kilometre east of Midway, Highway 3 leaves the Kettle River, which here flows southward into Washington, and veers sharply northeastward up Boundary Creek. For about the next 40 kilometres, it passes through hilly country that contains the remains of several old mines and many holes and tunnels dug to look for lodes of copper, gold, and silver. In the 1890s, this area became the Boundary mining camp. Most of the ore deposits were discovered by traditional prospecting methods helped by the relatively good exposures of bedrock, particularly where forest fires had removed the vegetation. Rail communication with the Castlegar area to the east, and thence by boat and train to the main line of the Canadian Pacific Railway, was provided in the late 1890s, and by 1904, a branch of the Great Northern Railway connected the area with Spokane, Washington. Smelters opened at Grand Forks in 1900 and at Greenwood in 1901, and mine production started. The town of Phoenix was located above its underground mine on the hilltop about 5 kilometres east of Greenwood and was also serviced by railway; remnants of the grade spiral up the hilltop to where Phoenix once was located. Production increased, and at its peak in 1913, production from the Boundary camp exceeded that of any other in British Columbia. The camp declined during World War I, and mining was interrupted in the early 1920s by exhaustion of many of the orebodies and by a shortage of coal for smelting caused by labor problems in the Crowsnest coal fields near the British Columbia–Alberta boundary. The rise of the price of gold in the mid-1930s caused a minor revival, but it was not until 1959 that the Phoenix Mine, like its mythical namesake, rose from the ashes of its predecessor and flourished again. The revival was prompted by improved metal prices, introduction of open-pit mining methods instead of underground tunnelling, and a better geological understanding of the ore deposits.

Pre-Eocene rocks in the Boundary camp area include serpentinite, diorite, amphibolite, greenstone, chert, argillite, and limestone, the latter containing fossils of Carboniferous or Permian age. In their range of rock types and ages, these are oceanic rocks like those in the Slide Mountain terrane. The rocks were folded and faulted at the end of Permian time, and

Shaded-relief image of topography along Highway 3 between Midway and Christina Lake. —Image constructed using Canadian Digital Elevation Data obtained from GeoBase

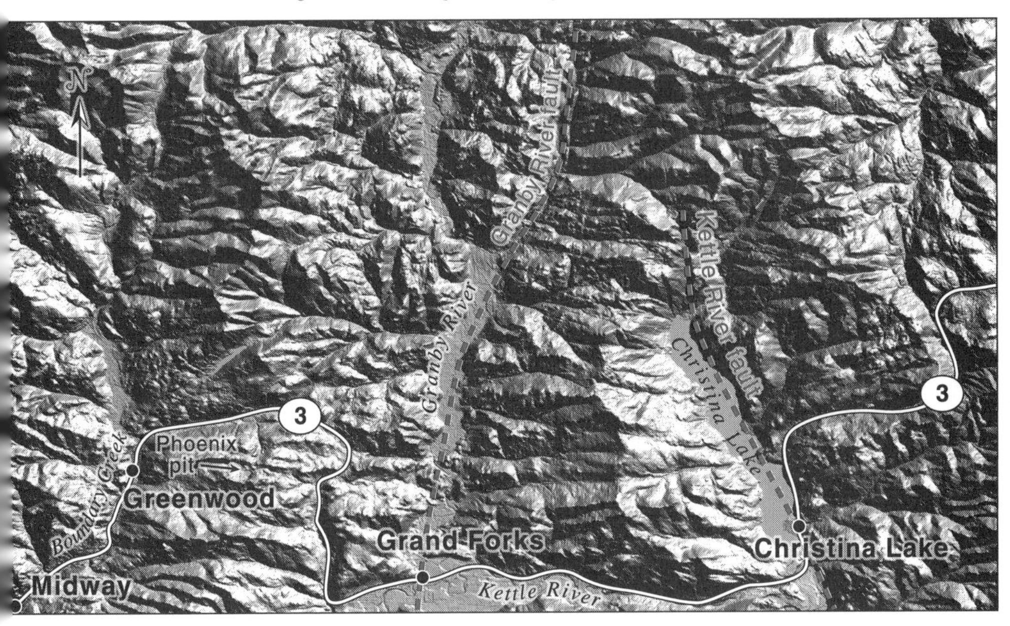

the Triassic strata deposited unconformably on top include chert breccia, locally referred to as "sharpstone conglomerate," with angular fragments of chert, limestone, chert-pebble conglomerate, sandstone, and volcanic rocks resembling some similarly aged rocks in the Quesnel terrane. All of these rocks are intruded by middle Jurassic to early Tertiary granitic bodies and cut by early Tertiary normal faults.

Mineralization is widespread in the area and includes skarn, a mineralized rock formed when limestone is altered. The skarn contains copper and gold, as well as sulfide-rich veins with the minerals chalcopyrite, pyrite, and pyrrhotite, and quartz-pyrite veins with gold. In every case, the rocks hosting these minerals are older than Eocene, so the mineralization predated Eocene time.

Smelter Slag at Greenwood

Greenwood was incorporated in 1897, housed three thousand people by 1899, and started smelting ore in 1901. Some of the larger buildings in town, such as the post office, reflect this prosperous and optimistic era. The present population is about six hundred people, leaving the town with the dubious distinction of being the smallest city in Canada.

At the turn of the century, the standard method of recovering copper, gold, and silver from sulfide ores was by smelting. The metal-bearing rock, plus additives such as quartz or limestone to reduce melting temperature and improve fluidity of the mix, was heated in a furnace. Heavy metals sank to the bottom of the furnace and were drawn off. The less-dense silicate melt, now free of nearly all its metal, was conveyed in large, 25-ton, bell-shaped cars to the slag pile where it chilled to a black glass, which is a kind of artificial lava. The smelter stack and slag pile are visible on the west side of Highway 3, 11 kilometres northeast of Midway and 1 kilometre south of Greenwood.

Phoenix Pit

For those with the willingness to drive a steep, winding road instead of continuing along Highway 3, which loops around to the north of the hills, we recommend a visit to the Phoenix Mine. Turn east in the centre of Greenwood, drive past the post office and out of town, and ascend 600 metres in a road distance of about 8 kilometres. The Phoenix open pit, visible south of the road, is the big hole that occupies the former site of the town of Phoenix, incorporated in 1900. All that remains of the town today is the war memorial, and even that has been moved from its original site. In the Phoenix pit, which is on private property, there is well-bedded, buff- to cream-coloured tuffaceous Eocene sandstone with woody material on

View of the abandoned open pit of the former Phoenix Mine. Tilted, buff to white, well-bedded Eocene sandstone with dark layers (in the left half of the pit) *overlies darker altered and mineralized Triassic strata that weathers to a rusty colour* (in the right half of the pit).

some bedding surfaces, overlying Triassic limestone that has been altered to skarn, as well as other sedimentary rocks.

From the mine site, continue east, crossing a hidden north-northeast-trending, west-side-down normal fault that brings together the upper, volcanic part of the Eocene sequence at Phoenix against late Paleozoic and early Mesozoic rocks to the east. This fault provides still more evidence that stretching during Eocene time probably pervaded the entire region, although such faults are generally hard to see in the pre-Eocene rocks. Continue down the east slope in a series of switchbacks for another 8 kilometres to rejoin Highway 3.

Limestone Cobble Conglomerate of the Quesnel Terrane

A spectacular conglomerate is exposed on the western side of Highway 3, 1 kilometre south of its junction with the eastern end of the Phoenix road. The conglomerate contains angular and rounded white limestone cobbles that are as much as 25 centimetres in diameter, as well as cobbles of volcanic rock and jasper, all embedded in a matrix of maroon and green

Triassic conglomerate with limestone cobbles in a matrix of maroon tuffaceous siltstone.

tuffaceous siltstone. Some of the limestone cobbles contain late Paleozoic fossils, whereas adjacent fine-grained beds of maroon tuff and white limestone in the matrix contain middle and late Triassic fossils. The Paleozoic rock fragments were eroded and incorporated into marine strata of Triassic age. The "sharpstone conglomerate" occurs near here beneath the limestone cobble conglomerate and is made up largely of angular fragments of chert and minor amounts of limestone and greenstone, all embedded in a matrix of chert grains, calcite, and chlorite. The rocks are probably related to the Nicola group of the Quesnel terrane. In this area and to the east, the Triassic rocks are largely sedimentary; the volcanic part of the Nicola group lies mostly west of a line from Kamloops south to between Princeton and Hedley.

For 20 kilometres west of Grand Forks, Highway 3 passes through green volcanic rocks including coarse volcanic breccia, unnamed here but equivalent to early Jurassic volcanic rocks in the Quesnel terrane crossed farther east along Highway 3 near Salmo and called the Rossland group.

Grand Forks Gneiss

The town of Grand Forks lies at the confluence of the Kettle River, flowing northeastward from Washington State, and the south-flowing Granby River. The north-south valley of the Granby River follows the trace of the Granby fault, a big Eocene normal fault that is down-dropped on its west side. Twenty kilometres to the east, the Kettle River fault, another Eocene normal fault, runs northward along Christina Lake. That fault is down-dropped on its east side. Between the two is an elongate, northward-trending, structurally high block, or *horst*, composed of gneiss.

The Grand Forks gneiss is a metamorphic rock derived from sedimentary rocks, lesser amounts of volcanic rock, and former plutonic rocks. Metamorphosed sediments with a relatively high alumina content, indicating the sediments contained a significant amount of mud, here commonly contain the aluminum silicate mineral sillimanite, which forms at temperatures of around 600 degrees Celsius. At such elevated temperatures, layered rocks become relatively plastic and can be easily warped and folded. Tight and overturned folds are common in the Grand Forks gneiss and folded rocks have been refolded. In places, partial rupture and pinching out of less-deformable rock layers form structures called *boudins*, which in two dimensions look like a string of linked sausages.

The Grand Forks gneiss contains well-cemented quartzite most likely derived from quartz sand brought in by streams from thoroughly weathered continental sources. Within the quartzite are former sand grains of isotopically dated zircon that crystallized about 650 million years ago in late Proterozoic time, so the rock must be younger than this. We can speculate that it is equivalent to the thick, latest Proterozoic to earliest Cambrian quartzite (the Hamill and Gog groups) exposed east of Rogers Pass along Highway 1 and in the Rocky Mountains. Other parts of the Grand Forks gneiss contain a lot of radioactive decay products, which hints that those rocks have been around a long time and may be as much as 2.5 to 1.6 billion years old, forming in early Proterozoic time. The rocks also give potassium-argon cooling ages of 60 to 50 million years, indicating that they were hot and cooled rapidly when the gneiss in the horst was uplifted and brought near the surface in early Tertiary time.

The Granby fault separates the Grand Forks gneiss to the east from the mildly metamorphosed late Paleozoic and early Mesozoic rocks, granitic intrusions, and unmetamorphosed Eocene rocks to the west. The Granby fault, and splays off of it, extend for about 70 kilometres north of the international boundary and nearly 100 kilometres south of it into Washington, where its continuation forms the eastern boundary of the Republic graben, a big down-dropped block of Eocene and older strata.

Grand Forks gneiss. Light layers are rich in quartz and feldspar; dark layers are rich in mica and amphibole.

The Granby fault surface is mostly hidden by alluvium along the Kettle and Granby Rivers but is exposed in a few places. You can see one exposure by following the paved road that runs north from the east end of the Granby River bridge in downtown Grand Forks. Follow the road about 6 kilometres up the east bank of the Granby River to Sand Creek. The exposure is half a kilometre up the Sand Creek road. The fault is marked by a crushed zone in greenstone, several tens of metres wide, that dips about 50 degrees to the west. Former chert lenses within the fault zone have been converted to mylonite, which can be mistaken for vein quartz. Gneissic granitic rocks of the Grand Forks gneiss in the uplifted block of the fault are shattered and contain the clay mineral chlorite, a low-grade metamorphic alteration mineral sometimes found along fault zones.

In the absence of a rock unit common to both sides of the fault, the fault displacement cannot be measured directly, although the difference in metamorphic grade of the rocks juxtaposed here by the fault suggests that it is considerable. Guesstimates of minimal amounts of displacement on the fault range from 5 to 10 kilometres.

The Kettle River fault, the east-side-down fault on the east side of the Grand Forks gneiss, begins about 40 kilometres north of the international boundary, runs southward beneath Christina Lake, and continues across the international boundary for about 60 kilometres, following the lower Kettle River valley.

Between Christina Lake and Castlegar

The rocks east of the Kettle River fault are exposed for a distance of about 10 kilometres in roadcuts on the long hill that ascends the valley of McRae Creek. They are far less metamorphosed than the Grand Forks gneiss but still contain foliation layers rather than bedding layers. They probably were derived from late Paleozoic calcareous siltstone and shale. The rocks are intruded by a variety of granitic and syenitic rocks of Jurassic and early Tertiary ages.

Between Christina Lake and Castlegar, Highway 3 crosses a high, rolling, 60-kilometre-wide plateau underlain by mainly granitic rocks of Jurassic and early Tertiary ages. Although exposures of bedrock are reasonably good on the lower slopes above Christina Lake and on the lower part of the descent to the Columbia River, glacial drift covers most of the bedrock on the plateau surface.

Rossland and Trail, south of Highway 3, have important mining connections. Rossland, a well-preserved and scenic old town now developed as the centre of a recreational area, at one time serviced mines of the Rossland mining camp. Production started there in 1894 and ended in 1974. Over that interval the camp was the second-largest producer of gold in British Columbia, after the Bridge River camp, northwest of Lillooet. The ore was in gold- and copper-bearing iron sulfide minerals in veins associated with granitic bodies that intruded volcanic rocks of the early Jurassic Rossland group of the Quesnel terrane. Trail, located on the Columbia River, hosts a major smelter that processes ore from regional mines, but in the past, principally from the former Sullivan Mine at Kimberley.

Castlegar Gneiss

Castlegar sits in a bend of the Columbia River at its confluence with the Kootenay River. A rest area on the west side of Highway 3 about 6 kilometres south of the bridge over the Columbia River provides a good view northward toward the confluence. The valley of the southward-flowing Columbia River is to the west and below the viewpoint; Castlegar is to the north, on its western bank; and the Kootenay River valley is to the northeast. The rock underfoot at the viewpoint is dark, deformed granitic rock of probable Jurassic age with a strong northeast-dipping foliation that is intruded by dark dikes of possible Tertiary age.

The broad, rounded mountain north-northwest of Castlegar is formed of Castlegar gneiss and Cretaceous and early Tertiary granitic rocks. The gneiss is derived from a Cretaceous intrusion from which uranium-lead crystallization dates of zircon give an age of about 110 million years. Potassium-argon cooling dates from biotite within the gneiss are about 50 million years old, which tells us it arrived near the surface at about that time. The shape of the mountain mimics the broad arch structure of the underlying gneiss, whose gently dipping layering is visible northwestward across the river from the viewpoint. In three dimensions, the structure is an elongate dome, about 60 kilometres long and 25 kilometres wide, rather than the arch seen in cross section near its southern end. The once-deeply

Shaded-relief image of topography in the Castlegar region. —Image constructed using Canadian Digital Elevation Data obtained from GeoBase

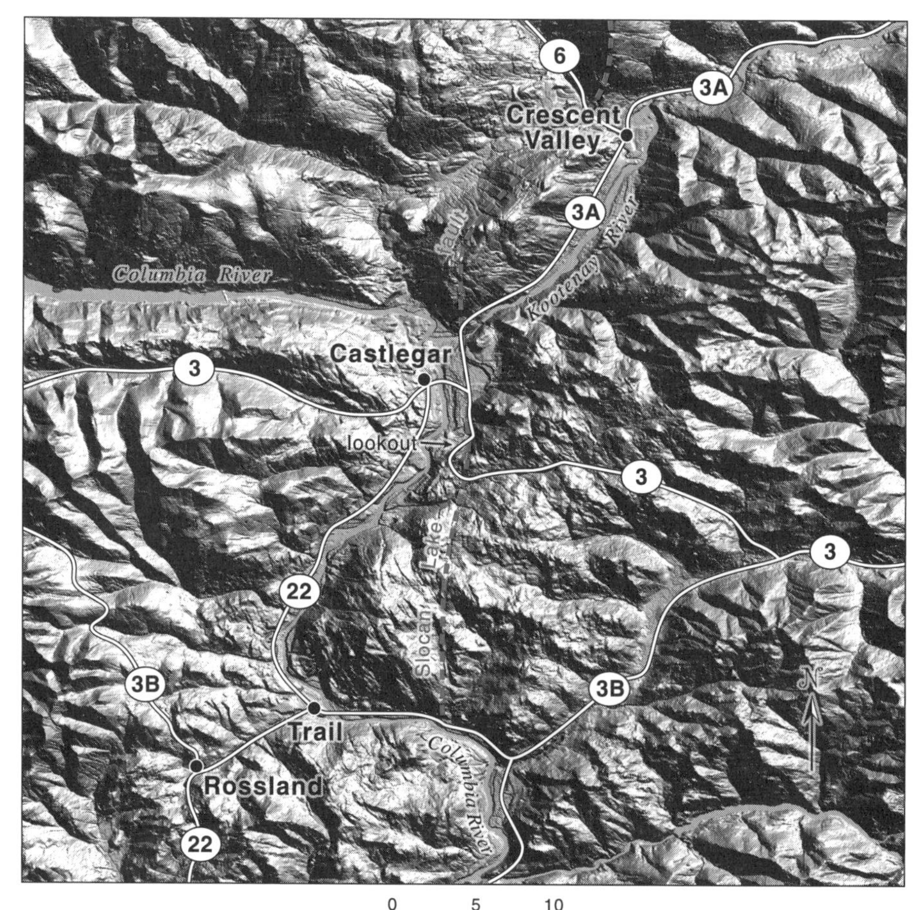

buried gneissic rocks are flanked on both sides by normal faults, somewhat similar to the Grand Forks horst, but here the faults are much less steep. To the west of the gneiss and associated rocks is the gently west-dipping, west-side-down Valkyr shear zone, a big detachment fault akin to the Okanagan Valley fault, above which is the northerly continuation of the Jurassic and Tertiary granitic rock crossed on the plateau between Christina Lake and Castlegar. To the east, the Slocan Lake fault, a big east-dipping, east-side-down normal fault exposed along Highway 6 east of Slocan Lake, separates the Castlegar gneiss and associated rocks to the west from Jurassic granitic rocks and early Mesozoic strata to the east. The southward continuation of the Slocan Lake fault lies about 1 kilometre east of the viewpoint but is not exposed along Highway 3.

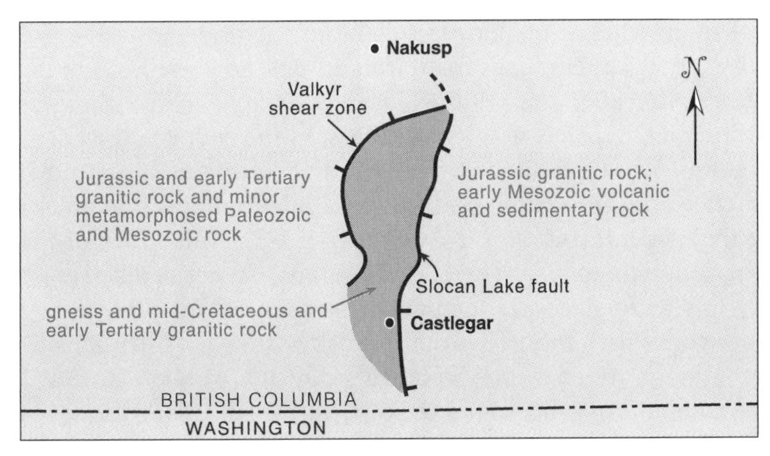

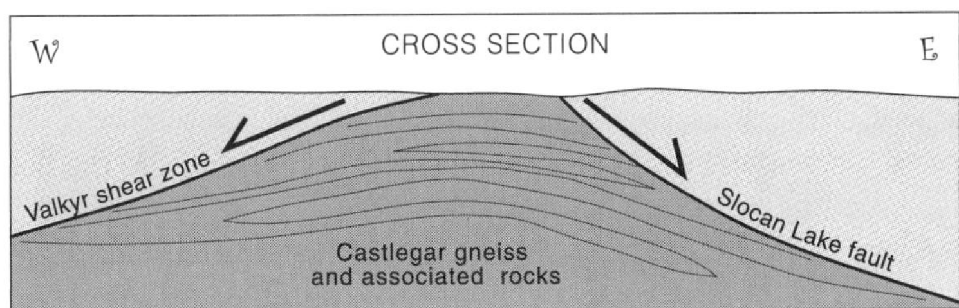

Simplified map and east-west cross section of Castlegar gneiss north of Castlegar.

Studies in the area north of Castlegar enabled geologists to closely pin down the time when the lengthy period of Cordilleran mountain building associated with lateral compression changed to the period of lateral extension, or stretching, that is associated with the normal faulting so widespread across this region. Gneiss with crystallization ages more than 60 million years old contain structures indicating that they formed during compression, whereas structures in rocks with ages less than 55 million years old formed during extension. The gneiss was originally deep in the crust but was brought to the surface when the crust was pulled apart during early Tertiary time. Formerly deep-seated rocks, once buried to about 20 kilometres, formed a series of broad swells and constrictions whereas the cold, brittle, near-surface rocks responded by normal faulting. A major challenge to geologists working in this area is that in order to understand the nature of the older compressional deformation that led to crustal thickening and Cordilleran mountain building, they must "see through" the effects of Eocene extensional deformation.

Between Castlegar and Salmo

East of the viewpoint 6 kilometres south of the bridge over the Columbia River, Highway 3 swings eastward and up and away from the valley, crosses the concealed trace of the Slocan Lake fault, and continues its ascent through the mainly granitic rocks east of the fault. On the north side of Highway 3 about 5 kilometres east of the viewpoint, a roadcut about 400 metres long reveals complex intrusive relationships. In a quick walk along the roadcut, you can see medium-grained granitic rock of Jurassic age, the dominant rock type here, cut by at least two sets of dikes: the oldest is very fine-grained, sugary-textured, light-coloured quartz-feldspar; and the youngest is dark rock. The country rock hosting the intrusions is fine-grained, thinly layered metamorphosed sedimentary rock that weathers to a rusty colour. It may be part of the early Jurassic Rossland group, which, exclusive of intrusions, is the main rock unit between Castlegar and Salmo. In its general nature, the Rossland group resembles the late Triassic rocks of the Nicola group, but it is the more eastern and younger part of the early Mesozoic Quesnel island arc.

East of 1,215-metre Bombi Summit, Highway 3 descends to the floor of the east-trending valley of Beaver Creek and follows this to Salmo. About 3 kilometres west of the junction with Highway 3B from Trail, and for the remaining 15 kilometres or so west of Salmo, the roadcuts are mainly in volcanic and sedimentary rocks of the Rossland group. A large roadcut in dark argillite and siltstone that weather to a rusty colour is about 11 kilometres west of Salmo.

In places, the Rossland group contains fossils of early Jurassic age, some of which are nearly as young as 180 million years old and the youngest marine fossils in the southern Omineca Belt. Shortly after this time, the rocks in this region were folded, thrust-faulted, intruded, and uplifted above sea level. The fossils are important because they provide a maximum age for the initiation of Cordilleran mountain building.

Salmo

Salmo, an old mining town, experienced its boom years between about 1935 and 1970, first with the gold-quartz mines of the Sheep Creek mining camp, about 12 kilometres east of Salmo, and then with lead and zinc mines southeast of town. The productive quartz veins of the Sheep Creek camp occupied steeply dipping fault surfaces in early Cambrian quartzite beds. The lead and zinc deposits were in late Cambrian carbonate. In 1941, one property was discovered to hold commercial quantities of tungsten, at that time of great importance for the war industry but in short supply because the Japanese had invaded China, a major source of this metal. The property was hastened into production, but nearly all of its production took place after World War II.

Shaded-relief image of topography along Highway 3 over the Selkirk Mountains between Salmo and Creston, as well as along Highway 3A between Nelson and Creston. —Image constructed using Canadian Digital Elevation Data obtained from GeoBase

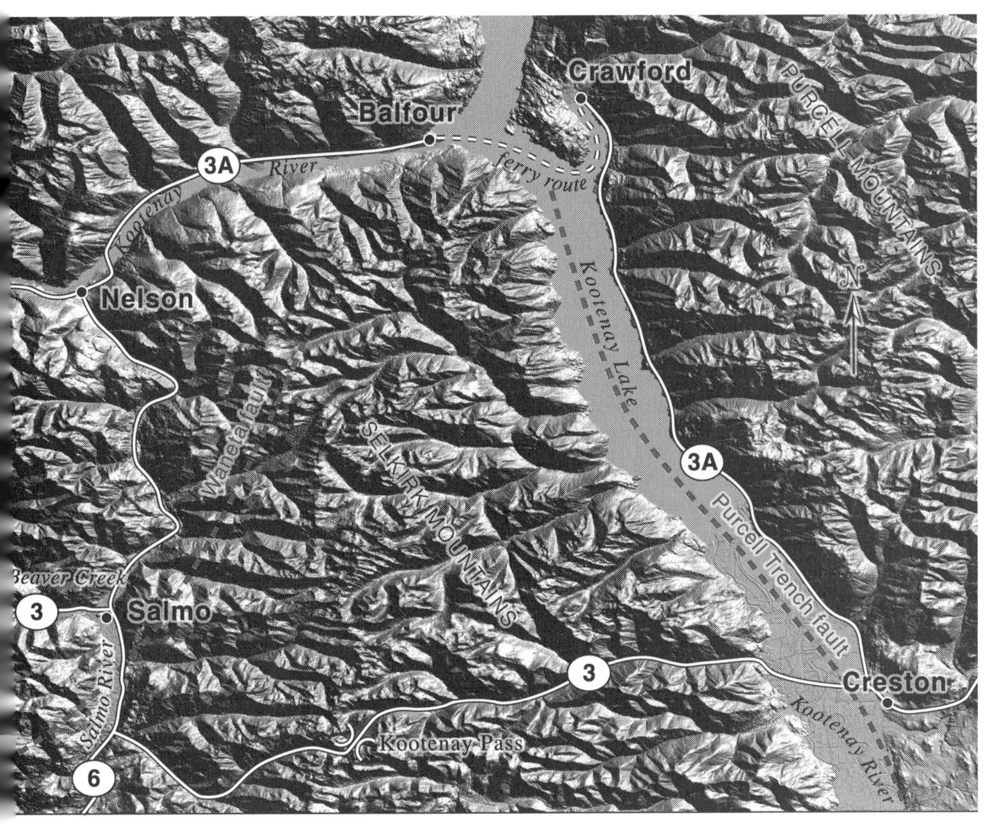

Though the mines are now closed, there is a new geologic attraction in Salmo. Early Cambrian quartzite, hard sandstone composed entirely of quartz grains, has been assembled in slabs of different colours to form mosaics on the exterior walls of buildings in the central part of town. The mosaics depict native animals and mining activities.

Most roadcuts along Highway 3 south of Salmo are in the Rossland group. About 5 kilometres south of Salmo, roadcuts just north of the bridge over the Salmo River are in massive, greyish green, fine-grained volcanic rocks that locally weather to a rusty colour. These are the easternmost exposures of rock along Highway 3 that are included in the Quesnel terrane.

About 4 kilometres south of the bridge over the Salmo River, Highway 3 passes an old tailings pile from one of the several mines in the area. Note the lack of vegetation on the finely crushed rock of the pile. Legislation introduced long after mining operations ceased here now requires that a vegetative cover be planted on such man-made deposits.

Waneta Fault and the Kootenay Arc

Although not exposed on Highway 3, the Waneta fault, a big, east-dipping thrust fault, separates the early Jurassic Rossland group of the Quesnel terrane to the west from overriding Proterozoic and early Paleozoic rocks to the east, which were deposited on and near the Laurentian continental margin. The fault brings the older rocks westward over the younger rocks. The Waneta fault involves rocks as young as early Jurassic in age and is intruded by early Cretaceous granitic rocks, so it is middle or late Jurassic in age. In its general nature, geological relationships, and east-over-west structure, the Waneta fault resembles the Louis Creek and Bolean Creek faults that separate rocks of the Intermontane Belt from those of the Omineca Belt near Pritchard on Highway 1.

The Waneta fault is a structure within the southern part of the Kootenay Arc, a long, narrow, westward-concave belt of very tightly folded and faulted rocks so named because of its arcuate map pattern, *not* its plate tectonic setting. The arc extends about 300 kilometres northward from northeastern Washington, along the northern part of Kootenay Lake, almost to Highway 1 near Revelstoke. For much of its length, the axis of the arc lies between the middle Proterozoic Belt-Purcell strata to the east and the late Paleozoic and early Mesozoic Quesnel and, locally, Slide Mountain terranes to the west. The tightly folded and faulted latest Proterozoic and early Paleozoic strata within the arc are similar in ages and rock types to those in the Selkirk fan, crossed on Highway 1 near Rogers Pass, but they are much more tightly compressed than those near Highway 1.

Why did the Kootenay Arc form? Our best guess is that the arcuate map pattern reflects the original shape of the old continental margin. It may be that the very tight folding characteristic of the arc has something to do with the stiff, strong Belt-Purcell rocks that flanked it on the east, for where they are absent to the north, the rocks are much less tightly folded. The rocks within the Kootenay Arc may have been deposited in or near an enormous bay on the old continental margin of North America that was created when the supercontinent Rodinia rifted apart in late Proterozoic time. The continental rift left stiff, strong middle Proterozoic Belt-Purcell rocks east of the bay (**see "Montania" section later in this roadlog**). During the early stages of Cordilleran mountain building, rocks in the Kootenay Arc were squeezed in the tectonic vise between the Belt-Purcell strata to the east and the suspect terranes to the west.

Between Highway 6 and Creston

East of the junction with Highway 6, Highway 3 crosses the southeastern Selkirk Mountains over 1,774-metre Kootenay Pass, then descends to the Purcell Trench, the great valley containing Kootenay Lake and Creston. This segment of Highway 3 crosses a folded and faulted stratigraphic section that becomes older from west to east, or as geologists might say, goes "down-section," from rocks laid down in early Paleozoic time to those deposited in middle Proterozoic time. This is a time span of over 1 billion years, ranging from about 450 to about 1,500 million years ago. Its western part, on the ascent to Kootenay Pass, crosses mostly Cambrian and Ordovician strata deposited along the newly formed Laurentian continental margin and tightly folded and faulted in the Kootenay Arc in Jurassic time. At Kootenay Pass, Cretaceous granitic rocks intrude late Proterozoic Windermere strata, which are exposed to the east along the upper part of the descent. The Windermere rocks formed during rifting of the supercontinent Rodinia. The lower part of the descent to the Purcell Trench is through middle Proterozoic Belt-Purcell strata, rocks deposited in a basin within an old continent.

It is worth making several stops along this part of Highway 3. About half a kilometre east of the junction with Highway 6 and for the next 2 kilometres are roadcuts in pale grey, white, and locally dark early Cambrian limestone. At about 5.5 kilometres east of the junction, Highway 3 crosses a fault that juxtaposes early Cambrian rocks on the southeast against Ordovician dark argillite of the Active formation to the northwest. Just over 7 kilometres east of the Highway 6 junction and 15 kilometres west of Kootenay Pass, the road crosses latest Proterozoic to earliest Cambrian quartzite, here called the Quartzite Range formation, but elsewhere referred to as the Hamill or Gog formation. The quartzite is exposed in the core of an anticline. In the Sheep Creek camp east of Salmo, about 15 kilometres due north-northwest, vein quartz deposits in these rocks host gold, but here the veins are barren. The highway continues to climb and pass roadcuts in mostly dark Cambrian argillite, slate, and limestone.

About 20 kilometres east of the junction with Highway 6 and 3 kilometres west of the rest stop at Summit Lake is white and pale grey latest Proterozoic to earliest Cambrian quartzite. Just west of the summit are roadcuts in grey and greenish "grit" made up of angular fragments of mainly quartz, a rock type typical of the Windermere supergroup. This rock is intruded at the summit by mid-Cretaceous granitic rock that interrupts the stratigraphic section for a road distance of about 3 kilometres.

The upper part of the downgrade east of the pass is in the Windermere supergroup, here exhibiting a wide variety of rock types. About 6 kilometres east of the summit, a distinctive exposure of late Proterozoic rocks passes from the calcium and magnesium carbonate called dolomite at the western end, through Monk conglomerate with blocks of dolomite in a greenish matrix, to green fragmental Irene volcanic rocks at its eastern end. The

Toby conglomerate, composed of flattened, mostly angular blocks and cobbles of carbonate and quartzite in a phyllitic matrix, here is at the base of the Windermere strata and may be ancient till.

volcanic rocks were extruded during rifting of the supercontinent Rodinia. On the downgrade about 5 to 9 kilometres east of this and 20 to 24 kilometres from the start of the ascent on the west side of the Purcell Trench, are exposures of the Toby conglomerate, a coarse conglomerate containing angular, rounded, and in places flattened fragments of quartzite and dolomite up to 50 centimetres and more across in a matrix with a slaty or phyllitic cleavage. The conglomerate is at the base of the Windermere supergroup and probably was deposited as till during an ancient glaciation, an event recognized in late Proterozoic rocks in many other parts of the world.

Though not visible along Highway 3, the major unconformity between the conglomerate at the base of the late Proterozoic Windermere supergroup and an erosional surface of quartzite and slate of the middle Proterozoic Belt-Purcell supergroup is about 16 kilometres east of Summit Lake. A sobering thought is that the unconformity here probably spans a time gap of over 500 million years, representing an interval equal to that between earliest Paleozoic time and the present.

About 14 kilometres east of the hidden contact, outcrops of metamorphosed Belt-Purcell quartzite and mica schist contain the aluminum silicate minerals sillimanite and kyanite and many small quartz-feldspar intrusions. The outcrops have weathered to a rusty colour. Here, the metamorphism obscures the original nature of the rocks, which are far better preserved to the east of the Purcell Trench.

Purcell Trench at Creston

Highway 3 crosses the floor of the southern part of the Purcell Trench, the large valley that separates the Selkirk Mountains to the west from the Purcell Mountains to the east. The flat floor of the trench, here about 8 kilometres across, is the floodplain of the northward-flowing Kootenay River before it enters Kootenay Lake. Today the flats are mostly rich farmland, with wetlands set aside for wildlife, but in their natural state they were subject to annual floods. The first successful reclamation for farming was in 1891, and today about 80 kilometres of dikes protect the farmland from flooding. The south end of Kootenay Lake is about 10 kilometres north of Highway 3, and the lake occupies the trench floor for a distance of 100 kilometres farther north. The outflow of the lake is about halfway along its western side, where the lower part of the Kootenay River drains southwestward to join the Columbia River at Castlegar.

For a distance of about 120 kilometres, from northern Idaho to about halfway along Kootenay Lake, the southern Purcell Trench follows the trace of a west-side-up normal fault, the Purcell Trench fault. This fault is unrelated to the Purcell fault, which is a thrust fault that parallels the Rocky

Mountain Trench near Golden. The Purcell Trench fault brings high-grade metamorphic Belt-Purcell strata west of the trench against less-metamorphosed Belt-Purcell rocks to the east. The fault cuts and offsets mid-Cretaceous granitic intrusions and appears to be yet another of the ubiquitous early Tertiary normal faults in the region.

Along the northern part of Kootenay Lake, the Purcell Trench follows rocks and structures of the Kootenay Arc. North of the lake, the trench is the valley of the south-flowing Duncan River and then crosses a drainage divide to align with the north-flowing Beaver River, which joins the Columbia River in the Rocky Mountain Trench north of Golden.

Purcell Mountains

The Purcell Mountains are made mostly of rocks of the Belt-Purcell supergroup, and this part of Highway 3 lies entirely within these rocks. The supergroup is enormously thick, over 11 kilometres in Canada and 20 kilometres in the United States. For comparison, stable continental crust is about 30 to 45 kilometres thick. Detailed mapping, section measuring, and correlation of rock units from place to place is needed to establish the stratigraphic order because the mainly sedimentary rocks within the supergroup contain no fossils apart from *stromatolites*, which are mounds formed by algal communities that trap sediment. Such detailed studies have been encouraged by the presence within the supergroup of many lead and zinc showings and the enormous Sullivan Mine at Kimberley, but to date no more Sullivan-sized deposits have been found. Isotopic dates from lava, related sills, and small intrusions that crystallized during deposition of the sedimentary rocks tell us the supergroup was laid down in middle Proterozoic time, between 1.5 to 1.4 billion years ago.

The oldest part of the Belt-Purcell supergroup is the most common rock unit along this segment of Highway 3. It consists of monotonous, dark to medium grey, thick to thinly bedded argillaceous quartzite and argillite that were deposited in deep water by turbidites, sediment-rich slurries that move downslope under the influence of gravity and settle out in the bottom of a basin. Modern analogues form on the continental slope, and on a smaller scale, within the Strait of Georgia where they originate from the Fraser River delta. In places, the sedimentary rocks host diorite and gabbro sills.

The abundance of quartz grains in these rocks indicates that they were eroded from a continent. Measurement of the orientation of sedimentary structures in them, such as ripple marks, allows us to determine that the directions of ancient currents were mostly from the east but locally from the southwest. This suggests that Belt-Purcell sediments were deposited in a trough that lay within a continent. Part of that old continent rifted

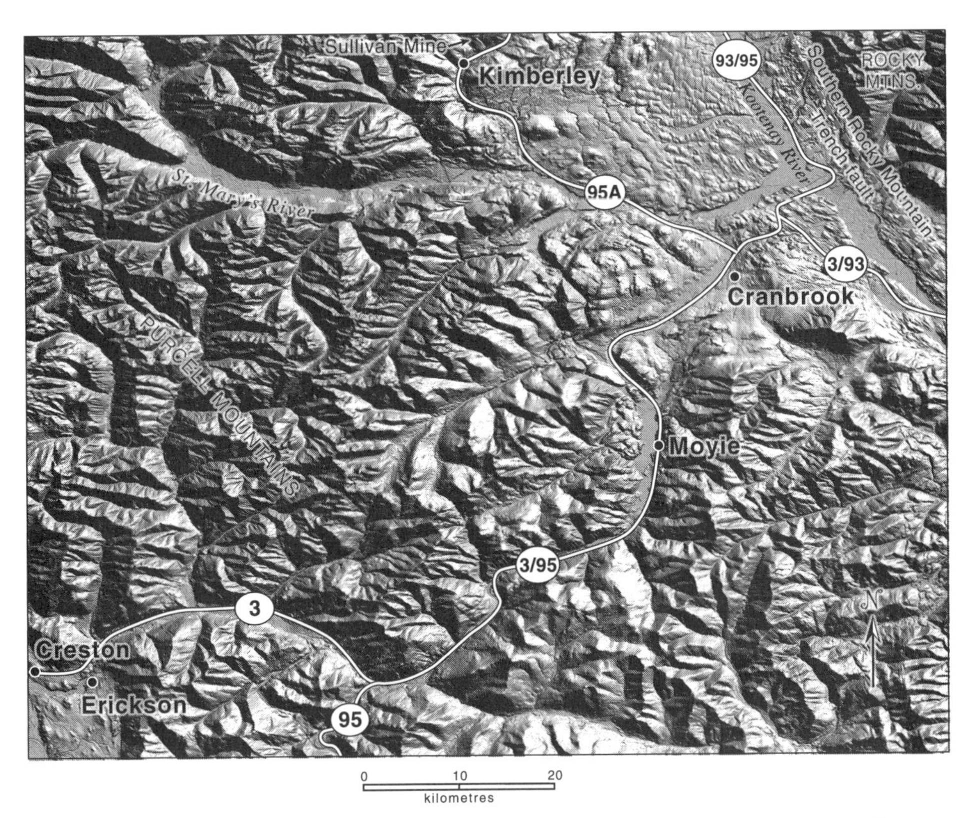

Shaded-relief image of topography along Highway 3 between Creston and Cranbrook. —Image constructed using Canadian Digital Elevation Data obtained from GeoBase

off and drifted away in latest Proterozoic to early Cambrian time. No one knows for certain where it is today, but rocks found today in both Australia and Siberia have been proposed as candidates.

About 4 kilometres east of Creston, Highway 3 passes the village of Erickson; from here you can look to the southeast and see the eastern wall of the Purcell Trench, appropriately called the Ramparts. These cliffs of the massively bedded older part of the Belt-Purcell sequence contain many sills of gabbro and diorite.

East of Erickson, scattered roadcuts are in rusty-coloured sandstone of the older part of the Belt-Purcell supergroup. The rusty colour comes from the weathering of finely disseminated iron sulfide within the rocks. A small quarry about 7 kilometres east of Creston exposes fine-grained gabbro that is part of one of the many sills within the Belt-Purcell rocks. About 54 kilometres east of Creston and 45 kilometres west of Cranbrook, a series of roadcuts exposes sedimentary rocks in the older part of the supergroup.

The Ramparts, on the east side of the Purcell Trench near Creston, are made of the older part of the Belt-Purcell supergroup and contain thick sills of diorite and gabbro.

The rocks here are generally hard, rusty-weathering quartzite, siltstone, and argillite that are well bedded and contain sedimentary structures, such as graded bedding in which each bed displays a progressive change in particle size, that indicate the sediments were deposited as turbidites in deep water. The metamorphic grade of these rocks is higher than one would expect from a casual glance. Fine-grained chlorite and biotite indicate that the rocks were heated to between 300 and 500 degrees Celsius.

About 55 kilometres east of Creston and 47 kilometres west of Cranbrook is a large sill of gabbro, one of many that probably were intruded into the hosting sedimentary rocks shortly after they were deposited. A uranium-lead date from this sill shows that it crystallized about 1.4 billion years ago and probably was intruded only slightly after deposition of the host rocks.

Between about 65 and 75 kilometres east of Creston, and 28 and 38 kilometres west of Cranbrook are the remains of several old mines. Mining started here in the 1890s, as in many other places in southern British Columbia, and lead, zinc, and silver were produced.

About 6 kilometres east of Moyie are brightly coloured rocks in the middle part of the Belt-Purcell sequence, whose lighter colours contrast with the typically somber and rusty-weathering rocks of the older part. They include green, grey, brown, and purple siltstone and argillite deposited in shallow water. The rocks here grade upward and eastward into dolomite-rich deposits containing algal mounds. The algae—which needed sunlight to grow—together with ripple marks, mud cracks, and in a few places salt casts, indicate that most of the rocks higher in the sequence were deposited in very shallow water, in intertidal or floodplain settings. Green and maroon basaltic flows are present in places.

Near the junction with Lumberton Road, about 10 kilometres west of Cranbrook, an outcrop of rocks belonging to the older part of the Belt-Purcell sequence features graded beds of quartz-rich sandstone interbedded with argillite that probably were deposited in deep water, near the toe of a submarine fan. A fault has carried these older beds up into the younger, shallow-water deposits.

The uppermost part of the Belt-Purcell section, exposed along Highway 3 mainly east of Cranbrook, is composed of deposits of grey, green, and red carbonate, argillite, and siltstone also deposited in shallow water, probably on tidal flats.

Well-bedded quartz-rich sandstone is interlayered with green to black slightly rusty argillite in the older part of the Belt-Purcell supergroup near the junction of Highway 3 with Lumberton Road.

SIDE TRIP TO KIMBERLEY

Highway 95A branches from Highway 3 about 3 kilometres east of Cranbrook and heads northwestward to Kimberley in what miners might call "elephant country." "Elephant" is the name sometimes used by economic geologists for giant mineral deposits, and the deposit exploited by the Sullivan Mine at Kimberley certainly qualifies for the name. The deposit was discovered in 1892, mining commenced in 1900, and after a shaky initial period, this became the largest single source of lead in the British Commonwealth during World War I. The mine, which was almost entirely underground, finally closed in 2002.

The mine exploited one of the largest lead, zinc, and silver deposits known on Earth. It contained 160 million tons of ore containing 6.5 per cent lead, 5.6 per cent zinc and 67 grams per tonne of silver. The orebody was about 1.6 by 2 kilometres in area and up to 100 metres thick. The extracted ore was moved by rail to the large smelter in Trail.

The orebody was in the older part of the Belt-Purcell supergroup, in rocks similar to those exposed along most of Highway 3 across the Purcell Mountains. Much of the ore is spectacular, with argillite layers interbedded with the lead sulfide mineral galena and the zinc sulfide mineral sphalerite. Careful study indicated that the sulfide was deposited in part from solution and in part as fine sulfide particles. Hot fluids, heated by the same magmas that formed the sills exposed along Highway 3, carried metals in solution up along faults that were active during sedimentation. Some cooled within the seafloor, while others were expelled into the seawater from hot springs emerging from the floor of the basin, analogous to the modern sulfide-rich "black smokers," or mounded hot spring deposits, found today on mid-ocean ridges.

Southern Rocky Mountain Trench Fault

About 12 kilometres east of Cranbrook, Highway 3 enters a wide part of the Rocky Mountain Trench that here contains the southward-flowing Kootenay River. East of the trench are the westernmost ranges of the southern Canadian Rocky Mountains, which are included in the Foreland Belt, whereas the Purcell Mountains to the west belong to the Omineca Belt.

The southernmost segment of the trench is controlled by a west-side-down normal fault, the Southern Rocky Mountain Trench fault. The structure of the trench here is a trapdoor-like structure called a *half-graben*. The hinge of the "trapdoor" is on the west side of the trench and the location of the bounding normal fault is near the foot of the prominent, 1,500-metre-high eastern wall of the trench called the Steeples. This is a classic *fault-line*

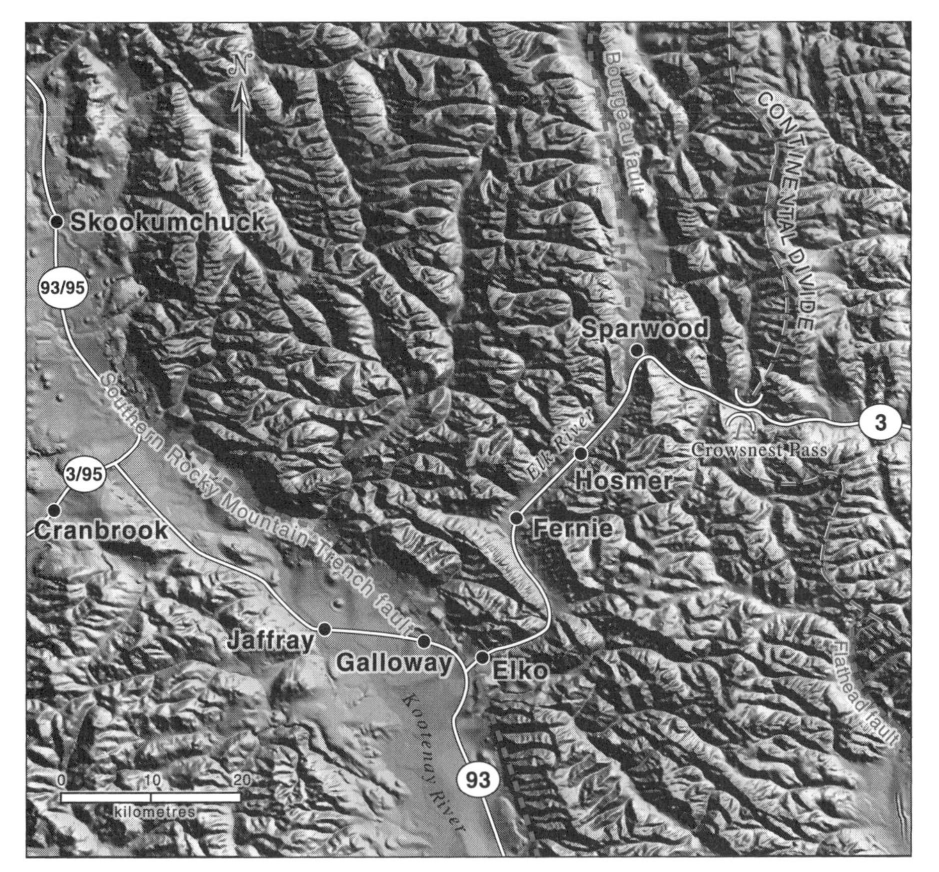

Shaded-relief image of topography along Highway 3 between Cranbrook and Crowsnest Pass. —Image using Canadian Digital Elevation Data obtained from GeoBase

scarp, an escarpment formed by differential erosion along a fault so that resistant rock on one side of the fault stands up and softer rock on the other side is eroded away. The youngest movement on the fault, amounting to about 600 metres, took place after Miocene time, that is, in the last 5 million years. Seismic reflection studies show that below the Pleistocene and recent sediments covering the trench floor, Tertiary strata dip eastward into the fault surface. At shallow depths, the fault surface dips westward at an angle of about 40 degrees, but deeper down it curves and becomes flatter. Such curved faults are called *listric* (shovel- or scoop-shaped) faults.

Looking north from near Jaffray, you can see the fault-line scarp of the snowcapped ridge of the Steeples, made of middle Proterozic Belt-Purcell rocks that are overlain on their eastern side by Paleozoic strata. The Southern Rocky Mountain Trench fault lies near the foot of the ridge. The rounded

View northward along the southern Rocky Mountain Trench from near Jaffray on Highway 3. Low hills (left centre) *are late Paleozoic limestone and dolomite down-dropped on the Southern Rocky Mountain Trench fault. The trace of the fault is at the base of the Steeples* (right), *which are middle Proterozoic Belt-Purcell strata overlain by late Paleozoic strata.*

hills low in the trench west of the ridge are late Paleozoic limestone and dolomite overlying the Belt-Purcell strata in the down-dropped block west of the fault.

Montania

Southeast of Cranbrook, the bedrock that forms low rounded hills in the trench on the western, down-dropped side of the Southern Rocky Mountain Trench fault is exposed in roadcuts along Highway 3 between 22 and 32 kilometres southeast of Cranbrook (12 to 22 kilometres west of Jaffray). Here it is early Carboniferous northeast-dipping limestone correlated with the Rundle formation, and just east of the bridge over the Kootenay River, Devonian dolomite that weathers to a brownish colour is correlated with the Palliser formation.

Along Highway 1 near Banff, the Rundle and Palliser formations lie east of the Bourgeau fault and were deposited, together with underlying Cambrian strata, on the edge of the Laurentian continent, above the late Archean and early Proterozoic rocks of the Canadian Shield. Here, in the southeastern Purcell Mountains and adjoining Rocky Mountains in Canada and nearby parts of the United States, the middle and late Paleozoic strata were deposited, not on the Canadian Shield, but on top of the enormously thick block of middle Proterozoic Belt-Purcell supergroup. This block, called Montania, acted physically just like the old high-standing continent. Only relatively thin sediments, if any at all, were deposited on it, and the entire late Proterozoic and early Paleozoic section, which is very thick in the western Rockies along Highway 1 west of Banff, is absent near this part of Highway 3.

In earliest Tertiary time, before about 60 million years ago, the Belt-Purcell rocks and their thin middle to late Paleozoic cover were carried eastward over the edge of the old continent on the large Lewis thrust fault. The fault is about 10 kilometres deep below here but surfaces about 25 kilometres east of the boundary between British Columbia and Alberta and crosses the international boundary.

Southeast of Galloway, the highway is very straight for a distance of about 8 kilometres and here follows the trace of the Southern Rocky Mountain Trench fault. The highway passes through green and grey siltstone and shale of the younger part of the Belt-Purcell strata, and east of the junction of Highways 3 and 93 near Elko, through still more of the younger part of the Belt-Purcell beds, some of which are bright red.

Northeast of Elko, Highway 3 follows the canyon of the southwest-draining Elk River, which cuts through the fault-line scarp of the Southern Rocky Mountain Trench fault to join the Kootenay River in the floor of the Rocky Mountain Trench about 20 kilometres north of the international boundary. The Elk River canyon cuts across the regional grain through middle Proterozoic, Paleozoic, and early Mesozoic strata that have been folded and thrust-faulted. The highway passes through a short tunnel in Carboniferous Rundle limestone and eventually emerges into a wider part of the Elk River valley that contains the town of Fernie.

Southern Foreland Belt Structures

The easternmost part of Highway 3 is the only place along the roads covered in this guidebook where the classical fold-and-thrust structure so characteristic of much of the Foreland Belt is encountered, although here the structure is quite complicated. Horizontal compression, acting on the originally flat-lying, well-bedded sedimentary deposits typical of the Foreland Belt, generated the folds and thrusts. In many places, the same layer of rock may be stacked upon itself again and again by repeated thrusts that become younger downward. Continued compression may also fold earlier thrust faults to produce still more structural complexity.

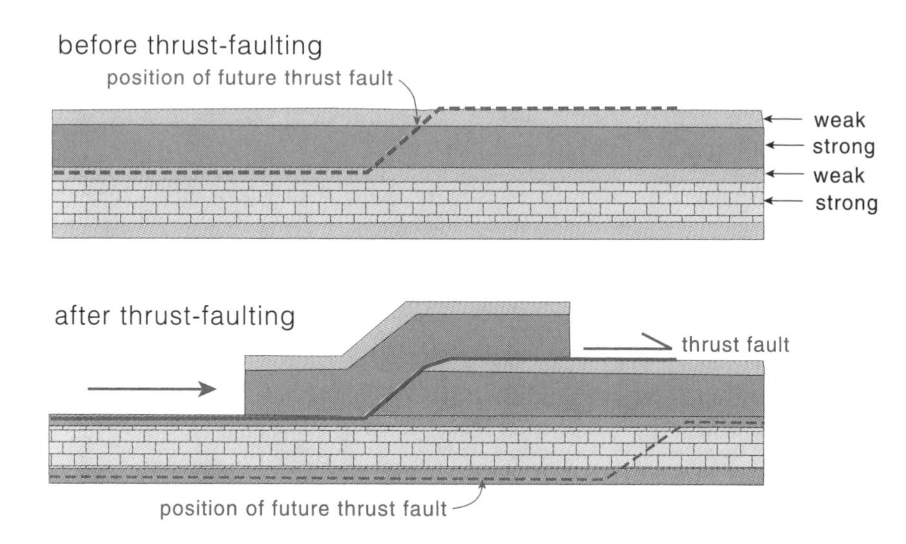

The Foreland Belt thrust faults tend to lie within weak layers, such as shale or shaley limestone, but in strong layers, such as massive limestone or quartzite, the faults cut obliquely across layers in the direction of thrusting. Once the thrust fault has formed, the next one to form will be beneath it.

The thrust faults carried rocks eastward and over the top of the western part of the continental margin. In general, the oldest thrust faults, of latest Jurassic or early Cretaceous age, are structurally high and to the west, and the youngest thrust faults and folds, of earliest Tertiary age, are structurally low and to the east. The easternmost folds and thrusts mark the eastern limit of Cordilleran deformation. Geologists have studied these structures for many years, not the least because the rocks affected by them potentially host deposits of oil, gas, and coal.

Fernie Basin and the Hosmer Fault

Fernie sits on the northwestern side of what is called the Fernie basin, an area about 60 kilometres long by 25 kilometres wide of Jurassic and Cretaceous shale and sandstone surrounded by harder Paleozoic limestone. The shale, sandstone, and local conglomerate belong to the Kootenay group and were deposited in a foreland basin that received the earliest sediment eroded from the emerging Cordilleran mountains to the west in Jurassic time.

Near Fernie, four major rock packages are involved in folding and thrust-faulting: the Proterozoic Belt-Purcell supergroup to the west; the relatively

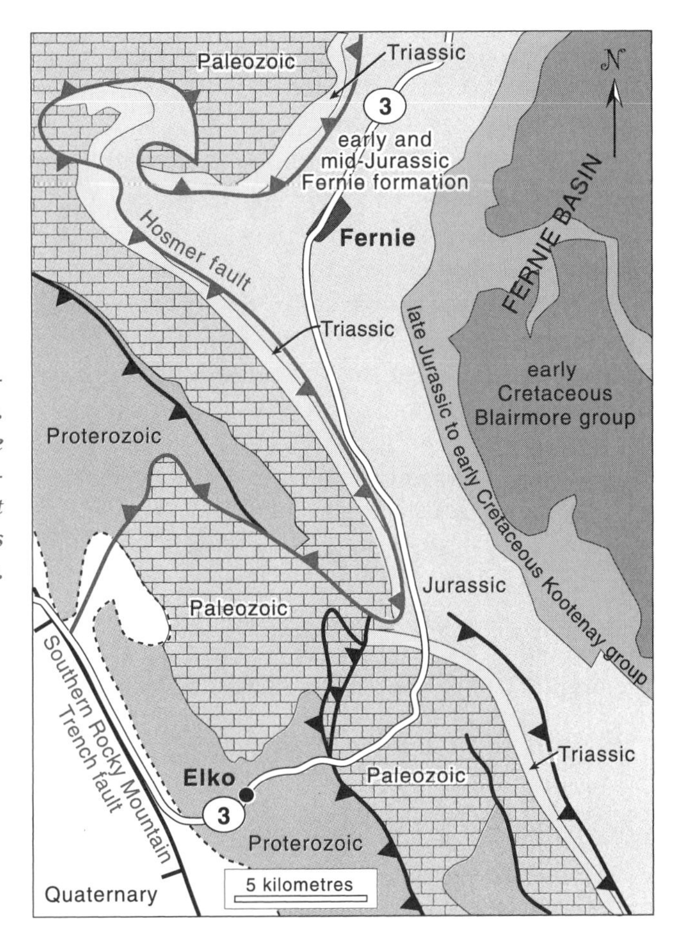

Rocks and structures near Fernie. Normal faults have bars on downthrown side; thrust faults have barbs on overriding side.

thin middle Paleozoic and Triassic sequence of limestone, dolomite, siltstone, and shale characteristic of Montania and the continental platform; early and middle Jurassic black shale of the Fernie formation; and the late Jurassic through early Cretaceous marine and nonmarine shale and sandstone of the Kootenay and Blairmore groups that collected in the foreland basin. The Kootenay group is economically important as it contains enormous coal deposits.

Scattered roadcuts along the western side of the Elk River valley southwest of Fernie are in soft, dark Jurassic Fernie shale, which is readily eroded by the river. Just above these rocks on the west side of the Elk Valley is the Hosmer fault, a big thrust fault deformed by later folding and faulting. Above the fault is an overturned section of rock: lowermost are dark Triassic and latest Paleozoic sandstone and shale; above is cliff-forming Carboniferous and Devonian limestone that caps the ridges; and uppermost and to the west are middle Proterozoic rocks. The overturned section above the thrust fault is best viewed by looking west from Highway 3 just west of Fernie.

The section underlying the treed mountains south and east of Fernie is right-side-up, with early and middle Jurassic Fernie marine shale in the floor of the valley and more massive Kootenay sandstone and conglomerate of latest Jurassic and earliest Cretaceous age higher up on the ridges that are capped by early Cretaceous Blairmore shale.

A good place to get a closer view of the Hosmer fault is near the bridge over the Elk River at the north end of Fernie. The dark rocks underlying the lower slopes of the valley north of town are Jurassic Fernie marine shale and siltstone. The break in slope is close to the trace of the Hosmer fault, above which are cliffs of light-coloured Paleozoic limestone.

Mount Hosmer, northwest of Highway 3, is a *klippe*, a piece of a once-continuous thrust sheet that has been isolated by erosion from the main part of the sheet. A good place to view the klippe is near the Highway 3 bridge over the Elk River at Hosmer, about 10 kilometres northeast of Fernie. The upper part of Mount Hosmer features cliffs of Devonian and Carboniferous limestone in which bedding surfaces dip steeply into the Hosmer fault, below which is Jurassic Fernie shale in the trees.

View of the south side of Mount Hosmer from near the bridge over the Elk River. The Hosmer fault is at the break in slope, below the cliffs and above the treed area, below which are Jurassic strata. The Devonian and Carboniferous limestone above the thrust fault moved from west **(left)** ***to east*** **(right)*****. The fault trace runs around the entire mountain, so rocks above the fault form a klippe.***

This entire package of rocks and structures has been carried eastward on the back of another big thrust fault, younger than the Hosmer fault, called the Lewis fault, which is of earliest Tertiary age, about 60 million years old. This fault is buried here but reaches the surface 25 kilometres east of the British Columbia–Alberta boundary. It carried the Belt-Purcell strata and their cover of younger rocks up and onto the old continent. The Lewis fault can be traced more or less continuously along its length for over 500 kilometres in a north-south direction.

Highway 3 northeast of Hosmer, and the north-south part of the Elk Valley north of Sparwood, runs close to the trace of the southernmost part of yet another big thrust fault, one that is sandwiched between the overlying Hosmer fault and the underlying Lewis fault. This thrust fault, the Bourgeau fault, runs northward from near here for about 400 kilometres and is crossed by Highway 1 near Banff, Alberta. In this area, the thrust fault is hidden within soft Jurassic shale, but its presence can be recognized from differences in the Paleozoic strata on either side of the strip of Jurassic shale that lies low in the valley.

Man-made Topography: Big Coal at Sparwood

At Sparwood, Highway 3 turns southeastward. At the bend, the Elk River continues northward along a valley carved in soft Jurassic shale along the trace of the Bourgeau fault. The Elkview coal mine lies mostly on the mountaintop northeast of Sparwood. The size of this enormous open-pit operation is difficult to comprehend from the level of the highway. The mine shapes the topography: the horizontal flat surfaces visible on the crest northeast of the town and the flat-topped hillside visible from Highway 3 about 5 kilometres east of town are both parts of the same operation.

The raw coal is conveyed from the mine to the preparation plant located just north of Sparwood, where in 2002 about 30,000 tonnes of raw coal per day were processed. The coal is distributed worldwide, and the mine is expected to have a future life of about forty years. Most of the coal is moved by rail from this area westward for a distance of over 1,100 kilometres, up the Rocky Mountain Trench to Golden, and then west across the mountains along the route followed by Highway 1 to the coal port near the ferry terminal at Tsawwassen. This is only one of five large open-pit mines within a radius of 80 kilometres of Sparwood. Three others are near the Elk River valley north of Sparwood, and one to the southeast. An information office in Sparwood, near the 350-ton capacity dump truck, provides details of the mining operations.

The coal is in numerous seams that range from 2 to 15 metres thick within a 1.2-kilometre-thick sequence of sandstone, shale, and conglomerate of

View looking north from Highway 3 southeast of Sparwood. The sloping hillside is reclaimed waste from the open-pit Elkview coal mine, located above on the mountain.

the Kootenay group of latest Jurassic and earliest Cretaceous age, about 145 million years old. The beds originated as sand, mud, and gravel deposited near sea level by rivers flowing from the southwest that carried detritus eroded from then-uplifting mountains on the site of the Omineca Belt. The coal is derived from peat deposited in nearshore swampy areas. The gradational contact between the older Jurassic Fernie marine sediments and the latest Jurassic to earliest Cretaceous Kootenay group is exposed in a roadcut on the south side of Highway 3 near the bridge over Michel Creek, about 2 kilometres east of Sparwood.

The Geological Survey of Canada reported the presence of coal in this area in 1884. After a railroad was built in the 1890s, underground mines were developed and were active from 1899 until about 1980. You can see entrances to some of the old underground mines and mine buildings along Highway 3 between Sparwood and Natal. Open-pit mining started in the late 1960s and, after some initial engineering problems related to the complex geologic structures in the area, developed into the present enormous operations.

About 12 kilometres southeast of Sparwood, Highway 3 crosses the trace of a west-side-down normal fault along which latest Jurassic to earliest

Cretaceous strata west of the fault have been dropped down against late Paleozoic and early Mesozoic limestone, sandstone, and shale. This fault may be the northern continuation of the big Flathead fault, which straddles the international boundary about 80 kilometres south-southeast of here. Like the Southern Rocky Mountain Trench fault, it is a normal fault bounding the eastern side of a half-graben that contains down-dropped sedimentary rocks as young as Oligocene, about 30 million years old. It is the easternmost of the many big Tertiary normal faults widespread in southeastern British Columbia.

Highway 3 crosses the Continental Divide and enters Alberta at 1,382-metre-high Crowsnest Pass. In this area, the eastern limit of Cordilleran deformation is about 50 kilometres east of the Alberta boundary.

HIGHWAY 3A

Castlegar—Creston via Nelson

153 kilometres

See the map of the Castlegar area on page 332 and the map of the Selkirk Mountains between Salmo and Creston on page 335.

The Kootenay River flows north into the south end of Kootenay Lake, a large natural lake that fills the Purcell Trench between the Selkirk and Purcell Mountains. The lake drains southwestward through an arm about midway along its western side, into another reach of the Kootenay River, which eventually joins the Columbia River at Castlegar. The Kootenay River between Nelson and Castlegar has no sign of a buried interglacial or preglacial channel, which suggests that the present course of the river did not become established until the end of the last ice age. The original drainage was to the south, but toward the end of the last glaciation, the Purcell Trench was dammed by a lobe of ice and contained Glacial Lake Kootenai. The glacial lake water spilled out of the trench, cutting a path across the Selkirk Mountains into the Columbia Valley. The valley sides between Nelson and Castlegar are mainly Jurassic granite and granitic gneiss, and this solid bedrock provides a stable foundation for hydroelectric dams.

The Kootenay River drops about 100 metres in the 40 kilometres between Nelson and its confluence with the Columbia River at Castlegar. Its power to produce electricity was harnessed early, by the city of Nelson and the mining industry at Trail. Generating capacity grew until about 1930, when it served much of the southern interior of British Columbia. Later hydroelectric developments on the Pend Oreille River, which joins the Columbia just north of the international boundary, and on the Duncan River, which flows into the north end of Kootenay Lake, have increased hydroelectric generating capacity in the region to almost 1,000 megawatts.

Nelson

Nelson was established at the end of the nineteenth century at the head of the navigable outflow of the west arm of Kootenay Lake and near an early, though short-lived, silver mine. It is now the main administrative centre for the region, as well as a flourishing recreational and retirement community. A series of buildings in town constructed in the early twentieth century are faced with building stone quarried locally.

Kokanee Glacier Provincial Park protects part of the rugged, glacially sculpted wilderness in the Selkirk Mountains north of Nelson and contains peaks up to 3,000 metres high. The Nelson batholith, about 170 to 160 million years old and approximately 60 by 40 kilometres in extent, underlies much of the region, and its southern part is crossed by Highway 3A along the Kootenay River near Nelson. The batholith intruded during the early stages of Cordilleran mountain building. Although it mostly intrudes metamorphosed volcanic and sedimentary, early Mesozoic rocks of the Quesnel terrane, its southeastern extremity, north of Salmo, also intrudes late Proterozoic and early Paleozoic strata deposited near the old continental margin, thus unequivocally establishing that the two were together by middle Jurassic time. Structurally, it is on the eastern, down-dropped side of the large early Tertiary Slocan Lake normal fault. In the late 1800s, discovery of silver- and gold-bearing veins within the granite attracted people to the region, and small mining operations were developed for a few years.

Highway 3A continues for 34 kilometres east of Nelson along the north side of the west arm of Kootenay Lake to the Balfour ferry landing. Boarding the ferry and crossing the lake to its eastern side at Crawford Bay takes about 20 minutes, though delays are possible.

Cody Caves Provincial Park

Cody Caves Provincial Park is on a steep, rough, 13-kilometre road off Highway 31, 3 kilometres north of Ainsworth Hot Springs and about 10 kilometres north of the Highway 3A ferry terminal at Balfour. Water flowing through fractures in early Cambrian limestone dissolved the rock, forming the caves. Water, probably originating in the region of the caves, percolates down to depths where it is heated to 40 degrees Celsius, thence emerging in artificially enlarged caves at Ainsworth Hot Springs.

Crawford Bay to Creston

On the east side of Kootenay Lake, south of Crawford Bay, the Purcell Trench follows the trace of the Purcell Trench fault, a west-side-up normal fault extending southward across the international boundary.

From the ferry landing to Crawford Bay, Highway 3A passes tightly folded metamorphic rocks containing the metamorphic mineral sillimanite, indicating the rocks once were heated to as much as 600 degrees Celsius at some depth. These rocks are part of the Kootenay Arc, the long, narrow, westward-concave structural feature of tightly folded rocks that extends from northeastern Washington south of Salmo and northward follows the northern part of Kootenay Lake. The arc is sandwiched between the Belt-Purcell strata to the east and the suspect terranes to the west, and it appears that rocks in the arc were squeezed between the strong, unforgiving Belt-Purcell strata and the suspect terranes during Cordilleran mountain building.

South of Crawford Bay, the metamorphic grade drops off steadily, and about 6 kilometres south, it is of the biotite-bearing greenschist grade that characterizes most rocks in the Purcell Mountains. Along this part of the route we also traverse the northeastward continuation of a section of rock crossed by Highway 3 between Salmo and Creston. A band of early Cambrian limestone reaches the east shore of Crawford Bay. South of this are latest Proterozoic to earliest Cambrian quartzite and late Proterozoic Windermere rocks, including the distinctive Toby conglomerate at the base of the Windermere supergroup, which may have originally been glacial till. The same conglomerate is found in the high country east of the lake, where it unconformably overlies beds of the middle Proterozoic Belt-Purcell strata. Unfortunately the unconformity is not obvious near the lake.

Bedrock exposures for the next 25 kilometres are mainly coarsely crystalline granitic rock of the Bayonne batholith of mid-Cretaceous age. The crystals of feldspar are noticeably larger than other minerals in the granite, a texture called *porphyritic.* The batholith is one of several plutons in the region that crystallized about 115 and 90 million years ago and is one of the many mid-Cretaceous intrusions in the eastern Omineca Belt that lie east of all other granitic rocks in the Canadian Cordillera. They intruded when Cordilleran mountain building probably was at its peak, and their isotope chemistry suggests they contain a large component of old continental crust, that may have been partly melted when it was pushed down to great depth during mountain building.

From 10 kilometres north of the Highway 3 junction to Creston, Highway 3A passes scattered exposures of somewhat-altered Belt-Purcell rocks.

HIGHWAY 6
Vernon—Highway 3A (North of Castlegar)
318 kilometres

Vernon sits near the boundary between the Intermontane and Omineca Belts near the intersection of two fault systems of different ages. The belt boundary south of Vernon is defined by the northern end of the early Tertiary west-side-down Okanagan Valley normal fault. Just south of Vernon, the fault lies about 5 kilometres east of Okanagan Lake, near the valley containing Kalamalka Lake. North of Vernon this fault continues north-northeastward, largely within metamorphic rocks, probably emerging as the Eagle River fault at Sicamous. The belt boundary northeast of Vernon is marked by the north-northwest-trending Louis–Bolean Creek fault system, probably a reverse-fault system formed in Jurassic time shortly after the Quesnel terrane was accreted. It separates little-metamorphosed Intermontane Belt rocks to the west from older and more-metamorphosed Omineca Belt rocks to the east.

Highway 6 east of Vernon follows valleys that lie near a highly irregular but roughly east-southeast-trending bedrock boundary. To the north of the boundary are high-grade metamorphic rocks probably derived from late Proterozoic and early Paleozoic mainly sedimentary rocks. To the south of the boundary are rocks of the Quesnel terrane, abundant Mesozoic and early Tertiary granitic intrusions, metamorphic rocks of uncertain origin, and patches of Eocene volcanic and sedimentary rock. The nature of the bedrock boundary near Highway 6 is poorly understood and probably reflects a combination of the northwest-trending Jurassic faults formed during lateral compression, such as that northwest of Vernon, and the early Tertiary normal faults that extended and disrupted the older structures.

Bedrock between Vernon and 1,190-metre Monashee Pass shows the variety of rocks along this irregular boundary: Eocene sedimentary and volcanic rocks; variably metamorphosed late Paleozoic and Triassic sedimentary and volcanic rocks of the Quesnel terrane; and, in the segment west of Cherryvale, highly metamorphosed rocks of possible late Proterozoic or early Paleozoic age. The Triassic and older rocks are intruded by Jurassic granitic rocks, which dominate the bedrock east of Monashee Pass.

Glacial Erratic

About 6 kilometres east of downtown Vernon, a block measured by G. M. Dawson in 1879 as "22 feet long, 16.5 feet wide, and 18 feet high" sits on the top of a hillock about 100 metres north of the highway. This limy block

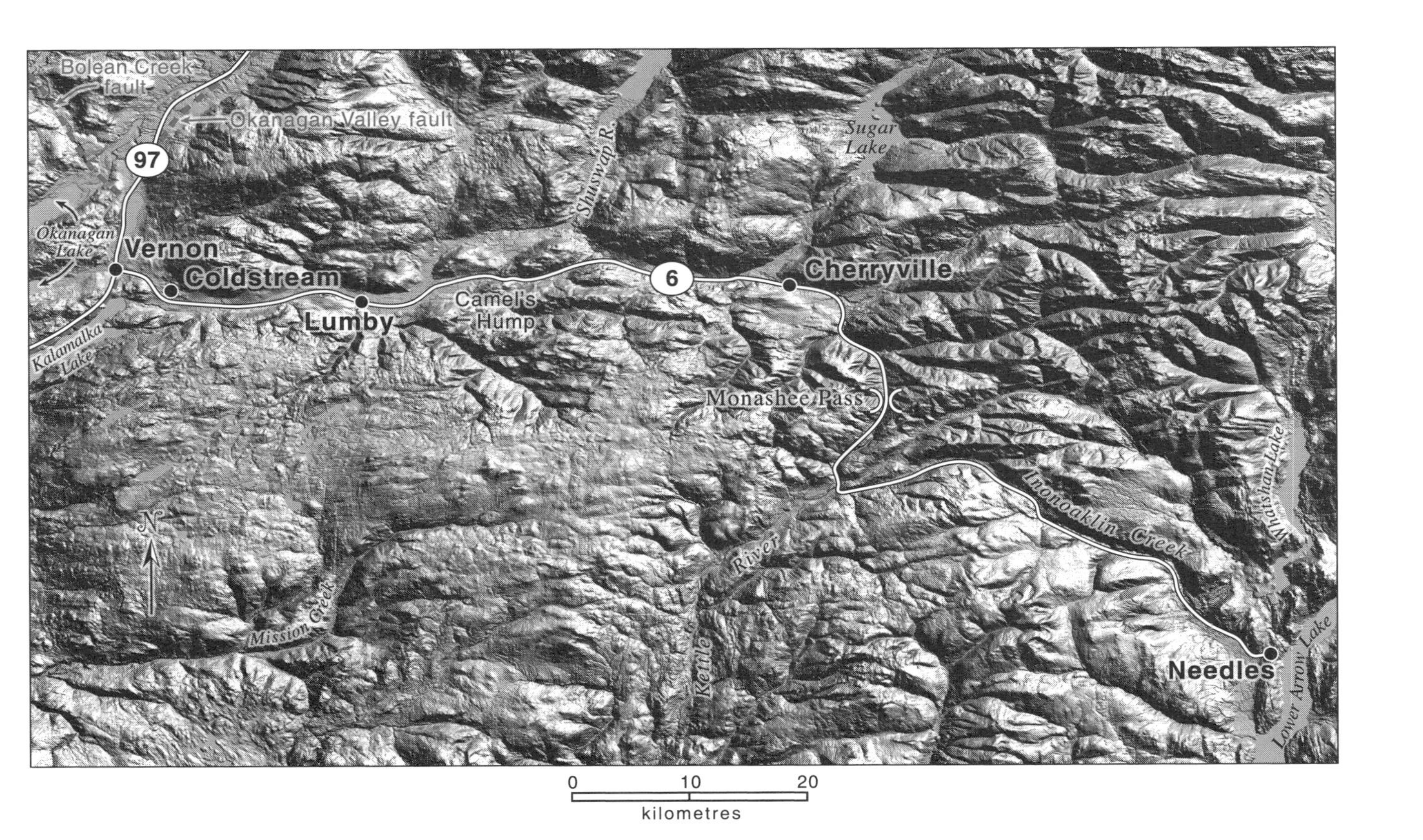

Shaded-relief image of topography along Highway 6 between Vernon and Needles, on Lower Arrow Lake.

—Image constructed using Canadian Digital Elevation Data obtained from GeoBase

A large glacial erratic of limy rock is 6 kilometres east of Vernon along Highway 6.

differs markedly in composition and character from the highly metamorphosed bedrock on which it stands. The great size of the block, its angularity, lack of a nearby source, and absence of nearby topography that could promote rolling or sliding to its present site led Dawson to the correct conclusion that it is a glacial erratic, a large rock carried by ice and then deposited during Quaternary glaciation.

Artesian Water at Coldstream

One kilometre east of the erratic and directly west of the Highway 6 junction with the road to Coldstream is a test hole drilled for the Geological Survey of Canada in the 1960s to determine the Quaternary stratigraphy. At a depth of about 50 metres, the drill broke through an impervious cap into a layer containing water under enough pressure to drive the drill back to the surface. Attempts to stem the artesian flow at a cost of several hundred thousand dollars were unsuccessful, and the borehole was left to drain into nearby Coldstream Creek. More than three decades later this still-flowing hole was looked on with favour as a potential source of potable water for the municipality of Coldstream.

Lumby to Monashee Pass

Highway 6 continues eastward through a gap, 525 metres in elevation, on the drainage divide between the Okanagan River and South Thompson River watersheds, and then through Lumby. The Camel's Hump, the steep-sided summit south of the highway and 13 kilometres east of Lumby, is the easternmost occurrence of Eocene volcanic rocks along this route. The bedrock mosaic of late Paleozoic and Triassic sedimentary and volcanic rocks, granitic intrusions, and metamorphic rocks continues east of here. For several kilometres, Highway 6 follows the south bank of the Shuswap River, here cut in metamorphic rocks containing widespread garnet and in places sillimanite. For 8 kilometres southeast of Cherryville, the highway climbs a terraced slope with almost no bedrock exposure and much cultivated land before entering the steep-sided, heavily wooded valley of Monashee Pass Creek. Here, bedrock exposures are more common and consist of mildly metamorphosed argillite and sandstone of probable late Paleozoic age.

Porphyritic Granitic Rock

Nearly all the bedrock between Monashee Pass and Lower Arrow Lake is Jurassic granitic rock. The highway south of Monashee Pass follows a meltwater channel southward for almost 8 kilometres, past picturesque ponds and elongated lakes, before making a sharp turn to the east into the southwestward-draining valley of the headwaters of the Kettle River. Five kilometres up the gravel-floored Kettle Valley, the highway turns southward for a little more than 1 kilometre and crosses over to the headwaters of southeastward-draining Inonoaklin Creek.

Near the crossing of Inonoaklin Creek is a rather different-looking granitic rock. Its striking feature is the large size of the rectangular feldspar crystals embedded in a matrix containing far smaller feldspar crystals. The association of the two markedly different sizes of one mineral in a rock is called *porphyritic texture*. This kind of texture is very common in volcanic rocks: the initial slow cooling of a molten body permits a period of slow crystallization and growth of one or more minerals. After eruption, the slow crystallization is followed by quick chilling of the lava into a fine-grained rock that encloses the larger crystals formed earlier. Porphyritic granitic rocks form in a similar way, except that the molten rock might have moved upward to a somewhat less-deep, cooler environment *within* Earth's crust rather than breaking out at the surface. It is also possible that a high content of dissolved gases in the molten granitic rock encourages growth of the large crystals. A sudden loss of these gases, for example, by fracturing of the roof above the intrusion, could then require the remaining liquid to crystallize to a significantly finer size. Alas, we are unable to observe these processes in action and must infer much of this behaviour from laboratory experiments carried out with unnatural compositions and at a small scale.

Highway 6 continues down the south side of Inonoaklin Creek as a narrow and winding paved road with numerous granitic rock exposures for 10 kilometres before the valley turns southward and widens. Here, the road keeps to the northeast of the stream and away from bedrock exposures.

Lower Arrow Lake

Travellers on Highway 6 must cross the valley of the Columbia River, here containing Lower Arrow Lake, on a ferry between Needles and Fauquier. Though the dam on the Columbia River just west of Castlegar has raised the water level, Lower Arrow Lake fills a natural lake basin on the Columbia. Highway 6 follows the eastern shore of the lake northward through granitic and metamorphic rocks for nearly 40 kilometres. There, at the big bend just south of Upper Arrow Lake, we cross the southern extension of the Columbia River fault. In the 20 kilometres south of Nakusp, the main bedrock is the south-southeastward continuation of the tract of Triassic sedimentary and volcanic rocks of the Quesnel terrane, that was crossed on Highway 6 west of Monashee Pass.

Remnants of an Early Mesozoic Basin near New Denver

Between Nakusp and New Denver, located at the north end of Slocan Lake 47 kilometres by road southeast of Nakusp, Highway 6 passes through early Mesozoic rocks in the northeastern margin of the Quesnel terrane, cut locally by Jurassic and Cretaceous granitic intrusions. They are more or less along a trend with those rocks west of Monashee Pass.

To see some good exposures of the early Mesozoic rocks, take a side trip along Highway 31A, which heads eastward from New Denver to end at Kaslo on Kootenay Lake. Rocks along this route include tightly folded dark grey shale and slate with very thin beds of fine-grained sandstone and siltstone of the late Triassic Slocan group. Geochemical studies on these rocks indicate that they had two sources: a continental source, which lay to the east, and a volcanic arc—presumably the Nicola group—which lay to the west. These rocks evidently represent the deepwater back-arc basin that in Triassic time separated the offshore Quesnel island arc from the old continental margin. South of here, these Triassic rocks are overlain by Jurassic volcanic and sedimentary rocks of the Rossland group. In the mountains northeast of Highway 31A, the Slocan group overlies a narrow sliver of late Paleozoic basalt, chert, and minor ultramafic rocks probably belonging to the Slide Mountain terrane.

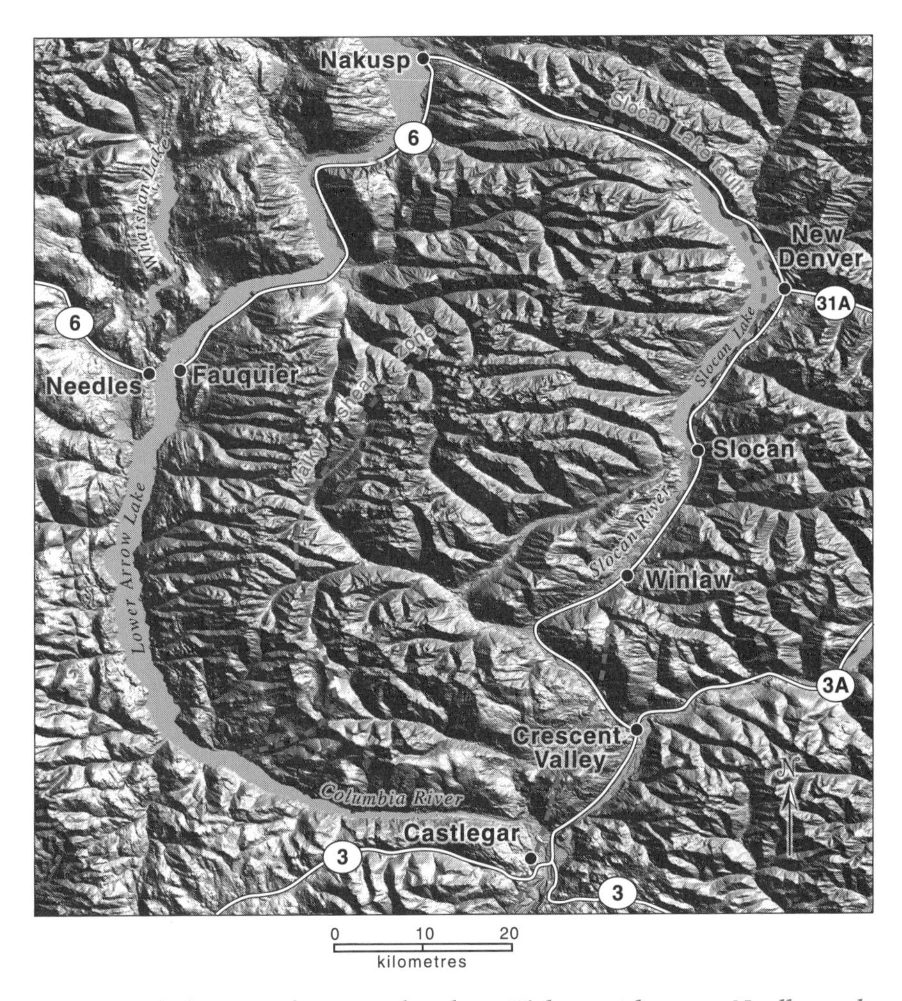

Shaded-relief image of topography along Highway 6 between Needles and Crescent Valley. —Image constructed using Canadian Digital Elevation Data obtained from GeoBase

The Slocan mining district in this area has a long history of activity that started in the early 1890s. The ore was in veins containing the lead and zinc sulfide minerals galena and sphalerite, which are related to intrusion of the middle Jurassic Nelson batholith, whose northern end is 5 to 10 kilometres south of the highway. Between 1894 and 1950, mines in the area produced 192,000 tonnes of lead, 140,000 tonnes of zinc, and—the commodity for which the region is known—1,780 tonnes of silver. Today, little remains of this activity, apart from some old buildings.

It is worth continuing on to the historic town of Kaslo on the west shore of Kootenay Lake, where one of the sternwheelers is preserved that provided transportation in the early days.

Slocan Lake Fault

South of New Denver, Highway 6 runs near the eastern side of Slocan Lake, a natural basin on the southward-flowing Slocan River. Starting about 10 kilometres south of New Denver magnificent exposures show the deformation in and near the Slocan Lake fault, a major north-northeast-trending, east-side-down normal fault that near the surface dips eastward at an angle of about 30 degrees. To the west across Slocan Lake, the uplifted fault block is composed of mid-Cretaceous to early Tertiary granitic rocks, ranging from about 100 to 50 million years old, and granitic gneiss that, before about 55 million years ago, were buried to depths of 20 kilometres. These hard, crystalline rocks form the high glaciated peaks and spires in Valhalla Provincial Park, a large protected area west of Slocan Lake. Careful isotopic dating in this area shows that rocks as young as 58 million years old were deformed at the end of the last stage of lateral compression, whereas slightly younger structures are related to extension. East of the Slocan Lake fault, in the down-dropped fault block, middle Jurassic granitic rocks intrude the relatively little-metamorphosed early Mesozoic black to dark argillite and phyllite of the Quesnel terrane. The fault was active in early Tertiary time, between about 55 and 47 million years ago, during the period of crustal stretching. Thus, isotopic dating from this area shows that the change from compression to extension took place between 58 and 55 million years ago.

At a viewpoint along Highway 6 about 10 kilometres south of New Denver and 20 kilometres north of Slocan, the trace of the fault lies directly beneath the highway. Broken, fractured, and brecciated, rusty-weathering middle Jurassic granitic rock in the down-dropped block of the fault is exposed east of the highway, and west of the highway in the picnic area are highly deformed early Tertiary granitic rocks.

About 8 kilometres south of the viewpoint and 6 kilometres north of Slocan, a pullout 250 metres above the lake provides a view northward of a low bench above the east side of lake. The break in slope above the bench marks the trace of the normal fault. The rocks near the pullout, and forming an extremely high roadcut to the south of it, are deformed granitic rocks of early Tertiary age in the uplifted block. They have strong foliation and an east-plunging lineation that is related to the fault.

About 20 kilometres south of Slocan, Highway 6 leaves the fault and heads southwestward through rocks in the uplifted fault block. In the large outcrop about 7 kilometres south of Winlaw, gneiss derived from sedimentary rock contains sillimanite and in places kyanite. The presence of kyanite indicates that the rocks once were buried to depths in excess of 20 kilometres. Highway 6 recrosses the Slocan Lake fault from the uplifted block to the down-dropped block just west of the Highway 3A junction, where the Slocan River joins the Columbia River.

HIGHWAY 23
Nakusp—Revelstoke
98 kilometres

For most of the distance between Revelstoke and Nakusp, Highway 23 follows Upper Arrow Lake. Prior to construction of the hydroelectric dam near Castlegar, Upper and Lower Arrow Lakes were two natural lakes connected by a 32-kilometre section of the Columbia River. The dam increased the water level so that now there is just one continuous, 230-kilometre-long lake, though the names Upper and Lower Arrow Lake have been retained. Upper Arrow Lake generally follows the north-south-trending Columbia River fault, the large, east-side-down normal fault that moved in early Tertiary time and can be traced along the Columbia River for about 140 kilometres. The fault mostly separates high-grade metamorphic rocks in the Monashee Mountains to the west from lower-grade metamorphic rocks and granitic intrusions in the Selkirk Mountains to the east.

North of Nakusp, Highway 23 follows the east side of the lake through black and dark green metamorphosed volcanic rock and dark grey, locally rusty-weathering phyllite and local conglomerate of late Paleozoic and early Mesozoic age, probably belonging to the Quesnel and Slide Mountain terranes. These rocks form a narrow band between the lake and the Kuskanax batholith, a large middle Jurassic granitic intrusion to the east. The highway cuts through the northwestern tip of the intrusion, which is exposed at both terminals of the Highway 23 ferry across the north end of Upper Arrow Lake between Galena Bay and Shelter Bay. It is worth examining the complex, convoluted contact between the granitic rocks and the black foliated amphibolite exposed on the south side of the eastern ferry terminal.

Along the western side of Upper Arrow Lake north of Shelter Bay, Highway 23 leaves the intrusion and crosses the concealed Columbia River fault into the high-grade metamorphic rock at the core of the Monashee Mountains. The blue metamorphic mineral kyanite can be found in some of the many roadcuts of gneiss along this part of the highway. Some of

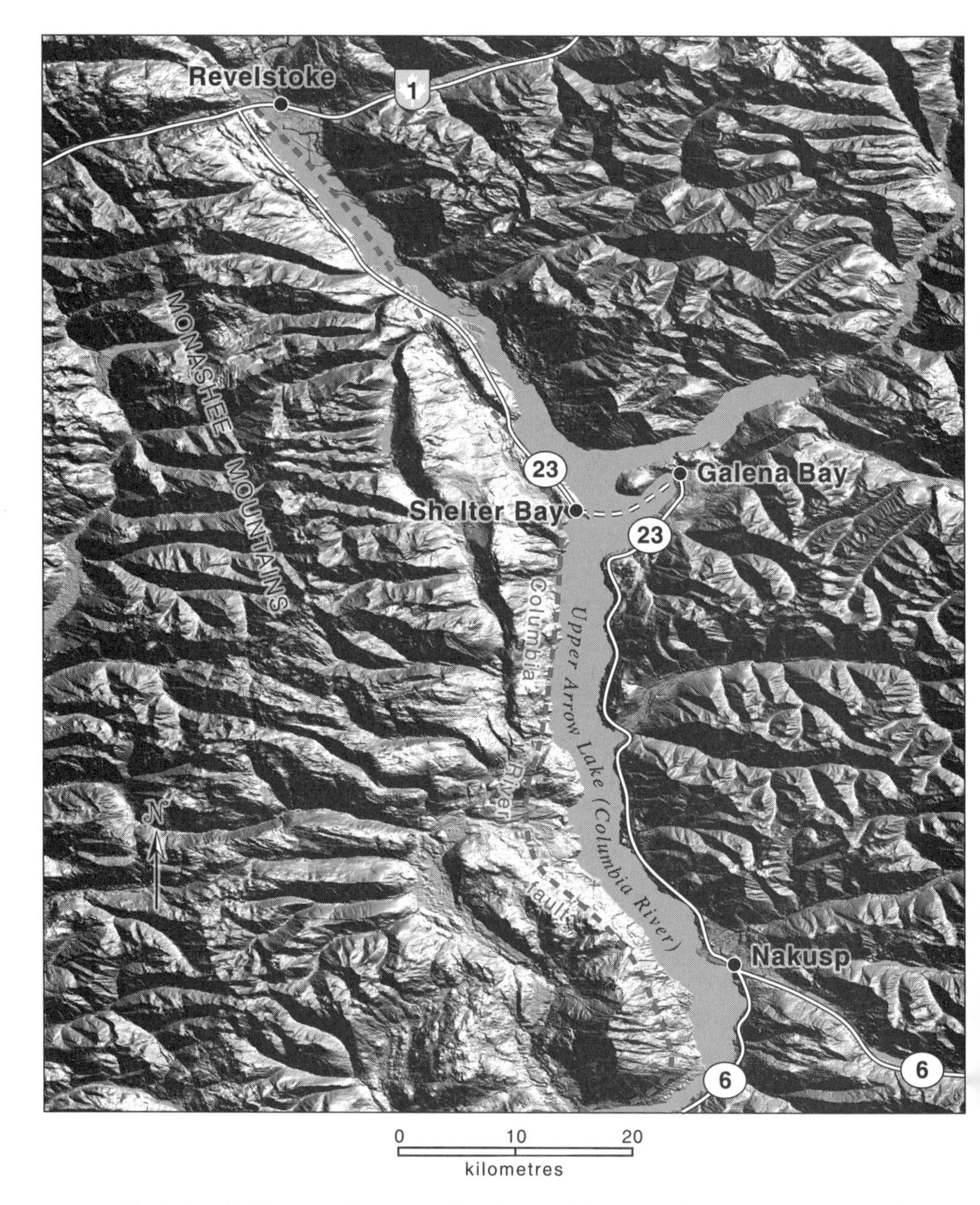

Shaded-relief image of topography along Highway 23 between Nakusp and Revelstoke. —Image constructed using Canadian Digital Elevation Data obtained from GeoBase

these metamorphic rocks may be as much as 2 billion years old, the oldest known in the Canadian Cordillera. They may be a piece of the Canadian Shield brought to the surface during west-side-up early Tertiary movement on the Columbia River fault.

HIGHWAY 33
Rock Creek—Kelowna

129 kilometres

See the map of Highway 3 between Osoyoos and Rock Creek on page 320.

Highway 33 passes through the Okanagan Highlands between Rock Creek and Kelowna, heading north along the west side of the Kettle and West Kettle Rivers. Most of the route parallels the Kettle Valley Railway Trail, a cycling and hiking trail along the abandoned railway that once connected Midway, on Highway 3, with Hope, east of Vancouver. The railway leaves the highway about 25 kilometres southeast of Kelowna and heads south down the Okanagan Valley, crossing the Okanagan River near Penticton. Unfortunately, big forest fires during the summer of 2003 burned some of the trestle bridges on this part of the route, but moves are afoot to have these repaired.

Serpentinite Soil near Rock Creek

Two to three kilometres north of Rock Creek, on the western side of the road, patches of dark soil contain chips of dark green to black rock with a greasy luster. The soil is the remnant of weathered serpentinite associated with late Paleozoic rocks probably belonging to the Slide Mountain terrane. Serpentinite, a soft slippery rock made largely of the mineral serpentine, forms when peridotite, made of the iron- and magnesium-rich minerals olivine and pyroxene and present in the upper mantle, reacts with hot water at an oceanic ridge. The dearth of vegetation on the soil surface here is notable. Few plants like serpentinite soils, although in places some species have adapted to them. It is not known whether these particular soils lack nutrients essential for plant growth, whether they carry some constituent in toxic concentrations, or whether the proportions of two or more constituents in the soil inhibit growth.

About 10 kilometres north of here, Highway 33 enters the valley of the West Kettle River.

Beaverdell Silver Mine

Beaverdell, a settlement 32 kilometres north of the confluence of the Kettle and West Kettle Rivers, is about 3 kilometres west of a former silver mine. The mine is not visible from Beaverdell, but you can see it from the highway about 1.5 kilometres north of the settlement and 7 kilometres south of Carmi, where the highway is aimed almost directly at the western mine portals to the southeast.

The deposit consists of about half a dozen ore-bearing quartz veins, or lodes, 0.9 to 1.5 metres wide, within Jurassic granitic rocks. The lodes are absent from nearby occurrences of Triassic or older, slightly metamorphosed sedimentary and volcanic rocks. Isotopic evidence indicates that the mineralization is related to another pluton, of early Tertiary age, also near the ore-bearing area. A complicated system of closely spaced faults, some no more than a few metres apart, and some with displacements of as much as 150 metres, has literally butchered the ore bodies, making the ore reserves very difficult to assess. The quartz veins and associated calcite contain pyrite, sphalerite, and arsenopyrite, plus silver-bearing minerals, notably galena, tetrahedrite, pyrargyrite, and native silver.

Like many other mineral deposits in this part of British Columbia, ore was discovered and staked in the late 1890s. There was limited production from 1900 to 1910, with hand-sorted ore hauled by wagon or stoneboat for 78 kilometres to Midway, the location of the nearest railway until the Kettle Valley Railway was completed by about 1916. From Midway, the ore went east by rail to the smelter at Trail. From 1916 until it closed in 1991, the mine was in production continuously, albeit at a limited rate for a long time. The annual yield of ore first passed 1,000 tons in 1927 and exceeded 10,000 tons only in 1951, after the first mill was built at the mine. From 1967 to 1990, the last full year of activity, production was between 30,000 and 40,000 tons per year. The mine produced metals with a gross value at 1985 prices of almost $300 million and employed twenty-eight miners in its later years. With the mine now closed, all underground workings are (or should be) sealed, but the casual visitor can scout the old mine dumps for mineral specimens, most notably pyrargyrite or "ruby silver," a metallic mineral black in bulk but brilliant red when light is transmitted through thin flakes.

Rock Canyon Carved by Meltwater

Thirty kilometres north of Beaverdell and about 15 kilometres south of McCulloch Road, Highway 33 enters a rock-walled channel that was cut by a short-lived meltwater stream in late-glacial time. Rockfalls from the channel walls in postglacial time partly dam the channel, creating attractive ponds.

Uranium Prospects

Geologists from Japan saw geological conditions in this area that resembled uranium occurrences in Japan, and in the 1960s they initiated a search for uranium-bearing deposits. Their efforts and those of North American geologists led to the discovery of numerous prospects between the head of Beaverdell Creek and the plateau southeast of Vernon.

The most promising of these deposits is in virtually unconsolidated conglomerate, sandstone, and shale of middle or late Tertiary age that accumulated in a channel eroded in Eocene or older bedrock. One of the properties, called the Blizzard, was extensively explored before a provincial decree banned development and mining of uranium ores from 1980 to 1987. The deposit lies on and near a hilltop about 9 kilometres east of Highway 33, near the northern entrance to the meltwater channel noted above. The potential ore is in a channel 1.5 kilometres long and up to 210 metres wide, defined with the help of more than 350 drill holes. The holes revealed, from bottom to top, granitic bedrock that is only locally mineralized, an erosional unconformity, a sequence of poorly consolidated sediments up to 35 metres thick that contains most of the uranium, and a capping of basalt flows about 5 million years old. The basalt cap is up to 75 metres thick and prevents erosion of the soft sediment.

The erosional unconformity on the granite surface records a stream-eroded landscape that sloped at angles of 10 to 35 degrees toward a low channel. These steep slopes do not occur more than 250 metres away from and 100 metres above the channel, so a former flat valley floor beyond these limits is possible and would be consistent with regional data. The channel itself has been traced for a distance of 2.4 kilometres along its meandering course. The elevation of the channel drops 30 metres southward in this distance, but whether the drop reflects a high original gradient or is a product of later tilting is undetermined.

Experience in uranium geology from elsewhere tells us how the concentrations of this element developed here. Eocene lava and granitic rock in the neighborhood held relatively high amounts of uranium, which was converted to a soluble form during weathering and carried in percolating groundwater into the mid-Tertiary valley fill. There, it came in contact with the carbonaceous material in mudstone within the sediments, deoxidized, and converted back into relatively insoluble uranium minerals.

Exploration drilling indicated 2.1 million tonnes of material containing 0.226 per cent uranium. At prices prevailing when the ban on mining uranium was introduced, the Blizzard property had a gross value of $400 million. When the ban expired, however, the price of uranium had dropped more than 50 per cent, and at this rate the property was not economically interesting.

HIGHWAY 95 AND 93/95

Golden—Cranbrook

248 kilometres

See the map on page 315 for the Golden area and the map on page 345 of Highway 3 between Cranbrook and Crowsnest Pass for the southern part of the route.

Highway 95, numbered Highway 93/95 south of Radium Hot Springs, runs along the floor of the Rocky Mountain Trench, the great valley that here separates the Purcell Mountains of the Omineca Belt to the west from the Rocky Mountains of the Foreland Belt to the east. The Rocky Mountain Trench and its northern counterpart, the Tintina Trench in the Yukon, a linear system of interconnected valleys extending for more than 2,500 kilometres from northwestern Montana to east-central Alaska, has several origins.

Erosion, acting along different but aligned geological features, formed the Rocky Mountain Trench. The segment of the trench north of about 54 degrees latitude (the latitude of Prince George) and the Tintina Trench in Yukon and Alaska follow the trace of a large right-lateral strike-slip fault. In Yukon, rock units and structures as young as earliest Tertiary time have been offset across the fault by about 425 kilometres. South of 54 degrees latitude, some of the right-lateral strike-slip movement may link up with movement on the Fraser fault in southwestern British Columbia. The part of the trench south of 54 degrees latitude veers more to the southeast and follow the traces of normal and thrust faults as far as Kinbasket Lake, just north of where Highway 1 enters the trench and where the presence of the same rock units on both sides of the trench precludes strike-slip movement. Between Golden and Canal Flats 160 kilometres to the south, which is the segment containing the headwaters of the northward-flowing Columbia River, the floor of the trench is in soft, calcareous shale of the early Paleozoic McKay group, and there the trench is largely erosional in origin. However, the Purcell fault, which runs parallel to the trench low on the mountain front west of Golden, probably controls the trend of the valley. South of Canal Flats, where the vigorous south-flowing Kootenay River enters the trench from its eastern side, the steep eastern wall of the valley is the fault-line scarp of the Southern Rocky Mountain Trench fault, on which sedimentary rocks as young as Miocene age have been down-dropped against middle Proterozoic rocks to the east.

Glacial Lake Invermere

Between Golden and about 30 kilometres south of Canal Flats, the lower parts of the trench contain terrace and delta deposits whose tops are up

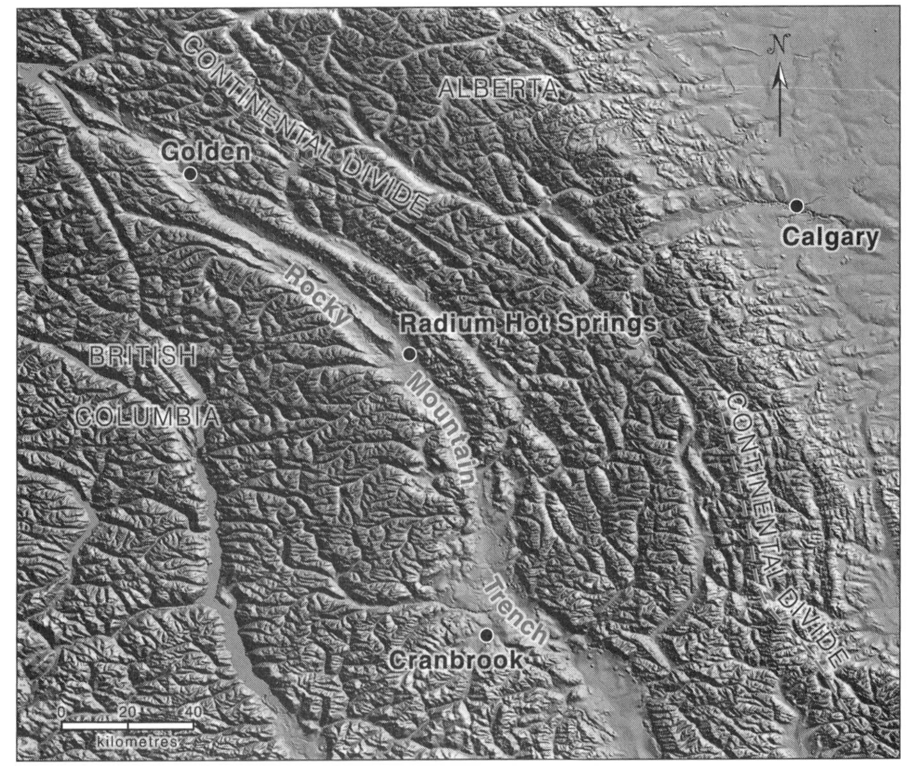

Shaded-relief image of the Rocky Mountain Trench between Golden and Cranbrook. —Image contructed using Canadian Digital Elevation Data obtained from GeoBase

to 100 metres above the valley bottom. At one time the trench contained Glacial Lake Invermere, a big lake that formed at the end of the last ice age. About 14,000 years ago, ice occupied the Rocky Mountain Trench, depositing glacial till on the valley floor as it melted. Prior to 10,000 years ago, the melting Cordilleran ice sheet retreated northward along the valley floor as far as the Beaver River, just north of where Highway 1 enters the trench, and also withdrew to the mountains on each side of the trench. Water ponded against the ice in the valley to the north because the valley floor north of Canal Flats slopes gently northward, dropping about 15 metres in 180 kilometres. Till, outwash gravel, and bedrock formed a dam north of Skookumchuck, about 30 kilometres south of Canal Flats. Thus, the trench was temporarily dammed north and south and received water from streams and rivers flowing into it from both sides. Glacial Lake Invermere, named from Invermere on the western side of the trench midway between Golden and Cranbrook, was 210 kilometres long, averaged 2.5 kilometres wide, and was up to 100 metres deep.

A cutbank above the Kicking Horse River in Golden exposes delta sediments deposited when the ancestral Kicking Horse River flowed into Glacial Lake Invermere.

Golden is on the floor of the trench at the confluence of the west-flowing Kicking Horse River and the Columbia River. East of Golden, Highway 1 climbs over sand and gravel deposited in an old delta before entering Kicking Horse Canyon. The delta formed where the ancestral Kicking Horse River, whose fast-running water was capable of carrying a heavy load of sediment, entered the quiet waters of Glacial Lake Invermere and deposited its load. A cross section of the old delta showing its west-dipping beds is beautifully exposed in the cutbank above the Kicking Horse River across from the municipal campground in southeastern Golden. You can gain a more distant view and better appreciation of the size of the delta complex by driving along Fourteenth Avenue South, ascending partway up Selkirk Hill, and looking back to the north.

South of Golden, Highway 95 crosses gravel terraces well above the present valley floor. These are the remains of former deltas formed where tributary streams entered Glacial Lake Invermere. In places, the terraces are made of finer-grained lake deposits, but these are more common in the trench about 100 kilometres south of Golden.

For the first 20 kilometres or so south of Golden, roadcuts through bedrock are in early Paleozoic calcareous shale and thinly bedded limestone.

Jubilee Mountain

Jubilee Mountain, about 44 kilometres south of Golden near Spillimacheen, is on the western side of the trench. A thin sequence of early Paleozoic carbonate and sandstone overlies latest Proterozoic to earliest Cambrian quartzite, in sharp contrast to the thick section of strata followed along the trench north of here and crossed to the east on Highway 1. The stratigraphic thicknesses of Paleozoic and late Proterozoic rocks in this region vary dramatically from place to place, and the thickness changes were controlled by old faults affecting the underlying rocks. The rocks on Jubilee Mountain were deposited on a high-standing normal fault block that was

Shaded-relief image of topography along Highway 93/95 between Spillimacheen and Skookumchuck. —Image constructed using Canadian Digital Elevation Data obtained from GeoBase

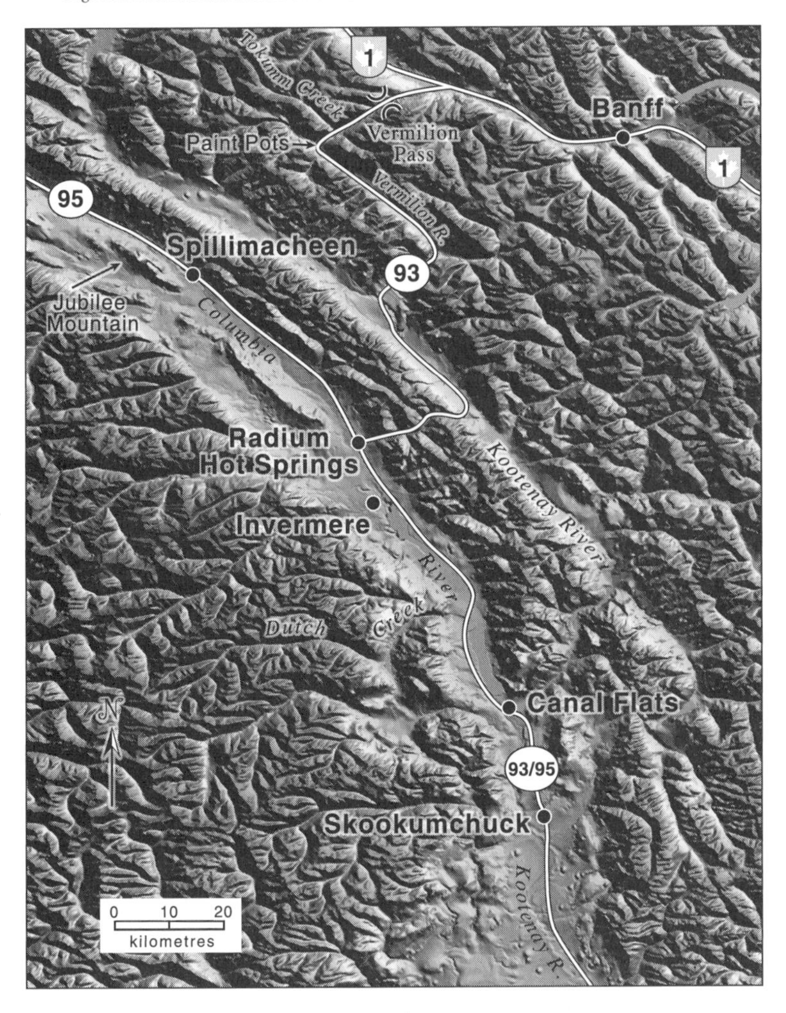

uplifted and tilted to the east during early Paleozoic time. It seems that normal faulting associated with crustal extension did not stop after the supercontinent Rodinia rifted apart in late Proterozoic time but continued through early Paleozoic time.

Bugaboo Provincial Park

About 40 kilometres west of Jubilee Mountain is Bugaboo Provincial Park in the northern Purcell Mountains. There, mid-Cretaceous granite intrudes strata belonging to the late Proterozoic Windermere supergroup. The granite forms spires protruding above glaciers and provides a challenging rock-and-ice environment that draws mountaineers from around the world. The park can be reached by a road that leaves the trench at Spillimacheen.

Radium Hot Springs

Travellers may wish to take a brief side trip up Highway 93, which joins Highway 95 at Radium Hot Springs, to visit the hot springs. It sits in a narrow canyon carved by glacial meltwater in early Paleozoic limestone, which is part of the same thin sequence of strata as that on Jubilee Mountain. The springs emerge from a breccia zone and have a constant temperature of 40 degrees Celsius. Though the water is mildly radioactive, levels are too low to be a health hazard.

Four kilometres east of its junction with Highway 95, near a parking lot, Highway 93 crosses the Redwall fault, a thrust fault that carried rocks above it to the northeast. Here, the fault is steeply dipping to vertical, and in part overturned. It is marked by red, iron-stained breccia. However, the breccia is probably not directly related to faulting but rather resulted from collapse of a cavity caused by solution of evaporite beds that lay below the fault. The Redwall fault defines the eastern limit of the thin Jubilee Mountain strata; east of it, strata of the same age and composition are thicker.

Kootenay National Park

Highway 93 northeast of Radium Hot Springs traverses Kootenay National Park, which includes sedimentary rocks deposited near the margin of the old Laurentian continent between late Proterozoic and early Mesozoic time. Several sites of geologic interest are along Highway 93 in the park. At the Paint Pots along the Vermilion River, cold mineral springs emerge and deposit iron oxides. Native peoples used the red soils and mud for paint. At Marble Canyon about 8 kilometres south of Vermilion Pass, Tokumm Creek has carved a narrow canyon along a fault in early Paleozoic dolomitic limestone.

Hoodoos at Dutch Creek

Buff-coloured to yellow cliffs of late-glacial outwash fan deposits near Dutch Creek, about 40 kilometres south of Radium Hot Springs, have been eroded into hoodoos, which are spectacular columnar sculptures. They are common in many old glacial and river deposits in the drier parts of British Columbia. Boulders in the deposits form caps that protect the underlying finer-grained material from erosion and eventually a spire or column forms.

Canal Flats and the Purcell-Selkirk Island

South of the hoodoos, Highway 93/95 climbs the western side of the trench. From a viewpoint you can see Columbia Lake, the headwaters of the gently northward-flowing and in places meandering Columbia River. The flat, swampy ground at the head of Columbia Lake is all that separates it from the vigorously southward-flowing Kootenay River, which emerges from the mountains on the eastern side of the trench just south of Canal Flats. Each spring when the snowpack in the mountains melts, the Kootenay River floods. In 1888, a canal was dug across this flat ground, diverting

Hoodoos are eroded into glacial outwash at Dutch Creek, about 40 kilometres south of Radium Hot Springs.

the floodwaters into Columbia Lake, but the canal was abandoned before it ever saw much use because of concerns from settlers along both rivers. Theoretically, and probably temporarily, the canal would have made the Purcell and Selkirk Mountains into an artificial island because the Kootenay River flows into the Columbia River at Castlegar, about 160 kilometres due southeast of here.

Late Proterozoic Windermere Rocks

South of the viewpoint over Canal Flats, Highway 93/95 descends a hill toward the flats. Just north of the bridge over the railway near the bottom of the hill, several roadcuts display a variety of sedimentary rocks very typical of the Windermere supergroup. These include dark grey slate containing thin, lighter-coloured sandier beds and grit. Well-developed cleavage in the slate cuts across the bedding and is crinkled. The coarser-grained rocks contain quartz veins that, like the slaty cleavage, formed when the rocks were folded and faulted.

Early Proterozoic shale and sandstone near the Highway 93/95 bridge over the railway at Canal Flats have been metamorphosed to slate. The lighter, subhorizontal layers are bedding. The subvertical slaty cleavage has been crinkled.

Grit, a characteristic rock type in the Windermere supergroup, is coarse-grained, greenish grey sandstone to microbreccia composed of angular grains of feldspar and quartz. The grains were washed off the deeply eroded supercontinent Rodinia, which was barren because a cover of vegetation would not appear for another 200 million years, and were deposited in troughs related to rifting of the supercontinent. Grains of zircon in the grit yield ages of about 2 billion years, which tells us they probably were eroded from the 2-billion-year-old Canadian Shield that now lies beneath southwestern Alberta. In places, the grit contains *rip-up clasts*, dark-coloured shale fragments that formed where a slurry of sand and water flowed down a submarine slope and tore up pieces of mud from the underlying layers. The deposits at the roadcuts north of the bridge are the remains of an ancient deepwater turbidite.

Southern Rocky Mountain Trench Fault

The eastern side of the Rocky Mountain Trench, from east of Skookumchuck for more than 200 kilometres south-southeastward as far as Flathead Lake in Montana, is the scarp of the Southern Rocky Mountain Trench fault, a west-side-down normal fault or system of connected faults. The fault was active at least into Miocene time. The eastern trench wall south of here consists of middle Proterozoic Belt-Purcell supergroup unconformably overlain by a relatively thin layer of early Paleozoic strata.

The eastern wall of the Rocky Mountain Trench near Cranbrook, called the Steeples, contains hard rocks, in contrast with the softer strata of the trench floor; it is a fault-line scarp formed by differential erosion across the Southern Rocky Mountain Trench fault.

GLOSSARY

accretionary complex. Ocean-floor rocks scraped off a subducting oceanic plate, complexly folded and faulted, and stuck on, or accreted to, the overriding plate during plate convergence.

alluvial fan. A body of sediment, fan-shaped when viewed from above and gently sloping to flat in profile, generally formed where a fast-flowing stream carrying a heavy load of sediment enters a broad valley and slows down and deposits much of its load.

alluvium. A body of clay, silt, sand, or gravel deposited by running water.

ammonite. A group of extinct cephalopod molluscs, relatives of the octopus and cuttlefish, but with an external shell mostly coiled in a plane, that lived in the Mesozoic seas.

amphibolite. A generally hard, dark rock made mostly of the group of minerals called amphiboles and formed when basalt and gabbro are metamorphosed.

andesite. A volcanic rock whose silica content is intermediate between silica-poor basalt and silica-rich rhyolite, whose colour may range from light to dark, and which commonly contains small feldspar crystals in a fine-grained matrix; the extrusive equivalent of diorite. The rock type is common in continental volcanic arcs, such as those in the Andes.

anticline. A fold in which the bedding or layering has been bent upwards to make an arch in profile; the opposite of syncline.

argillite. A weakly metamorphosed rock made mostly from clay, but lacking the "splittability" or fissility of shale or the cleavage of slate.

asthenosphere. A term based on rock strength for the weak layer in the upper mantle that underlies the rigid lithosphere (or rock sphere). The layer behaves plastically and flows laterally when, for example, a great load such as a continental ice sheet is imposed above it, and flows back again when the ice melts and load is removed.

back-arc basin. A basin that develops behind a chain of subduction-related volcanic islands, or island arc, in an ocean basin and is located on the opposite side of the arc from the subducting oceanic plate.

basalt. A fine-grained, heavy, dark volcanic rock composed mostly of the minerals calcium plagioclase and pyroxene; the extrusive equivalent of gabbro.

basement. That part of the crust made of old metamorphic and igneous rock that lies beneath the sedimentary cover.

batholith. A large body of intrusive rock; generally defined as one whose surface area is more than 100 square kilometres.

bedding. Layering in a sedimentary rock, or volcanic rock made of fragments, that results from deposition of the sediment.

bedrock. The consolidated rock in a region, as opposed to unconsolidated soil, sand, gravel, or clay.

belt. A general term used geologically to describe a linear tract of rocks. In the Canadian Cordillera, the term *morphogeological belt* is used for major geological features that combine distinctive bedrock with topography; for example the Coast Belt is a rugged mountainous region made mainly of granitic rock.

biotite. See **mica**

breccia. A coarse-grained rock made of angular fragments generally in a finer-grained matrix; may be of sedimentary or volcanic origin or material crushed and fragmented by a fault; the opposite of conglomerate, which contains rounded fragments.

calcareous. An adjective used generally for sedimentary rock that is rich in calcium carbonate; for example, calcareous sandstone.

chert. A very hard, splintery, extremely fine-grained rock made entirely of silica either deposited as beds (for example, radiolarian chert derived from silica-rich ooze deposited on the deep ocean floor) or else formed as secondary concretions in limestone (flint nodules).

chlorite. A platy, soft green silicate mineral rich in iron and magnesium that is common in low-grade metamorphic rocks.

clastic. An adjective used for a rock, such as sandstone, that is made largely of grains or clasts derived from pre-existing rocks and minerals.

clay. Very fine-grained sediment or soft rock composed of extremely fine-grained fragments, including clay minerals; typically the unconsolidated precursor of shale, argillite, and slate.

cleavage. A property of a rock (or a mineral such as mica) that allows it to be split along parallel surfaces. In the case of a rock such as slate or schist, this is the result of deformation accompanied by the growth of aligned minerals.

coke. A nearly smokeless fuel produced by heating coal and driving off the volatile material.

conglomerate. A coarse-grained rock made of rounded fragments generally in a finer-grained matrix; generally of sedimentary origin; the opposite of breccia, which contains angular fragments.

continental arc. A subduction-related chain of volcanoes formed on a continental margin. A local example is the Cascade magmatic arc that extends from southwestern British Columbia to northern California.

continental platform. As used here, in western Canada, it refers to the stable, interior part of the continent, where a relatively thin, flat-lying, blanket of

Paleozoic through Tertiary sedimentary rocks overlies the continental basement composed of old Precambrian metamorphic and igneous rocks.

continental shelf. That part of the continental margin between the shore and the continental slope and submerged to depths of generally up to about 200 metres.

continental slope. The oceanward-sloping part of the continental margin between the continental shelf and the ocean bottom; submerged to depths between about 200 to 2,000 metres off western Canada.

convergent margin. The boundary region between two tectonic plates that are moving towards one another on which the lower oceanic plate descends into the mantle along a subduction zone, and a magmatic arc is generated in the upper, overriding plate.

country rock. The rock intruded by and hosting an igneous intrusion or a mineral deposit.

crossbedding. Layers within a bed of sedimentary rock, typically sandstone, that are inclined at an angle to the main bedding layer. Small examples form on the leeward face of current ripples and are inclined in the direction of water flow; large examples form on the lee side of sand dunes and are inclined in the direction of wind flow.

debris flow. A mass of rock fragments of all sizes that moves downhill as a water-rich slurry; intermediate between a mud flow and a rock slide.

delta front. The frontal part of a delta where much of the initial deposition of sediment occurs. It lies within and below the zone of wave erosion.

dike. A sheetlike intrusion that cuts across bedding, foliation, or layering of the hosting country rock.

diorite. An intrusive igneous rock typically composed of white crystals of feldspar (sodium and calcium aluminosilicate minerals) and black crystals of amphiboles (iron and magnesium silicate minerals) with little if any visible quartz; the intrusive equivalent of andesite.

distributary channel. A channel that carries water away from the main river channel crossing a delta or alluvial plain.

divergent margin. The boundary region between two tectonic plates that are moving away from one another. This typically occurs at mid-ocean ridges, which is where basalt wells up from the underlying mantle and cools to form new oceanic crust.

drumlin. A streamlined hill formed beneath a glacier and made of debris carried by the glacier. Its long axis indicates the flow direction of the glacier.

erratics. Rocks, transported and deposited by a glacier, that do not come from the local area.

fault. A fracture in a rock body along which there has been displacement of the rock on one side relative to the other. **Normal faults** form when the rock is extended so that one side drops down relative to the other. **Thrust and reverse**

faults form when the rock is compressed horizontally, so that one side overrides the other and the rock body is shortened; thrust faults feature mainly horizontal shortening and may be flat-lying; reverse faults feature more vertical movement and so are steeper. In **strike-slip faults** the movement is parallel with the trend of the fault so that the rocks on each side move sideways relative to one another and no great vertical displacement is involved.

feldspars. This group of aluminosilicate minerals are the most common minerals in Earth's crust and important components of most igneous rocks. **Plagioclase feldspar** has a range of compositions from sodium rich to calcium rich, and is generally white; **potassium feldspar** may be pink.

felsic. An adjective applied to light-coloured igneous rock such as granite or rhyolite; derived from *fel*dspar and *si*li*c*a.

foliation, foliated. These terms refer to the planar texture of a rock, especially that produced by the growth of oriented, flat minerals such as micas in metamorphic rocks, or by shearing and flattening of pre-existing rock grains during deformation and metamorphism.

fusulinids. Members of the fossil family Fusulinidae, these were single-celled animals that secreted calcareous shells, called tests, mostly less than 1 centimetre in maximum dimension, shaped like grains of wheat, cigars, or eggs, and with a complex internal structure. They lived during late Paleozoic time.

gabbro. A generally dark-coloured intrusive rock made mainly of calcium-rich plagioclase feldspar and iron- and magnesium-rich pyroxene, and generally with no or little quartz; the intrusive equivalent of basalt.

glaciation. A general term for the formation, movement, and retreat of ice sheets or flows.

gneiss. A generally layered or foliated, coarse-grained rock formed during high-grade metamorphism. It may be segregated into light-coloured layers rich in the minerals quartz and feldspar and dark layers rich in the minerals biotite and amphibole.

graded bedding. A sedimentary bed or layer in which the bottom has a sharp contact with the underlying layers, and within which the grain size grades from coarser near the base to fine at the top; formed by settling from a water-rich slurry that moved downslope as a turbidity current.

granite. An intrusive crystalline rock with abundant quartz and potassium and sodium feldspar and minor amounts of dark, iron- and magnesium-rich silicate minerals such as biotite. True granite with abundant potassium feldspar is relatively rare in the Canadian Cordillera. **Granitic rock** is the general name for crystalline intrusive rock with visible quartz, feldspar, and iron and magnesium minerals and includes granite and the far more common granodiorite and quartz-diorite.

granodiorite. An intrusive crystalline rock that is intermediate in composition between granite and quartz diorite, with a higher ratio of plagioclase feldspar to potassium feldspar than true granite, and biotite and hornblende as the mafic minerals; together with quartz diorite, the most common kind of granitic rock in the western Canadian Cordillera.

greenschist. A foliated metamorphic rock that is green because it contains the green metamorphic minerals chlorite, epidote, and actinolite. **Greenschist facies** refers to the level of low-grade regional metamorphism characterized by the metamorphic minerals noted above, plus others such as white mica and biotite, that form at temperatures between about 300 and 500 degrees Celsius.

greenstone. A general name for a nonfoliated rock formed by the low-grade metamorphism of basalt or gabbro.

greywacke. A generally dark grey variety of sandstone composed of poorly sorted angular grains of quartz, feldspar, and rock fragments in a finer-grained, clayey matrix.

hornblende. A black or dark green mineral that typically occurs as prismatic elongate crystals. The most common mineral of the amphibole family, it has a complex and variable composition featuring calcium, sodium, magnesium, iron, aluminum, oxygen, and hydrogen. A fine-grained fibrous form is **actinolite**.

igneous rock. A rock that solidified from molten rock, or magma, such as a lava flow on the surface or a granitic rock at depth.

intrusion. A body emplaced as molten rock that subsequently cooled to form a pluton, batholith, dike, or sill within the country rock.

island arc. A chain of volcanic islands within an ocean basin that is arcuate in map pattern, formed above a subduction zone, and is separated from the closest continent by a back-arc basin. The nearest example is the Aleutian Island chain in Alaska.

isotope. One of several possible varieties of a chemical element, each of which has the same number of protons in the nucleus but a different number of neutrons. **Radioactive isotopes** decay to other isotopes, called daughter products, at known rates, and the quantity of daughter product in a rock gives the age, in years, when the rock formed or cooled below a certain temperature.

kettle. A bowl-shaped depression found in sediments deposited during glaciation, which formed when a detached buried mass of ice melted and caused overlying sediments to subside.

late-glacial. Features that formed during the period when the last continental (and Cordilleran) ice sheet retreated.

lava. Molten rock extruded at Earth's surface; also applied to the cooled solidified rock.

limestone. A sedimentary rock made mostly of calcium carbonate generally derived from the calcareous remains, such as shells, of marine animals or biochemically precipitated by plants. With post-depositional alteration, magnesium may be introduced to make **dolomite**, a rock composed of magnesium and calcium carbonate. A general term for this group of rocks is **carbonate**.

lithosphere. The solid, and relatively rigid, outermost shell of Earth; above is the hydrosphere and atmosphere, and below is the plastic asthenosphere. The lithosphere forms the tectonic plates and in most places includes the crust and the uppermost part of the underlying mantle.

mafic. A geological adjective applied to dark igneous rocks such as basalt and gabbro. The term is derived from *ma*gnesium and *f*err*ic*.

magma. Molten rock that may be extruded on the surface as lava or intruded within the crust.

magmatic arc. A general term that includes the extrusive and intrusive rocks of both island arcs and continental that mostly exhibit arcuate map patterns.

mantle plume. An upwelling of hot material from the deep mantle that generally is columnar in form and thought to have a mushroom-shaped head. Its surface manifestation may be called a **hot spot**. Such upwellings from beneath a tectonic plate have generated the basaltic rocks that make up Hawaii.

marble. Metamorphosed limestone.

megathrust. Term applied to a subduction zone, such as the Cascadia megathrust, where the coupling between converging plates is very strong and creates the potential for great earthquakes. The young, hot, and therefore relatively buoyant subducting oceanic plate is not sinking passively into the mantle but rather is being overridden on a giant thrust fault by the continental plate.

mélange. Mélange is the French word for "mixture" and used geologically for a rock unit that is chaotic or highly disrupted internally, with little or no continuity of beds. It contains blocks of all sizes of a variety of rock types, some locally derived and some far travelled, in a matrix of finer-grained material.

metamorphism. The mineralogical, chemical, and textural changes brought about in a pre-existing rock by the application of conditions of temperature and pressure that are different from those under which the rock formed. Under the new conditions, a set of metamorphic minerals will grow whose nature depends both on the composition of the rock and the new temperature and pressure to which it has been subjected. **High-grade metamorphism** results when a rock is subjected to high temperatures and medium to high pressures corresponding to depths of burial in the crust of 15 to 25 kilometres. **Low-grade metamorphism** results from lower temperatures and pressures, corresponding to depths of burial of up to about 15 kilometres.

mica. A family of aluminosilicate minerals characterized by the ability to split into thin, slightly flexible, sheets. The white mica **muscovite** and the dark brown to black mica **biotite** are common in metamorphic rocks; biotite is found in most granitic rocks.

mid-ocean ridge. A continuous submerged ridge where tectonic plates pull apart, basalt wells up from the mantle, and cools to form new crust; found in the middle part of the Atlantic Ocean, less symmetrically in the Indian Ocean, and something of a misnomer in the Pacific Ocean where the ridge lies in the southeastern Pacific and is largely absent from the North Pacific.

mudstone. A blocky, fine-grained sedimentary rock of similar composition to shale but lacking the "splittability" or fissility of the latter. Argillite is a harder, slightly more-metamorphosed equivalent.

muscovite. *See* **mica**

mylonite. A dense, hard, fine-grained rock that has streaky or laminated structure and formed from a pre-existing rock by shearing and growth of new fine-grained minerals.

olivine. A typically yellowish green magnesium-iron silicate mineral that is present as equidimensional crystals within fine-grained basalt. With heat and water, olivine metamorphoses to green to black serpentine. Together with other magnesium and iron silicate minerals it is a major component of the mantle.

outwash. Sand and gravel deposited by a stream emerging from or under a glacier.

paleomagnetic. Refers to the ancient remnant magnetism in a rock that records the direction and intensity of Earth's magnetic field at the time the rock formed. Measurement of the direction may establish the paleolatitude at which the rock formed.

passive continental margin. A continent-ocean boundary within a plate formed by rifting and separation of a larger continent. A modern example is the eastern seaboard of North America within the North American Plate that includes the floor of the eastern Atlantic Ocean as well as the continent. An **active margin** is a plate boundary, such as the present western margin of the North American plate, where different plates interact with one another.

phyllite. A low-grade metamorphic rock derived from shale that is intermediate between slate and schist. It has shiny cleavage surfaces produced by the growth of minute mica crystals aligned on the cleavage surface.

pillow basalt. Basalt extruded underwater contains pillowlike masses that formed as the surface of the molten lava cooled instantly to make a glassy skin enclosing a bubble of still-molten lava.

pluton. An igneous intrusion; the term is generally applied to bodies of granitic rock.

pyrite. An iron sulphide that forms golden, cube-shaped crystals and is sometimes called "fool's gold." Fine-grained pyrite is common in sedimentary rocks, such as shale, and typically weathers to a rusty colour.

pyroclastic. Fragmental (or clastic) material expelled from a volcano, such as volcanic ash and volcanic breccia.

pyroxene. A family of magnesium- and iron-rich silicate minerals, typically dark and forming stubby crystals, that are common in some volcanic rocks.

pyrrhotite. An iron sulphide mineral, generally softer and darker than pyrite, that may be magnetic.

quartz. Hard, commonly colourless glassy silicon dioxide that next to feldspar is the most common mineral in Earth's crust; it may occur as hexagonal or bipyramidal crystals, but more typically is fine grained; common in sandstone and granitic rocks.

quartz diorite. An intrusive igneous rock, compositionally intermediate between granodiorite, which contains some potassium feldspar, and diorite, which has little or no visible quartz. Together with granodiorite, it is the most common variety of granitic rock in the western Canadian Cordillera.

quartzite. A type of sandstone composed largely of rounded quartz grains firmly stuck together by silica cement, thus making a very hard, tough rock.

radiolarian. Minute single-celled animals with siliceous skeletons, called tests, that lived from the Cambrian to the present in marine environments and are extremely important in providing the ages of rocks scraped off the ocean floors during plate convergence. **Radiolarian oozes** are deposits on the deep ocean floors that contain abundant radiolarians.

rebound. Uplift of Earth's surface to its original level after it was depressed by a great load, such as the continental ice sheet, that subsequently melted so that the load was removed. The underlying asthenosphere flows laterally away from the region beneath the load to flow back again after the load is removed. The depression containing Hudson Bay in eastern Canada has not fully returned to its preglacial level.

rhyolite. Extrusive rock, typically silica-rich, with crystals of quartz and/or feldspar in a glassy matrix and commonly exhibiting bands or flow structure; the extrusive equivalent of granite.

rift. A valley or trough bounded by normal faults that forms when the lithosphere is pulled apart. Rift valleys occur along the crest of mid-ocean ridges, where plates are pulling apart, and also within continents as precursors to the separation of plates.

sandstone. A sedimentary rock deposited by water or wind and composed of rounded or angular grains of sand that have been cemented together, commonly either by silica or calcium carbonate.

scarp. A cliff or steep slope formed by uplift along a fault. A **fault-line scarp** is formed by erosion; resistant rock on one side of a fault stands high, and soft rock on the other side is eroded.

schist. A medium- to coarse-grained metamorphic rock, typically with a foliation formed by aligned platy minerals such as mica or elongate minerals such as hornblende.

serpentinite. A yellowish green to black rock that typically breaks to give shiny surfaces with a lustrous feel. It is composed mainly of the serpentine group of iron- and magnesium-rich hydrated silicates and formed by alteration by hot water of such minerals as olivine.

shale. A fine-grained sedimentary rock, originally deposited as mud, that splits into thin layers parallel with bedding.

shear zone. A zone in rock caused by separate parts of the rock body moving past one another when the rock was hot and relatively plastic. Shear zones may merge with faults that were formed in cooler, more brittle rocks.

shield. The exposed continental basement composed of Precambrian metamorphic and granitic rocks. The Canadian Shield is largely surrounded by, and buried beneath, younger sedimentary cover; together, they form the continental platform.

siliceous. Adjective denoting "silica-rich," for example, siliceous shale.

sill. A tabular igneous body that was intruded parallel with the layering or foliation in the hosting country rock.

silt. Very fine-grained sediment whose grains are just visible to the naked eye; intermediate between sand and mud.

siltstone. A sedimentary rock made of cemented silt-sized grains.

skarn. A distinctive metamorphic rock, common around certain types of mineral deposits, and made of calcium-bearing silicate minerals, including garnet. It forms when limestone and dolomite are intruded and heated, and altered by addition of silica, iron, and magnesium.

stock. A small igneous intrusion that crosscuts the country rock; horizontally it is roughly equidimensional rather than tabular.

strata; stratified. Strata, the plural of stratum, are layers formed when sediments or volcanic ash are deposited one upon another; the layers may vary slightly in composition; for example sandstone interlayered with shale. A **stratified** rock is one with layers formed in this way.

subduction zone. The place where oceanic lithosphere descends into the mantle. The surface expression of the subduction zone may be a deep oceanic trench, such as south of the Aleutian Islands in Alaska, although elsewhere, such as west of Vancouver Island, there is no trench.

supercontinent. All continents were amalgamated into one supercontinent at least twice in Earth's history; about 1 billion years ago the supercontinent Rodinia formed and then broke up into continent-sized fragments, and about 300 million years ago the supercontinent Pangea formed from these fragments and subsequently broke up into the present continents.

syenite. A medium-grained to coarsely crystalline intrusive rock that resembles granite but lacks visible quartz.

syncline. A fold in which the bedding or layering has been bent downwards to make a trough in profile; the opposite of anticline.

talc schist. A generally light-coloured schist in which the mineral talc, a hydrated magnesium silicate, is the prominent aligned, or foliated, mineral; derived from the alteration of ultramafic rock.

talus. An outward sloping pile of rock fragments of any size accumulated by falling or sliding at the base of a cliff or steep slope. A **talus cone** generally accumulates below the mouth of a gully that opens on a steep slope or cliff and concentrates the rock fragments in one place.

terrane. A fault-bounded body of regional extent, within which there is some degree of compositional or, ideally, stratigraphic uniformity, and whose geological record is different from those of adjoining terranes. Some terranes contain paleontological and/or paleomagnetic records different from those of adjoining terranes and from the nearest continental platform, which raises suspicion that they were geographically separated from one another at the time they formed and has led to the term **suspect terrane**.

till. An unsorted and generally unlayered mixture of rock fragments of all sizes down to clay that were deposited beneath a moving glacier or continental ice sheet.

trench. The term may be used geologically in two ways: (1) it generally refers to narrow, elongate, deep depressions on the ocean floor that mark places at convergent plate margins where subducting oceanic lithosphere disappears from Earth's surface to descend into the mantle; (2) in the Canadian Cordillera *trench* is used for some big, continuous, relatively straight valleys within the mountains, such as the Rocky Mountain Trench.

tuff. A general term for particularly finer-grained, consolidated pyroclastic rocks, such as volcanic ash.

turbidite. A layered rock deposited on the ocean floor from a slurry carrying suspended sediment that moves downslope under the influence of gravity.

ultramafic. A generally dark, heavy rock made of iron and magnesium silicates, such as olivine and/or pyroxene.

unconformity. A major break in a succession of stratified rocks that represents cessation of deposition for a considerable time and/or erosion of the rocks below the break. A **disconformity** is where strata on both sides of the break are parallel. An **angular unconformity** is where strata below the unconformity were deformed and then eroded prior to deposition of the overlying strata.

veins. The filling of a crack or fracture in country rock, commonly with quartz, calcium carbonate, and/or an ore mineral.

volcanic ash. Fine-grained pyroclastic material ejected from a volcano; when consolidated it forms tuff.

volcanic cone. A hill or mountain of lava and/or pyroclastic material extruded and ejected from a volcanic vent and built up around the vent.

zircon. A mineral in igneous and metamorphic rocks that often contains small amounts of uranium and thorium, elements that can be used in isotopic age dating.

ADDITIONAL READING

denotes popular publications

*Canning, S., and R. Canning. 1999. *Geology of British Columbia: A Journey through Time.* Vancouver, B.C.: Greystone Books.

*Clague, J., and B. Turner. 2003. *Vancouver, City on the Edge.* Vancouver, B.C.: Tricouni Press.

Coney, P. J., D. L. Jones, and J. W. H. Monger. 1980. Cordilleran suspect terranes. *Nature* 228:329–33.

Engebretson, D. C., K. P. Kelley, H. P. Cashman, and M. A. Richards. 1992. 180 million years of subduction. *GSA Today* 2:93–100.

Gabrielse, H., and C. J. Yorath, editors. 1991. *Geology of the Cordilleran Orogen in Canada.* Geological Survey of Canada, Geology of Canada 4, and Geological Society of America. *The Geology of North America* G-2.

*Gadd, B. 1986. *Handbook of the Canadian Rockies.* Jasper, Alberta: Corax Press.

*Gould, S. J. 1989. *Wonderful Life: The Burgess Shale and the Nature of History.* New York: W. W. Norton and Company.

*Goward, T., and C. J. Hickson. 1995. *Nature Wells Gray: A Visitor's Guide to the Park.* Edmonton, Alberta: Lone Pine Publishing and Kamloops, British Columbia: Friends of Wells Gray Park.

Grand, S. P., R. D. van der Hilst, and S. Widiyantoro. 1997. Global seismic tomography: a snapshot of convection in the Earth. *GSA Today* 7(4):1–7.

Hyndman, R. D. 1972. Plate motions relative to the deep mantle and the development of subduction zones. *Nature* 238:263–65.

*Ludvigsen, R. (editor) 1996. *Life in Stone: A Natural History of British Columbia's Fossils.* Vancouver: University of British Columbia Press.

*Mathews, W. H. 1975. *Garibaldi Geology: A Popular Guide to the Geology of the Garibaldi Lake Area.* Geological Association of Canada, Cordilleran Section, Vancouver, B.C.

Mathews, W. H. 1986. *Physiography of the Canadian Cordillera.* Geological Survey of Canada, Map 1701A, scale 1:5,000,000.

Monger, J. W. H., R. A. Price, and D. J. Tempelman Kluit. 1982. Tectonic accretion and the origin of the two major metamorphic and plutonic welts in the Canadian Cordillera. *Geology* 10:70–75.

Monger, J. W. H., R. A. Price, and W. J. Nokleberg. 2005. Northern Cordillera. *The Encyclopedia of Geology*. Elsevier.

Price, R. A., and J. W. H. Monger. 2000. *A Transect of the Southern Canadian Cordillera from Calgary to Vancouver.* Geological Association of Canada, Cordilleran Section Field Trip Guidebook, p.164.

Smith, P. L., editor. 1991. *A Field Guide to the Paleontology of Southwestern Canada.* The first Canadian paleontology conference, University of British Columbia, Vancouver.

Wheeler, J. O., and P. McFeely. 1991. *Tectonic Assemblage Map of the Canadian Cordillera and Adjacent Parts of the United States of America.* Geological Survey of Canada, Map 1721 A, scale 1:2,000,000.

*Yorath, C. J. 1990. *Where Terranes Collide.* Victoria, B.C.: Orca Book Publishers.

*Yorath, C. J., and H. W. Nasmith. 1995. *The Geology of Southern Vancouver Island: A Field Guide.* Victoria, B.C.: Orca Book Publishers.

Many of the popular books and maps listed above are available from the sales office of the Geological Survey of Canada in Vancouver. Technical articles can be read in the library of the Geological Survey of Canada.

Geological Survey of Canada
101–605 Robson Street
Vancouver, B.C. V6B 5J3
(604) 666-0529
www.nrcan.gc.ca/gsc/pacific/vancouver

INDEX

Page numbers in italics refer to photographs

ABOUT THE AUTHORS

W. H. (Bill) Mathews was born in Vancouver in 1919 and passed away in 2003. He completed his Master of Applied Science degree at the University of British Columbia in 1941 and then became a mining engineer for the British Columbia Department of Mines. He obtained his Ph.D. from the University of California at Berkeley, where he met his wife, Laura Lou, and taught there from 1949 to 1951. In 1951, he returned to Vancouver to join the faculty in the Department of Geography and Geology at the University of British Columbia. He retired in 1984 and began work on this book as a part-time project. Bill's professional interests embraced a wide range of topics, particularly those in the late Cenozoic and Holocene record, such as volcanic activity and its relationship to glaciations, glacial processes, regional geomorphology and the evolution of landscapes, landslides, hydrogeology, Tertiary stratigraphy, coal geology, and mineral deposits. He was noted for his attempts to quantify geological processes.

J. W. H. (Jim) Monger was born in England in 1937 and received his geological education at the University of Reading and the University of Kansas, where he met his wife, Jackie. He then obtained his Ph.D. from the University of British Columbia in 1966. He joined the Vancouver office of the Geological Survey of Canada (GSC) in 1965, remained there until 1995, and currently is an emeritus scientist of the GSC and teaches part-time at Simon Fraser University. With the GSC, he studied the stratigraphy of late Paleozoic and early Mesozoic rocks in the western Cordillera and did regional mapping in southwestern British Columbia. Beginning in the early 1970s, following the emergence of plate tectonic concepts, he has attempted to unravel and explain the tectonic evolution of the Cordillera.